Handbook of Marine Biotechnology

Handbook of Marine Biotechnology

Edited by Gilbert King

SYRAWOOD
PUBLISHING HOUSE
New York

Published by Syrawood Publishing House,
750 Third Avenue, 9th Floor,
New York, NY 10017, USA
www.syrawoodpublishinghouse.com

Handbook of Marine Biotechnology
Edited by Gilbert King

International Standard Book Number: 978-1-64740-433-8 (Hardback)

Cataloging-in-publication Data

Handbook of marine biotechnology / edited by Gilbert King.
p. cm.
Includes bibliographical references and index.
ISBN 978-1-64740-433-8
1. Marine biotechnology. 2. Marine biology. 3. Biotechnology. I. King, Gilbert.
TP248.27.M37 H36 2023
660.6--dc23

TABLE OF CONTENTS

PREFACE

This book aims to highlight the current researches and provides a platform to further the scope of innovations in this area. This book is a product of the combined efforts of many researchers and scientists, after going through thorough studies and analysis from different parts of the world. The objective of this book is to provide the readers with the latest information of the field.

Biotechnology refers to a field of applied science that creates products and processes from living organisms and their derivatives. It entails the understanding of the functioning of living organisms at the molecular level. It draws from several academic fields such as chemistry, biology, physics, and mathematics. Marine biotechnology is the use of bioinformatics, biotechnology, and molecular and cell biology to produce products and processes from marine organisms. It is concerned with ocean exploration for creating novel chemical products, enzymes, medicines, and other products and processes. It also contributes significantly to biomaterials, seafood and aquaculture safety, health care diagnostics, and bioremediation. This book unravels the recent studies in the field of marine biotechnology. It will also provide interesting topics for research, which interested readers can take up. This book is a vital tool for all researching or studying marine biotechnology as it gives incredible insights into the emerging trends and concepts.

I would like to express my sincere thanks to the authors for their dedicated efforts in the completion of this book. I acknowledge the efforts of the publisher for providing constant support. Lastly, I would like to thank my family for their support in all academic endeavors.

Editor

1

Marine Sponge Derived Natural Products between 2001 and 2010: Trends and Opportunities for Discovery of Bioactives

Mohammad Ferdous Mehbub [1,2,3], Jie Lei [1,2], Christopher Franco [1,2,*] and Wei Zhang [1,2,*]

[1] Centre for Marine Bioproducts Development, Flinders University, Adelaide, SA 5042, Australia; E-Mails: mohammad.mehbub@flinders.edu.au (M.F.M.); etuferdous@yahoo.com (J.L.)

[2] Department of Medical Biotechnology, School of Medicine, Flinders University, Adelaide, SA 5042, Australia

[3] Department of Fisheries Technology, Faculty of Fisheries, Hajee Mohammad Danesh Science and Technology University, Dinajpur 5200, Bangladesh

* Authors to whom all correspondence should be addressed: E-Mails: chris.franco@flinders.edu.au (C.F.); wei.zhang@flinders.edu.au (W.Z.);

Abstract: Marine sponges belonging to the phylum Porifera (Metazoa), evolutionarily the oldest animals are the single best source of marine natural products. The present review presents a comprehensive overview of the source, taxonomy, country of origin or geographical position, chemical class, and biological activity of sponge-derived new natural products discovered between 2001 and 2010. The data has been analyzed with a view to gaining an outlook on the future trends and opportunities in the search for new compounds and their sources from marine sponges.

Keywords: marine sponges; porifera; marine natural products; anticancer; sponge-associated bacteria; drug; bioactive compounds

---

1. Introduction

Discovery of marine derived natural products is a promising, but comparatively new field, which started with the discovery of unusual nucleoside derivatives in the sponge *Tethya crypta* in the 1950s by Bergmann and Feeney [1,2]. In the early 1960s, research on marine natural products was driven by chemical studies and few compounds were tested for any relevant bioactivity [3] such as production of a pyrrole antibiotic by a marine bacterium *Pseudomonas bromoutilis* [4]. However, utilization of marine organisms as sources of bioactive metabolites started seriously at the end of 1960s [5] with the isolation of prostaglandin derivatives from the Caribbean Gorgonian *Plexaura homomalla* [6]. In the 1980s effective collaborations were established between marine chemists and pharmacologists and the investigations were focused on central nervous system membrane active toxins, ion channel effectors, anticancer and anti-viral agents, tumor promoters and anti-inflammatory agents [7]. In the 1990s pharmaceutical and biotechnological industries focused their screens on chemical libraries of both natural products, as well as synthetic compounds produced by combinatorial methods [8]. Invertebrates, mainly sponges, tunicates, bryozoans or molluscs provided the majority of the marine natural products involved in clinical or preclinical trials [9].

The discovery of marine natural products has accelerated over the last two decades with the number of new compounds discovered annually increasing from 20 to more than 200 [10]. It has been estimated that by 2010 more than 15,000 new marine natural products (NMNP) had been discovered [11–13] with 8368 new compounds recorded for the decade between 2001 and 2010. This constitutes over half of all the compounds discovered since 1951.

Among all the marine organisms investigated, marine sponges (Porifera) are recognized as the richest sources of NMNP, with about 4851 compounds to date, contributing to nearly 30% of all marine natural products discovered so far. It should be noted that of these, 1499 new compounds were isolated in the five years from 2008 to 2012 [14–18]. This makes sponges the most prolific marine producers of compounds with more than 200 new compounds reported each year for the last decade [19]. With this myriad of NMNP available, numerous studies have revealed a broad spectrum of biological activities for these compounds, including anticancer, antiviral, antibacterial, antifungal, antiprotozoal, anthelmintic, anti-inflammatory, immunosuppressive, neurosuppressive, neuroprotective, antifouling and a range of other bioactivities [20]. In addition, as infectious microorganisms evolve and develop resistance to existing pharmaceuticals, marine sponges provide novel leads against bacterial, fungal and viral diseases [19,21]. Figure 1a shows the almost linear growth of new compounds over the last three decades. It is predicted that if this rate can be sustained, the discovery of marine natural products from sponges, in particular, and as well as other major marine organisms will bring about new and effective therapies against human diseases [22–24].

Figure 1b shows the trends of novel marine natural products discovered from different phyla of marine organisms during 2001–2010. The annual discovery of marine natural products remained at a constant level of about 500 products in the late 1990s [10] but this number has increased from 600 to over 1000 compounds from 2008 to 2010, a significant increase which was partly driven by new developments in modern analytical technology and instruments, especially the development of the high resolution nuclear magnetic resonance spectrometer (NMR) and mass spectrometry (MS) coupled with high-performance LC and GC [10].

Figure 1. (**a**) Number of new compounds isolated from marine organisms per decade from 1970 to 2010; (**b**) Total number of new compounds isolated from different marine organisms from 2001 to 2010.

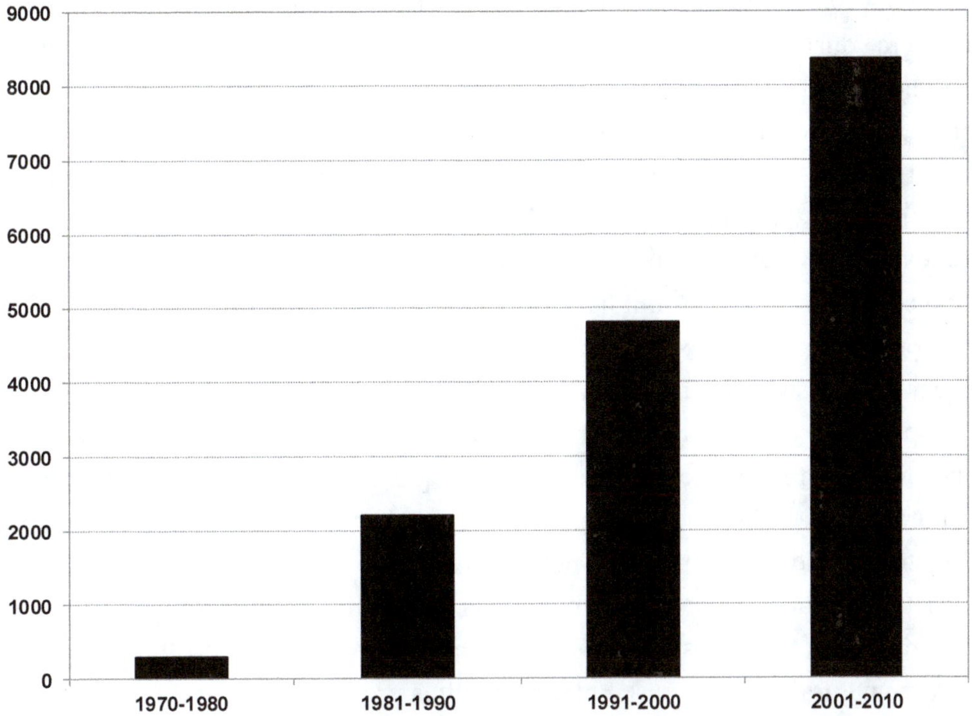

(**a**)

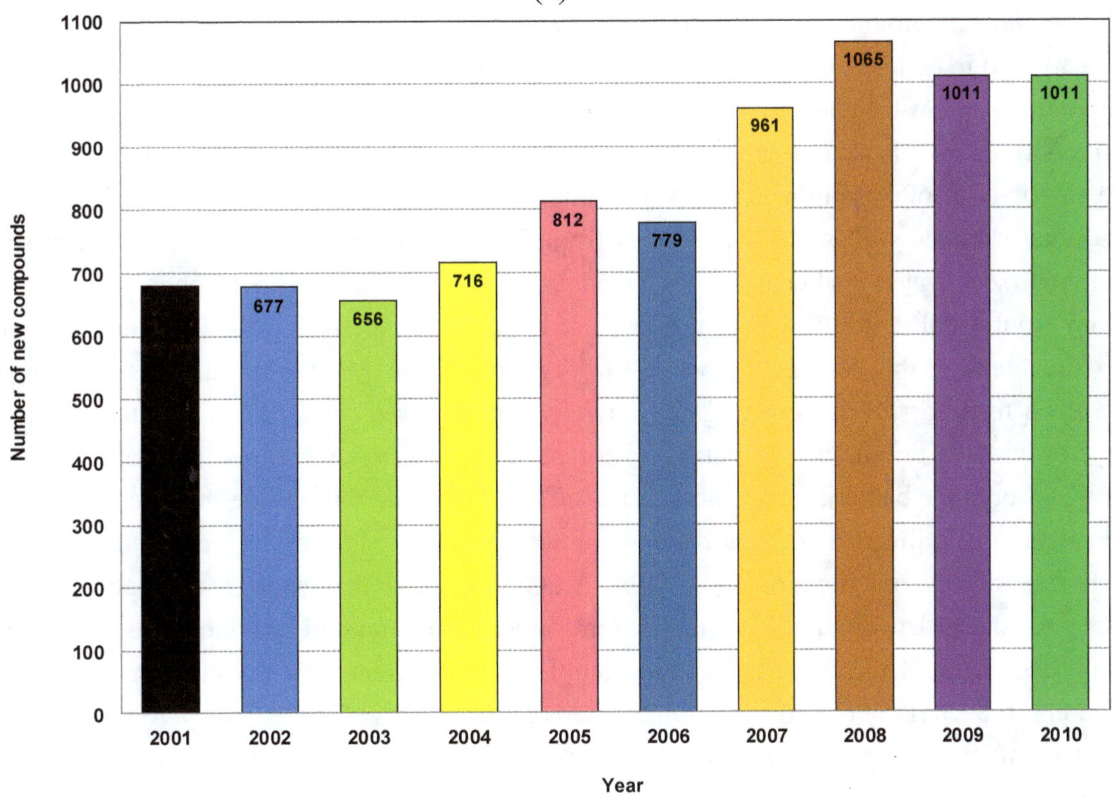

(**b**)

Although sponges have shown the highest potential for natural product discovery, no comprehensive reviews have been published that focus only on compounds from sponges in terms of their source areas, modes of action, chemical class and taxonomy. Two review papers written by Hu *et al.* 2011 and Leal *et al.* 2012 [10,25] described the overall trends in marine natural products, including those from Porifera, during the last two decades. Therefore, we prepared this manuscript based on sponge-derived new natural products from 2001 to 2010 and all the graphs and tables generated for this paper are based on the data reviewed by Blunt *et al.* from the *Natural Product Reports* of 2003–2012 [13–16,20,26–30]. Information was collected from each individual published paper during this time period and data was generated and analyzed accordingly.

Sponges are exclusively aquatic animals that dominate in many benthic habitats. They are sessile and do not have tissues or sensory organs but have different types of cells which conduct all forms of bodily function. They consume food and excrete waste products within cells without a body cavity [31].

Numerous ecological studies have shown that secondary metabolites produced by sponges often serve defensive purposes to protect them from threats such as predator attacks, microbial infections, biofouling, and overgrowth by other sessile organisms [32,33]. For this reason Porifera are attractive subjects for natural product chemists due to the sheer number of metabolites produced, the novelty of structure encountered, and the therapeutic potential of these compounds in the treatment of human diseases [7]. There is evidence that some compounds originally found in sponge cells are synthesized by microorganisms associated with sponges, since the mesohyl of sponges is often inhabited by microbes and many poriferan natural products resemble metabolites produced by marine microbes [34].

However, these natural products have interesting biomedical potential, pharmaceutical relevance and diverse biotechnological applications [5,35–39]. Moreover, sponge-derived antifouling molecules have been found to be less toxic, environmentally friendly biocides that are often very effective [40].

It is of both scientific and industrial interest as to why and how marine sponges possess such a high diversity of novel marine natural products. As the oldest metazoan, sponges have survived in the ocean for over 600 million years [41] throughout the vast changes experienced by the ocean. The fact that sponges still exist in all waters from fresh to saline, from intertidal to deep-sea, from tropical to frozen waters indicates the tremendous ability of sponges to respond and adapt to the varied environmental conditions over this period. In addition, sponges are one of the most efficient sessile filter feeders: they can filter up to 24 $m^3 \cdot kg^{-1}$ day^{-1} [42]. Bacterial numbers in sponge tissue often exceed those of the surrounding seawater by two to three orders of magnitude as the sponge mesohyl provides a unique ecological niche for particular bacterial species. In many cases, sponge mesohyl harbours the bacterial symbionts (30%–60%) [43]. Bacteria provide their hosts with products of their metabolism, thereby granting the sponge access to bacteria-specific traits such as autotrophy, nitrogen fixation and nitrification [44]. Other examples show that sponge-associated bacteria can process metabolic waste compounds, stabilize the sponge skeleton and provide protection against UV radiation [35,45,46]. The most prominent example of sponge bacterial symbiosis, however, is the involvement of bacteria in the production of bioactive metabolites [47] that have a role in defense [48].

These highly intensive, constant interactions with the environment have given sponges a unique biochemistry to produce the high diversity of metabolites that can either help them survive or prompt them to evolve. Being attached to a solid surface, a sponge is unable to escape when confronted with a predator, and so, when threatened they release stored secondary metabolites that have cytotoxic, antibiotic and feeding deterrent properties [48]. Some chemicals prevent settlement of fouling organisms on the sponge surface and restrict competition for space with neighbors. The sponge bacterial associations and interactions have been widely studied, with evidence that the sponge-associated bacteria can help the sponges to produce secondary metabolites to protect them against their predators [49]. In an ecological context, sponges have developed special mechanisms to protect themselves from pathogenic bacteria, viruses, parasites, fungus and other predators that include both chemical defense mechanisms and physiological responses. Chemical defense mechanisms help to protect sponges against certain deleterious bacteria [33,50,51]. In this way, sponges provide novel leads against viral, fungal and parasitic diseases [52]. By producing different types of toxins or malevolent tastes and odors, sponges protect themselves against predators or inhibit coral overgrowth that could threaten the sponge osculum or other systems. As a physical defense they have spicules and collagen. Sponges may also succumb to microbial and fungal infections which could result in the disintegration of the sponge fibers/tissue and ultimately lead to sponge death [53]. The fact that sponges are susceptible to microbial infection suggests that they should also possess mechanisms to prevent these types of diseases [54]. Maldonado and co-workers [55] showed how sponges recover from a bacterial infection: their ultrastructural study revealed that the sponges secrete successive collagen barriers at the diseased area and abandon decaying body parts external to the barrier.

Recently, the ubiquitous defense enzyme, phospholipase A2 (PLA2) detected in a sponge associated bacterium envisaged the possible functional role in the ecological succession of the host sponge against predatory/fouling pressure in the habitat [56]. In response to predators and pathogens, sponges have engineered complex secondary metabolites from a diverse set of biological precursors. Secondary metabolites are organic compounds that are not directly involved in the normal growth, development or reproduction of organisms. These metabolites produced by sponges and their associated microflora can be classified chemically as alkaloids, terpenoids, glycosides, phenols, phenazines, polyketides, fatty acid products and peptides, amino acid analogues, nucleosides, porphyrins, aliphatic cyclic peroxides and sterols [57,58]. Many of these compounds are very potent because the diluting effect of the ocean drives the construction of molecules that are highly active and stable in saline conditions [59].

Given the significance of sponges in marine natural product discovery, the aim of this review is to present a comprehensive overview of sponge-derived natural product discovery during the recent decade from 2001 to 2010, in order to understand the defining trends and provide insights into avenues for further compound discovery. The temporal trends of the discovery of sponge-derived marine natural products and their biological activities, the sources of discovery in terms of sponge taxonomy, the chemical classes of these natural products, and the countries of collection have been categorized. Our analysis also includes a short description of the relative distribution and contribution of these discoveries with reference to governmental funding, policies and known national priorities given to marine natural products. Finally, the opportunities and challenges have been identified for future R and D in this fast growing field.

The new compounds isolated during the last decade were classified into 18 chemical classes including acid, alkaloid, ester, fatty acid, glycoside, ketone, lipid, macrolide, alcohol, peptide, peroxide, polyketide, quinone, steroid, sterol, terpene, terpenoid and unclassified, based on the data reviewed by Blunt *et al.* from *Natural Product Reports* of 2003 to 2012 [13–16,20,26–30]. The World Porifera Database [60] was used for the taxonomic classification of the sponges mentioned in the Natural Product

Reports [13–16,20,26–30]. The World Register of Marine Species (WoRMS) database [60] was also used to cross check detailed taxonomical information (order and family) for each surveyed species and to validate and/or update their scientific names [61].

During the recent decade the sponges collected were from 19 known orders as well as 12 sponges of unknown identity which provided new compounds. These included Agelasida, Astrophorida, Axinellida, Chondrosida, Choristida, Clathrinida, Dendroceratida, Dictyoceratida, Hadromerida, Halichondrida, Haplosclerida, Homosclerophorida, Leucosolenida, Lithistida, Lyssacinosida, Ocilosclerida, Poecilosclerida, Spirophorida and Verongida. Sixty two countries (with Antarctica labeled as a country for reporting purposes) were the sources for the sponge samples studied. The bioactivities were mainly classified as anti-Alzheimer's, antibacterial, antituberculosis, anticancer, antifungal, anti-inflammatory, antimalarial, antiviral and anti-HIV.

2. New Compounds and Their Distribution 2001–2010

2.1. Yearly Distribution of Phyla that Produce Natural Products Discovered from 2001 to 2010

To investigate the distribution of sources of NMNP, 12 different categories including a separate group comprising marine microorganisms and phytoplankton were used for this review (Figure 2). Invertebrates comprise approximately 60% of all marine animal diversity [62]. Most belong to the phyla Porifera (sponges), Annelida, Arthropoda, Bryozoa, Cnidaria, Echinodermata, Mollusca and Chordata. Several studies addressing marine invertebrates also include these groups of organisms [32,33,63].

Figure 2. Total number of new compounds isolated from different types of marine sources, 2001–2010.

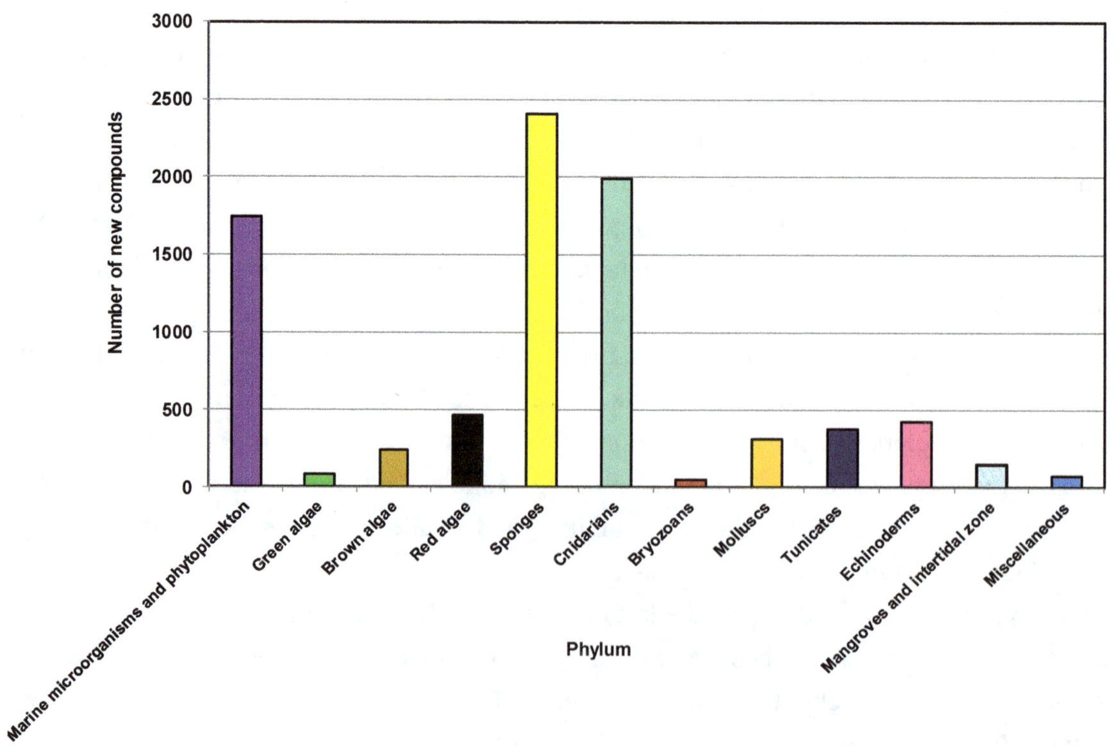

As highlighted in other reviews [10,64], the phyla Porifera and Cnidaria have been the two main sources of NMNP. During the last 10 years from 2001 to 2010, more than 2400 new natural products have been discovered from 542 genera and 671 species of sponges. These sponges belong to 19 known orders and 74 families, contributing about 29% of the marine natural products discovered during this decade, making it the largest source among all marine organisms [48,65].

2.2. Sponges (Porifera) as a Source of New Natural Products and Drugs for the Future

A review in 2003 collected the most important marine natural compounds which were undergoing preclinical and clinical trials (I, II, III) for anticancer activity. Among those, compounds from sponges were the following: Discodermolide, Hemiasterlins A & B, modified Halichondrin B, KRN-7000, Alipkinidine (alkaloid), Fascaphysins (alkaloid), Isohomohalichondrin B, Halichondrin B, Laulimalide/Fijianolide, 5-Methoxyamphimedine (alkaloid) and Variolin (alkaloid) [66]. A review paper was published by Sipkema *et al.* 2004 about the drugs from marine sponges [67].

Marine natural sources as potential anticancer agents were reviewed in 2011 which mentioned 39 marine-derived potential anticancer agents and among them 18 compounds from sponges with different mechanisms of action [68]. Interestingly, from the 16 marine natural products that are currently under preclinical trials as new drug candidates, most are derived from invertebrates. Of these, Porifera remain the most important phylum, with six of the 16 compounds [69–71]. A review paper published in 2013 classified anticancer molecules according to their current status in the clinical phase trials (approved/phase IV/phase III/phase II/phase I) and updated the data to April 2012 [72]. A very recent review, published in 2014, showed the compounds derived from marine sources currently in clinical trials against cancer with more updated information on clinical and late preclinical developments [73]. This paper also mentioned that although many compounds showed potential against cancer and entered clinical trials in cancer, to date, only Cytarabine, Yondelis® (ET743), Eribulin (a synthetic derivative based on the structure of Halichondrin B), and the Dolastatin 10 derivative, monomethylauristatin E (MMAE or vedotin) as a warhead, have been approved for use in humans (Adcetris®) [73].

Although a number of compounds from sponges showed promising activity to be potential drug candidates over the last few decades they are generally not ready for further development due to the challenge of obtaining continuous and larger supplies of the compounds, unless they can be chemically synthesized. Considerable quantities of a drug candidate are vital for clinical trials, but only a few milligrams of most natural products can be isolated from marine samples [74]. One solution is farming sponges to source bioactive metabolites [75]. On the other hand, sponge compounds that are produced by sponge-associated microorganisms can be scaled up as the microorganisms are able to flourish independently of the sponge. Another way to ensure supply is using sponge cell culture, although this is still a growing area. Muller *et al.* in 2000 described the production of bioactive compounds by sponge cell culture [76] and Zhang *et al.* in 2003 and Cao *et al.* in 2007 showed improved cell proliferation and spiculogenesis from primmorphs of sponges and dynamics of spicule production in *in vitro* sponge cell culture systems [77]. The sustainable production of bioactive compounds from sponges was reviewed in 2004 [78] and again in 2009 where the advantage and disadvantage of sponge cell culture was discussed, with the conclusion that the understanding of the metabolic pathways is one of the potential advantages of sponge cell culture systems [79]. However, the most promising solution was answered by Wilson *et al.* in 2014; his findings illuminated two promising approaches for addressing the supply problem [80]: firstly, large-scale cultivation of the microorganisms that produce interesting metabolites, and secondly, expressing the biosynthetic pathway of interest in an easily cultivable surrogate host. The discovery of Wilson and Piel and their colleagues identified *Entotheonella* and members of the newly proposed phylum Tectomicrobia as a "biochemically talented" phylum on a par with the actinomycetes [80–82]. Thus, these results could facilitate a new era of drug discovery.

2.3. The Distribution of New Marine Natural Products from Sponges

To date, about 11,000 species of sponges have been formally described of which approximately 8500 are considered valid, but as many as twice those numbers are thought to exist [83]. Well known sponge fauna in the Caribbean, Mediterranean, and the British Isles each contain 500–800 species, whereas less well characterized sponge fauna in Australia, Papua New Guinea and Indonesia possess a high biodiversity [84]. Sponges are currently divided into four distinct classes, 25 orders, 128 families and 680 genera [59,60].

The sponges reported to produce new compounds in the last ten years were from 19 known orders although a number of sponges were not identified. A careful analysis of the trends of the discovery of new bioactive compounds from different orders of sponges is presented here to guide scientists in future discoveries.

As shown in Figure 3a, 504, 355, 337, 274 and 201 new compounds were found from the five orders Dictyoceratida, Haplosclerida, Poecilosclerida, Halichondrida and Astrophorida, respectively. Notably, these five orders contributed more than 70% of the new compounds. The highest numbers of new compounds were found from Dictyoceratida with 72 found in 2008, 66 in 2004, 63 in 2009, 58 in 2001 and 51 in 2005.

Table 1 shows that some orders have been found to be productive sources of NMNP, many with a large chemical diversity. Astrophorida, Dictyoceratida, Halichondrida, Haplosclerida and Poecilosclerida are the orders from which more than 50 species were studied [83].

Dictyoceratida contributed at least 16% of new compounds each year except in 2010 (11.9%), reaching a peak of 28.5% in 2005 (Figure 3b). Haplosclerida contributed at least 13.58% of new compounds each year except 2004 and 2008 with values of 9.4% and 7.7% respectively. This order yielded the highest number of compounds in 2001 at 24.6%. In the case of Poecilosclerida, a gradual increase was observed until 2004 and reached a peak in 2009 at 22.1%. From 2004 to 2008 Halichondrida contributed at least 11.4% new compounds each year.

This corresponds with a recent review on clinically active compounds from sponges which had Astrophorida, Chondrosida, Dendroceratida, Dictyoceratida, Hadromerida, Halichondrida, Haplosclerida, Lithistida, Poecilosclerida, Spirophorida and Verongida as the main orders from which clinically active compounds were found [85]. Leal *et al.* 2012, covering 1990–2009, found that NMNP were recorded in 17 orders of Demospongiae, and about 89% of the natural products were derived from only eight of those orders, namely Astrophorida, Dictyoceratida, Halichondrida, Haplosclerida, Homosclerophorida, Lithistida, Poecilosclerida and Verongida [25].

Figure 3. (**a**) Total number of new compounds isolated from different orders of marine sponges 2001–2010; (**b**) Distribution of new compounds isolated from different orders of marine sponges as a percentage found within the year, 2001–2010.

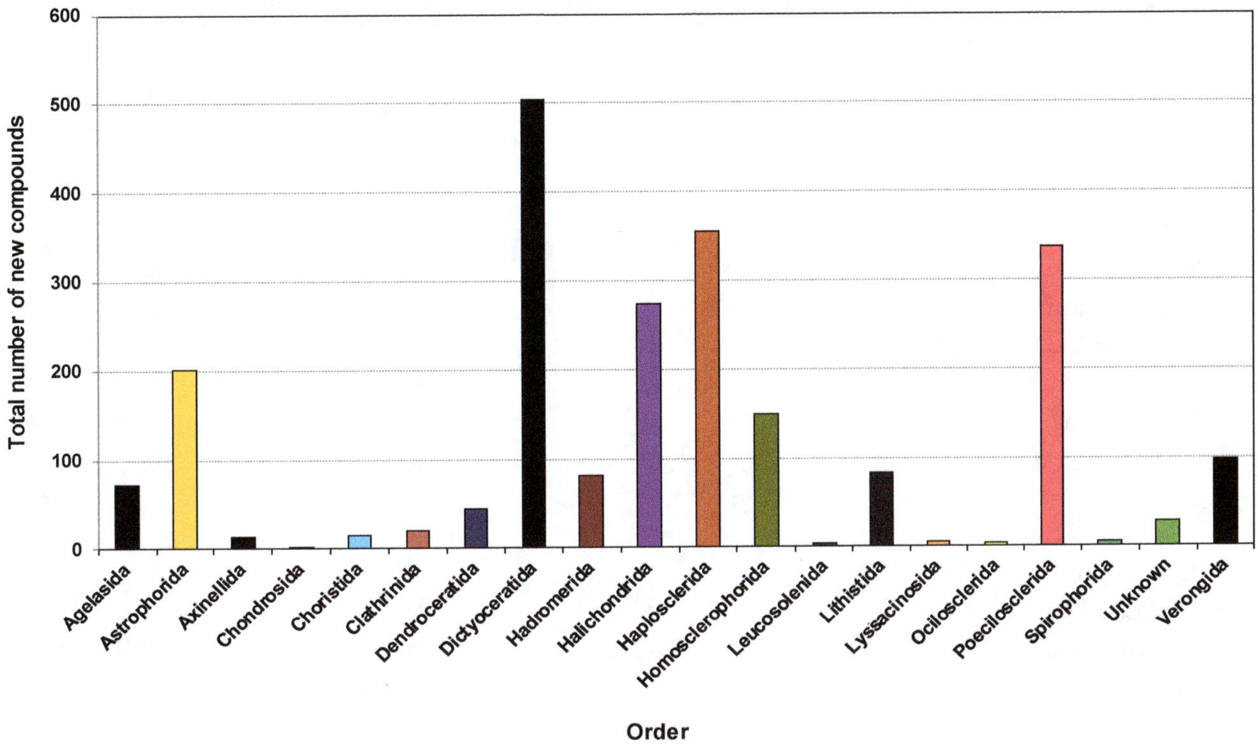

(**a**)

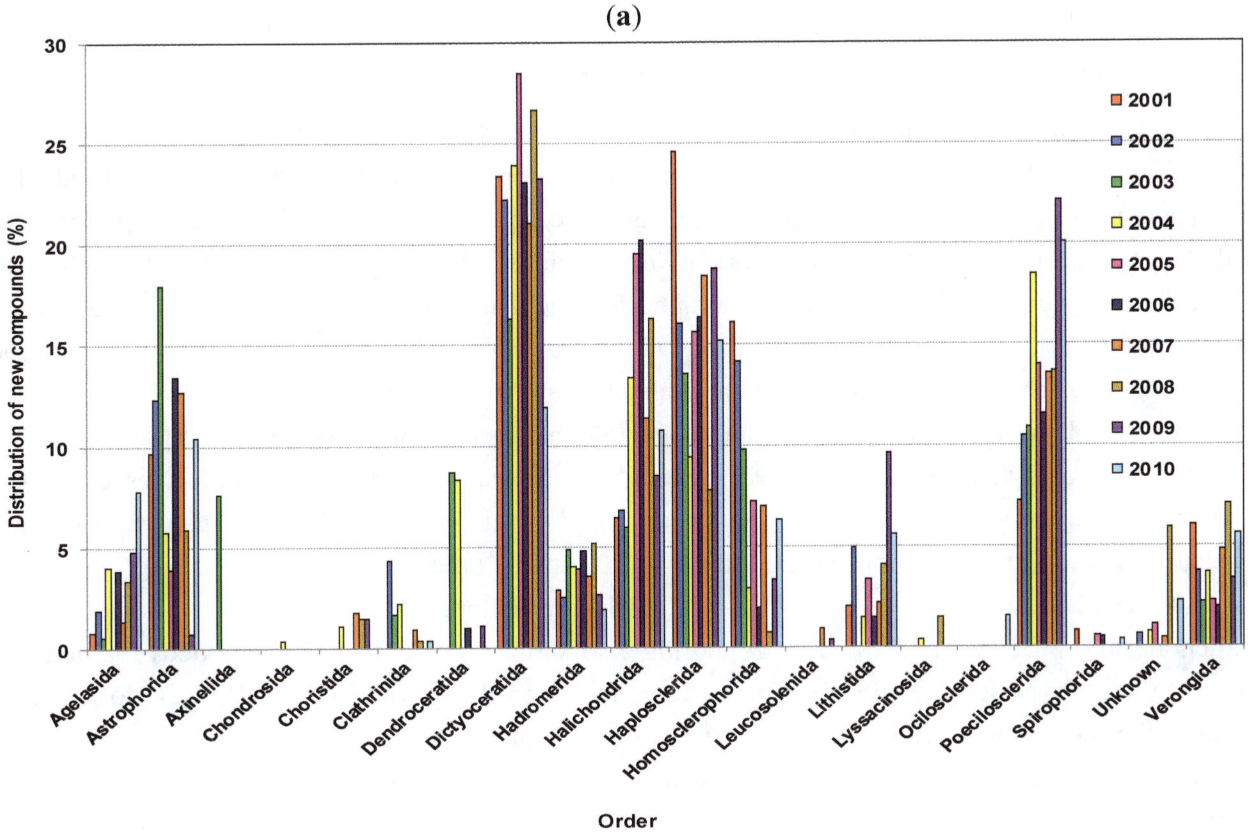

(**b**)

Table 1. Total number of new compounds isolated from different orders of marine sponges with number of families, genera, species and number of references published from 2001 to 2010.

Order	Number of Families	Number of Genera	Number of Species	Number of References
Agelasida	9	9	21	24
Astrophorida	26	58	63	62
Axinellida	2	2	2	2
Chondrosida	1	1	1	1
Choristida	3	3	3	3
Clathrinida	5	5	7	7
Dendroceratida	7	10	11	13
Dictyoceratida	40	117	145	161
Hadromerida	24	31	32	33
Halichondrida	31	69	86	84
Haplosclerida	52	80	100	120
Homosclerophorida	10	20	39	50
Leucosolenida	1	1	1	1
Lithistida	14	20	23	32
Lyssacinosida	2	2	2	2
Ocilosclerida	1	1	1	2
Poecilosclerida	67	68	81	83
Spirophorida	4	4	4	5
Unknown	8	12	12	10
Verongida	27	29	37	46
Total	334	542	671	741

All NMNP discovered since 1990 were recorded in 64 families belonging to the phylum Porifera. However, about 51% of these products were derived from only nine families: Spongiidae, Dysideidae and Thorectidae in the order Dictyoceratida; Chalinidae and Petrosiidae in the order Haplosclerida; Halichondriidae in the order Halichondrida; Ancorinidae belonging to the order Astrophorida; Plakinidae belonging to the order Homosclerophorida and Theonellidae of the order Lithistida. Here, the highest increase in the number of NMNP annually discovered was recorded for the families Chalinidae and Spongiidae. The family Theonellidae has yielded a number of unique compounds [86] with a broad spectrum of biological activities, including antifungal [87] and cancer cell growth inhibitors [88,89].

Figure 4 shows that between 9 and 16 genera were found to produce new molecules each year from the order Dictyoceratida which revealed a high availability of different genera from this particular order, with the highest number found in 2009. This order will be examined in greater detail in a subsequent review covering 2001–2010. Genera belonging to the orders Haplosclerida, Halichondrida, Poecilosclerida, and Astrophorida also showed more than five productive genera for most of the last decade. The number of new compounds correlated with the high diversity of sponges because the higher the diversity the higher the possibility of getting more novel compounds. Another possible reason that makes Dictyoceratida the most prolific producers of NMNP as well as Astrophorida, Haplosclerida, Homosclerophorida and Halichondrida is because these orders harbor high densities of microorganisms [25,85].

Figure 4. Number of different types of genera used for isolation of new compounds from different orders of marine sponges from 2001 to 2010.

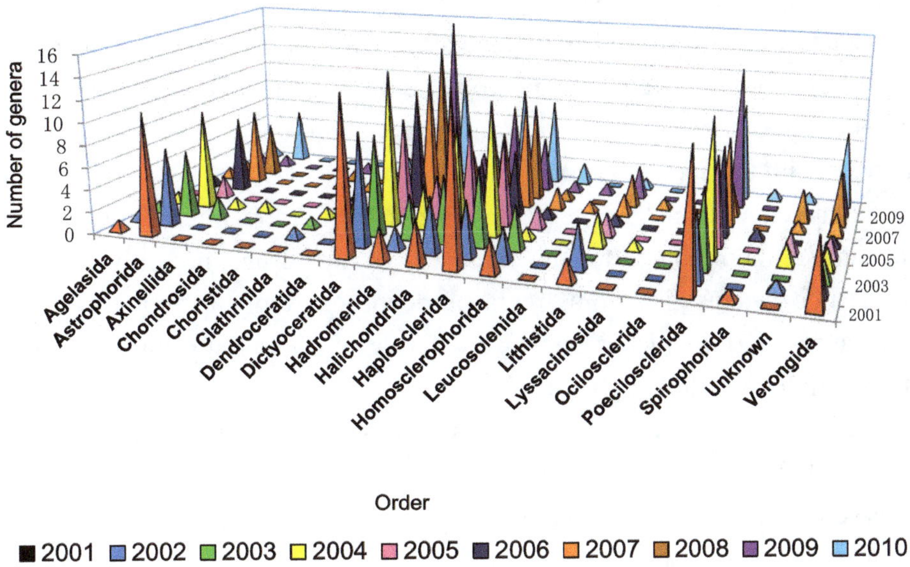

2.4. Distribution of New Compounds per Species from Different Orders

During the last decade each individual species from different sponges contributed on an average four new compounds under different orders (Figure S1, Supplementary Information). Although very few species have been studied from the orders Axinellida and Choristida, on an average they also produce almost the same number of compounds. This finding permits the inference that each individual species has the potential for contributing similar numbers of new compounds regardless of the order.

2.5. Symbiotic Relationships: Sponge Associated Microorganisms

Hentchel *et al.* 2002 reported that although it is generally believed that symbiotic interactions exist between sponges and specific microorganisms, consideration of alternative explanations such as the selective enrichment of ubiquitous seawater bacteria is also important [42]. Moreover, they stated that very specific type of selective pressures, perhaps the hostility to digestion, must exist to establish the uniform composition arrangements of the microbial communities existing in sponges that have otherwise few commonalities [42].

However, it was consequently observed that marine sponges host abundant and diverse communities of symbiotic microorganisms [90]. Webster *et al.* 2012 mentioned that microbial symbionts are undoubtedly important to sponge health, and therefore it is likely that interruptions to symbiosis as a result of climate change/environmental stress will influence sponge health, growth rates or their ability to defend themselves from predation, fouling and disease [91]. Symbiotic interactions between sponges and microorganisms could contribute to sponge nutrition as well [92]. Important roles of the symbionts include photosynthetic carbon fixation [93], nitrification [94,95], nitrogen fixation [44,96,97], and anaerobic metabolism [98]. Another important role of sponge-associated bacteria is the production of potential secondary metabolites, such as antibiotics, antifungal compounds and anti-predation or antifouling compounds [39]. Whilst the microbes associated with the sponges produce the secondary metabolites, possibly all of these sponges have particular microbial associations [99]. More research is necessary to explore the relationship between microorganisms and phytoplankton associated with sponges because it has been suggested that (at least) some of the bioactive secondary metabolites isolated from sponges are produced by functional enzyme clusters which originated from these microorganisms. The role of these microorganisms in sponge biology varies from source of nutrition to mutualistic symbiosis with the sponge [100].

It has been recognized that metabolites synthesized by marine microorganisms associated with sponges could become a major source for the discovery of new drugs, not only because the biological diversity in marine ecosystems like coral reefs or deep sea floors is probably higher than in the rainforest, but because marine microorganisms offer a renewable resource for the scale-up and development of potentially new drugs [101,102].

A number of examples of the functions of the sponge-associated microorganisms are provided to signify the diversity of functions. Dudler and Eberl 2006 conducted a study on interactions between bacteria and eukaryotes which showed increasing evidence to support the hypothesis that secondary metabolites produced by symbiotic bacteria are a result of bacterial cell-to-cell signaling [103]. In a similar vein, Schmidt *et al.* 2008 looked at how organisms cooperate in the synthesis of natural products. They found that partners may exchange and modify the natural products produced by each other and also explained how these secondary metabolites are utilized by the host organisms [104].

Many sponge-derived metabolites resemble bacterial natural products or belong to substance classes typical for these microorganisms [82]. In a recent paper, the Piel group demonstrated beyond doubt that almost all bioactive polyketides and peptides known from the marine sponge *Theonella swinhoei* were attributed to a single phylotype, *Entotheonella* spp. and are extensively distributed in sponges [80].

The diversity in the locations (Okinawa, the Philippines, Indonesia, the Red Sea, Italy, South Africa, and Papua New Guinea) and genera of sponges (*Amphimedon* sp. and *Acanthostrongylophora*) responsible for the production of manzamine alkaloids are widely believed to be a result of a symbiotic relationship between these sponges with common or closely related microorganism(s), which may account for the generation of manzamine enantiomers [105].

Crambe crambe (Schmidt, 1862) (Poecilosclerida) is a red incrusting marine sponge present in the Mediterranean Sea and reported to produce diverse PGAs, namely crambescidins 800, 816, 830, 844, as well as isocrambescidin [106,107]. Using 16S rRNA gene pyrosequencing it has been found that the associated bacterial community of *C. crambe* is dominated by a single bacterial species affiliated to the Betaproteobacteria [108].

2.6. The Distribution of Chemical Classes

Although acid, ester, ketone, peroxide do not directly fall into chemical classes, we considered these as chemical classes just to show the distribution as such described in the original paper as well as mentioned by Blunt *et al.* [13–16,20,26–30] in their annual review of marine natural products that were used in preparing this manuscript. In fact, classes of natural products should be classified based on their biosynthesis. However, it is important to remember that a macrolide is often a polyketide and a sterol a steroid and fatty acid belongs to lipid.

A wide range of chemical and functional diversity has been observed among new compounds during the last decade. The analyzed data (Figure 5a) has shown that in the last ten years 450, 331, 188 and 155 new compounds were classified as alkaloids, terpenes, terpenoids and peptides, respectively, and these four classes make up approximately 50% of compounds discovered. While alkaloids contributed 20% of the new molecules the terpene/terpenoid classification which are often not readily distinguishable, together made up 23% of the total within the decade.

Figure 5. (**a**) Chemical classes of new compounds isolated from marine sponges from 2001 to 2010; (**b**) The distribution of different chemical classes of new compounds isolated from marine sponges from 2001 to 2010.

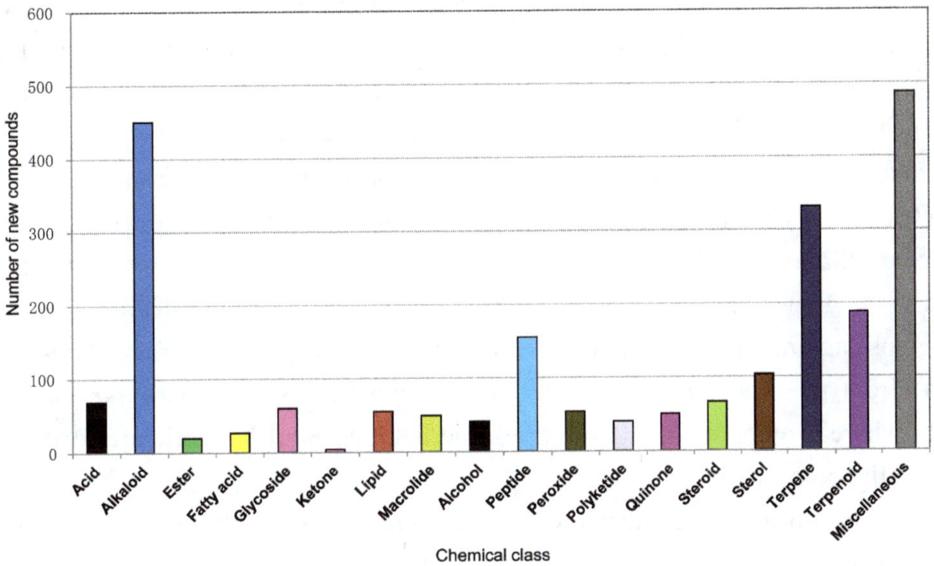

(a)

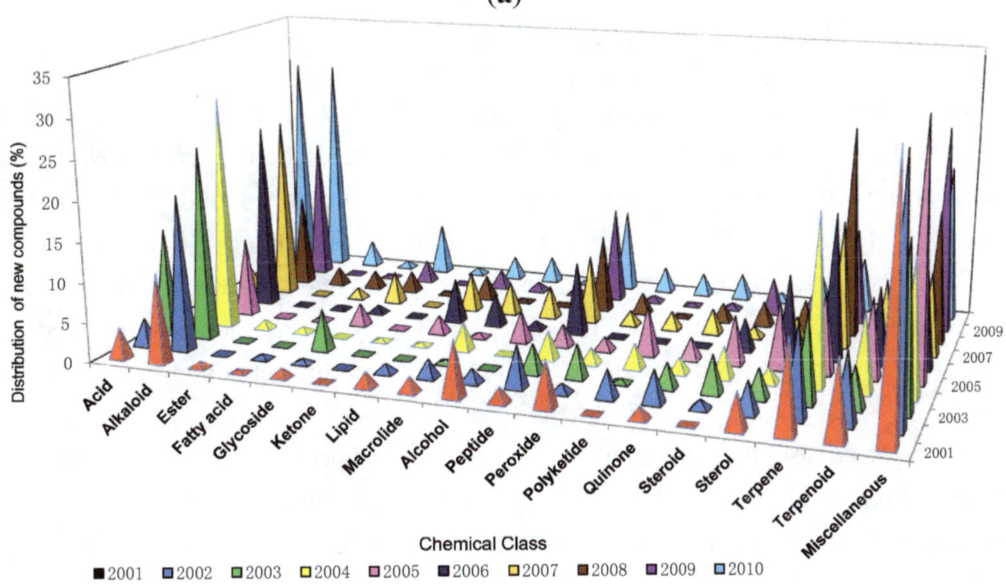

(b)

The chemical diversity of bioactive compounds reportedly produced by sponge-microbe associations showed that certain chemical classes such as quinones, steroids, fatty acids, diketopiperazines, alkaloids, terpenes, terpenoids, trichoverroids and prodigiosin derivatives, diglucosyl-glycerol, polyketides, cyclopeptides, glycoglycerolipid, benzoic acid derivatives are responsible for anticancer or antitumor activity; quinolone derivatives for anti HIV activity; fatty acid esters and fatty acids for anti-inflammatory activity; alkaloids and quinolone derivatives for antimalarial activity; polyketides glycopeptides, α-pyrone derivatives, peptides, proteins, antimycin, lipopeptides, polybrominated biphenyl ether, cyclic depsipetide, terpenes, pentaketides, furan carboxylic acid, alkaloid, diketopiperazine, anthraquinone, chromones, steroid, lactone, quinolone derivative, trisindole derivative, macrolactam, ethers, phenol derivative for antimicrobial activity; and dihydropyridine for neuroprotective activity [85]. In this review it was also evident that anticancer, antimicrobial, anti-HIV, antiflammatory, antimalarial, and neuroprotective disease and antituberculosis were the main classes activity exhibited by most of the new compounds during the recent decade. Terpenoid quinones and hydroquinones, the chemical classes found mainly from Dictyoceratid sponges [11,12,20], have been continuously reviewed and updated. Sesquiterpenoid quinones and hydroquinones showed versatile activities and even a hydro-quinone displays multiple activities [109].

The reason that most sponges produce alkaloids could be for protection from predators such as fish as alkaloids act as good toxicants against predators. Assman *et al.* 2000 presented data suggesting that bromopyrrole alkaloids fulfill multiple ecological functions in the defense mechanisms of the common and diverse genus *Agelas* [110]. A striking example of the significance of marine alkaloids for chemical defense against fish is provided by the red coloured sponge *Latrunculia magnificea* (order Poecilosclerida) from the Red Sea. Even though *L. magnificea* growth is exposed, it is apparently avoided by fish, whereas other sponges from the same habitat that are cryptic are readily consumed by fish when artificially exposed [111].

Further information on the different classes of secondary metabolites of marine sponges and their bioactivities are the subject of previous reviews [14,93,112,113].

2.7. The Distribution of Bioactive Compounds

Table 2 shows that during the decade until 2010, 332, 229, 227, 173 and 149 new bioactive compounds were found from Dictyoceratida, Haplosclerida, Poecilosclerida, Halichondrida and Astrophorida, respectively, the highest number of new bioactive compounds isolated from different orders of sponges in this period. Dictyoceratida alone contributed 20.6% of all bioactives, Haplosclerida and Poecilosclerida both contributed 14%, whereas Halichondrida and Astrophorida contributed 10.8% and 9.3% respectively. The detail of yearly distribution is presented in Figure 6. Therefore, these five orders contributed 69% of bioactive compounds among all the different orders of sponges.

In sponges the role of the chemical constituents is clouded by the complexity of the sponge-symbiont relationship [114]. The current body of evidence is too limited to make broad generalizations, but it suggests complex chemical and biological interactions that have not yet been resolved [115]. This knowledge will feed into strategies to relieve a major bottleneck for sponge metabolite production, namely: understanding metabolites production in the sponge [79].

Table 2. Total number of new compounds isolated from different orders of marine sponges with different bioactivities from 2001 to 2010.

Orders of Sponges	Anti-Alzheimer's	Antibacterial	Antituberculosis	Anticancer/ Cytotoxicity	Antifungal	Anti-inflammatory	Antimalarial	Anti-HIV	Antiviral	Miscellaneous	Total
Agelasida	0	17	0	11	6	0	6	0	0	14	54
Astrophorida	0	8	6	97	7	0	1	5	3	22	149
Axinellida	0	0	0	0	0	0	0	0	0	8	8
Chondrosida	0	0	0	3	3	0	0	0	0	0	6
Choristida	0	0	0	12	0	0	0	0	0	12	24
Clathrinida	0	4	3	3	2	0	0	0	0	0	12
Dendroceratida	0	4	0	14	3	3	0	0	0	14	38
Dictyoceratida	0	38	3	182	11	2	1	5	0	90	332
Hadromerida	0	2	3	45	1	0	0	5	0	18	74
Halichondrida	1	18	4	99	16	1	2	1	0	31	173
Haplosclerida	8	15	2	100	20	0	7	4	0	73	229
Homosclerophorida	0	2	3	55	10	0	9	1	0	34	114
Leucosolenida	0	1	0	2	0	0	0	0	0	0	3
Lithistida	0	5	2	38	9	2	0	16	0	16	88
Lyssacinosida	0	0	0	1	0	0	0	0	0	0	1
Ocilosclerida	0	0	0	0	0	0	0	0	0	0	0
Poecilosclerida	0	17	5	143	21	1	1	4	1	34	227
Spirophorida	0	0	0	4	0	0	0	0	0	0	4
Unknown	0	0	0	2	0	0	0	0	0	7	9
Verongida	0	14	0	17	5	0	0	0	0	34	70
Total	9	145	31	828	114	9	27	41	4	407	1615

The origin and role of a number of compounds such as bioactive peptides within the sponges have yet to be clarified, as many of these compounds have potent activities not always clearly related to their *in situ* role [116]. However, it was subsequently found that antimicrobial peptides (AMPs) are components of innate immunity, forming the first-line of defense used by any organisms against the invading pathogens [117]. Pasupuleti *et al.* 2012 termed AMPs as the key component of the innate immune system [118]. Two good examples of AMPs produced by Porifera are Stylisin and Discodermin A [119,120], although many AMPs have been produced since the first production of Discodermin from sponges [121]. In 2013 a review paper was published on antimicrobial peptides with versatile biological activities which included a few sponges producing AMPs [122].

There is a sign that the quick evolution of molecules related to cell adhesion and pathogen killing (AMP precursors) has been acute in the successful alteration of sponges [123]. In their 2010 review Otero-González *et al.* discussed the new frontier for microbial infection control by antimicrobial peptides including Porifera [121]. A review published by Brogden *et al.* 2005 commented that translocated peptides can modify cytoplasmic membrane septum creation, hinder cell-wall synthesis, constrain nucleic-acid synthesis, inhibit protein synthesis, or obstruct enzymatic activity [124].

Figure 6. Distribution of bioactive compounds isolated from various marine sponge orders 2001–2010.

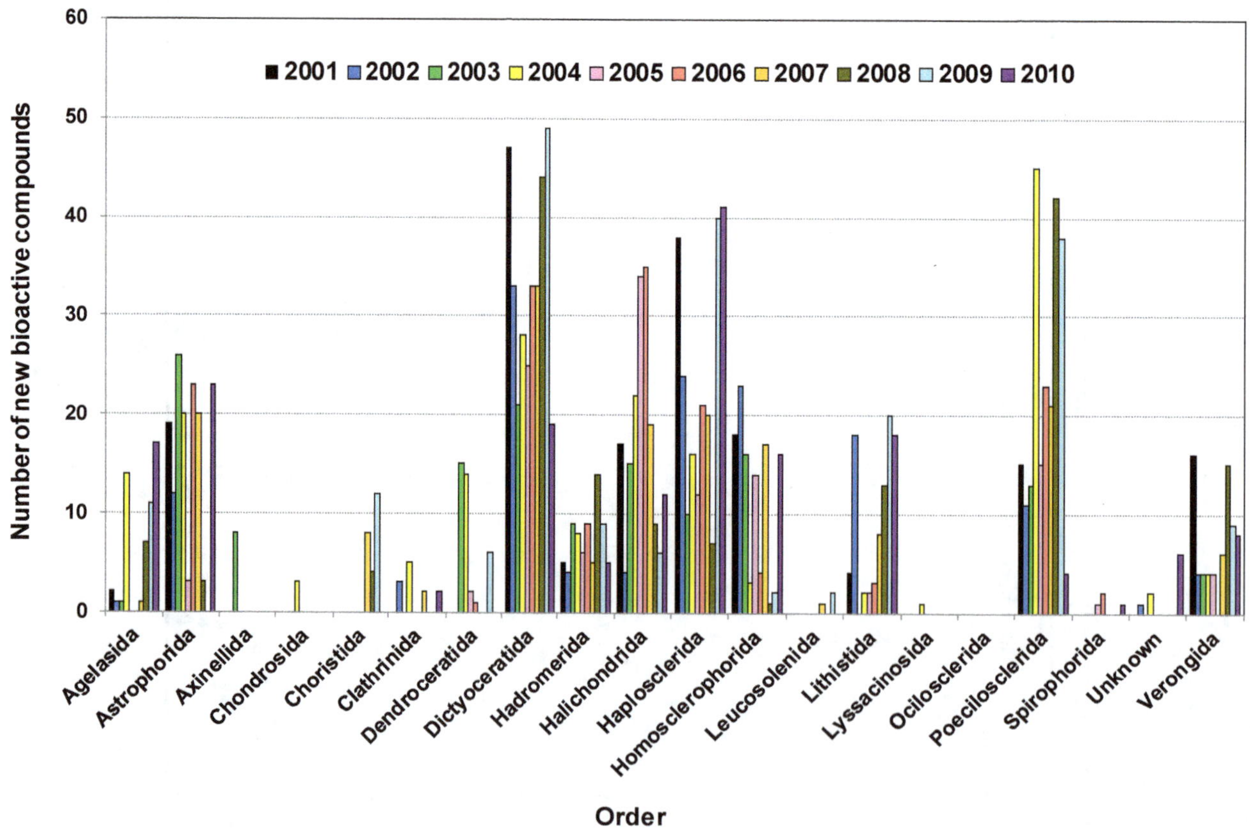

Some external activities of these peptides are as antitumorals, antivirals, immunosuppressive and antimicrobial agents, as well as neurotoxins, hepatotoxin, and cardiac stimulants. The various functional roles of some terpenoids are considered as hormones (gibberellins), photosynthetic pigments (phytol, carotenoids), electron carriers (ubiquinone, plastoquinone), and mediators of polysaccharide assembly, as well as communication and defense mechanisms [125]. Terpenes may also act as safeguard for a variety of organisms in the marine world, including algae, sponges, corals, mollusks and fish [33].

It is important to note that most of the orders showed cytotoxicity or anticancer activity although only a few have been tested for *in vivo* antitumor activity. In the five year period from 1986 to 1991, more than 400 novel marine natural products with cytotoxic activity were reported in the literature [126] with the majority of compounds only tested for cytotoxicity in cell culture assays. Although cytotoxic activity is regarded as the first indicator in identifying anticancer drugs [127], we have considered these compounds to be cytotoxic unless further experimentation indicates their potential role as an anticancer drug. Increasing evidence has shown that cell death can be induced via three different mechanisms: apoptosis, autophagy and oncosis [128]. Most of these sponge-derived novel compounds have been screened for cytotoxic activity but not for apoptosis [129–131], although analysis should focus on a number of cancer relevant targets associated with the cell cycle, signal transduction, angiogenesis or apoptosis [132–135]. A recent overview (2011) retrieved scientific papers identifying 39 compounds from marine sponges with apoptosis-inducing anticancer properties [136]. In another example that distinguishes cytotoxicity from antitumor activity, a recent study used a novel *in vitro* assay to screen 2036 extracts from 683 individual sponges that led to the identification of bioactive compounds (which were prepared in pure form and in sufficient quantities) that could treat solid tumors [137].

Figure 7a,b show that the number of compounds with reported inhibition of cancer cell lines (or cytotoxicity) was highest with the number of 825, antibacterial activity at 145, and antifungal and anti-HIV activities at 111 and 41, respectively. Thus, cytotoxicity or anticancer activity contributed at least half of the reported activity. Progress towards marine anticancer drugs dominates, with the prime source phylum being sponges, followed by microorganisms, tunicates and mollusks [22]. Antiviral and anti-HIV activities have been observed from samples of Astrophorida, Lithistida and Poecilosclerida. The most promising antiviral substances from sponges appear to be 4-methylaaptamine, manzamines [19], besides Papuamides C and D [138], haplosamates A and B [139] and avarol [140] which are examples of HIV-inhibiting compounds from different sponges.

In a recent overview of 132 natural products from marine sources obtained during the period 2002–2011, which exhibited anti-HIV activity, it was reported that sponges contributed more than half of all anti-HIV natural products from marine organisms. These were mainly alkaloids and cyclic depsipeptides [141], such as Cortistatin A (CA), a recently discovered natural steroidal alkaloid isolated from the marine sponge *Corticium simplex* [142]. It has been reported to display anti-proliferative properties towards human umbilical vein endothelial cells (HUVECs) with an average half-maximal inhibitory concentration (IC_{50}) of 0.35 μM [142,143]. A recent study showed that Cortistatin A potently suppresses Tat-dependent HIV transcription [144]. Further reading on the structural characteristics of sponge derived cyclodepsipeptides can be found in a recent review [145]. It is important to note that Halichondrida and Haplosclerida are the only orders from which anti-Alzheimer's activity was reported during the last decade.

There have been many interesting compounds with unusual structures with potential activity which have been observed during the last decade. Some of the compounds are illustrated in Table 3.

Figure 7. (**a**) Total number of new compounds isolated from different marine sponges with various bioactivities from 2001 to 2010; (**b**) The distribution of new compounds isolated from different orders of marine sponges with various bioactivities from 2001 to 2010.

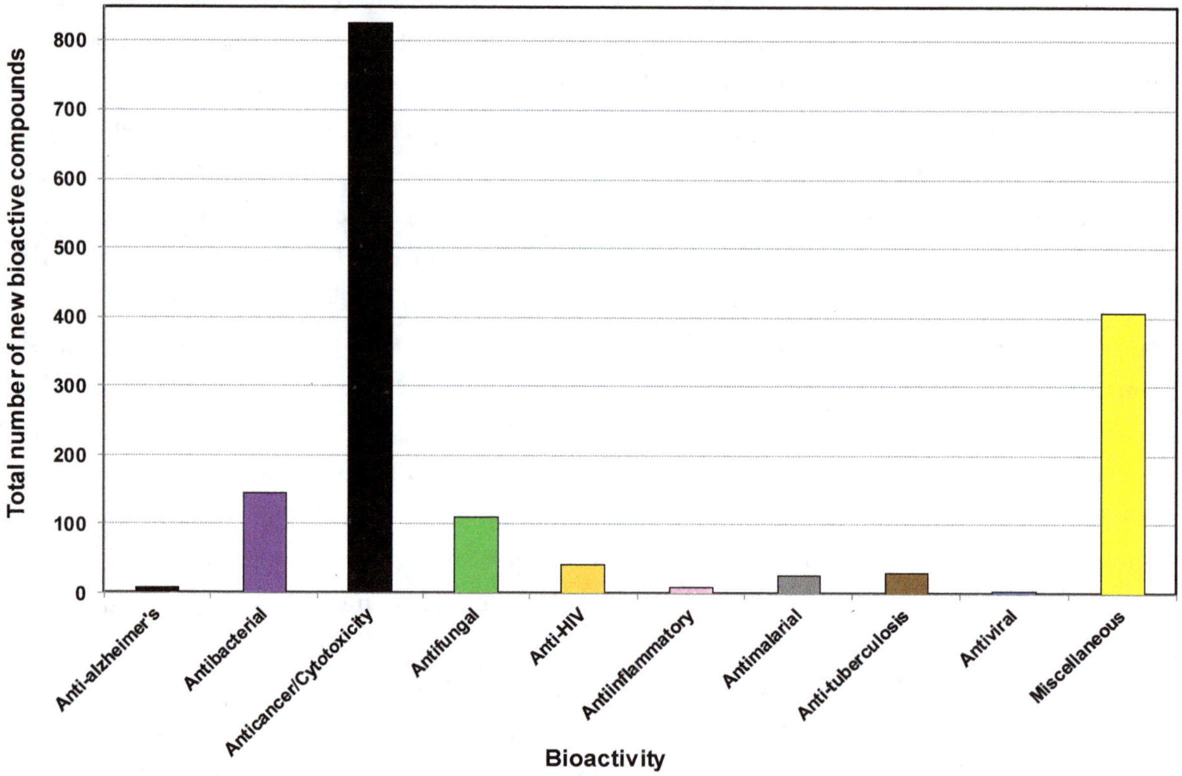

(a)

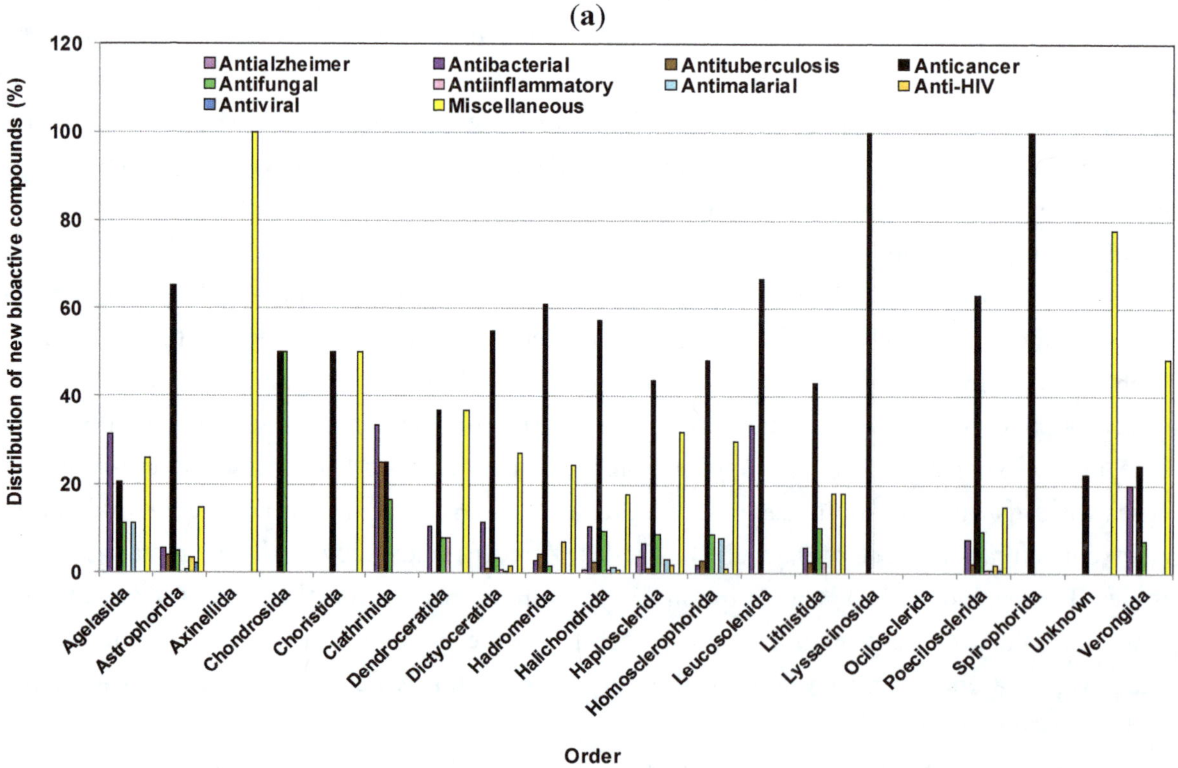

(b)

Table 3. Selected compounds with unusual structure and significant activity from sponges.

Organism	Order	Compound Name	Chemical Class	Special Feature/Activity	Source, Country, Year,/Depth	Reference
Sarcotragus sp.	Dictyoceratida	Sarcotragin A, & B	Trisnorsesterterpenoid lactam	Showed moderate cytotoxicity (LC_{50} 207 μg/mL) toward the leukemia cell-line K562	Seoguipo, Jaeju Island, Korea, 2001	[146]
Polymastia tenax	Hadromerida	5α,6α-epoxy-24*R**-ethylcholest-8(14)-en-3β,7α-diol and 5α,6α-epoxy-24*R**-ethylcholest-8-en-3β, 7α-diol	Sterol	Exhibited significant cytotoxic activity *vs.* human lung carcinoma (A-549), human colon carcinomas (HT-29 and H-116), and human prostate carcinoma (PC-3) cell lines with the LC_{50} (μg/mL) value of 5–10, 1–5, 1–5, 0.5–1 and 1–5	Punta de Betín, Bahía de Santa Marta, in the Colombian Caribbean, Colombia, 2002	[147]
Crella spinulata	Poecilosclerida	Benzylthiocrellidone	Bis-dimedone thioether	First report of a natural product containing a dimedone moiety. No activity reported	Davies and Bowden Reefs Australia, 2002	[148]
Ectyoplasia ferox	Poecilosclerida	Ectyoceramide	Galactofuranosylceramide (GSL)	The first example of a monohexofuranosylceramide and the first natural GSL with its first sugar in the furanose form. No activity reported	Island of Rum Cay, Bahamas, 2000	[149]
Cribrochalina olemda	Haplosclerida	Kapakahine E	Peptide (cyclic)	Kapakahine E showed moderate cytotoxicity against P388 murine leukemia cells at IC_{50} of 5.0 μg/mL	Pohnpei, Micronesia, 2003	[150]
Haliclona Viscosa	Haplosclerida	Viscosamine	Trimeric 3-alkyl pyridinium alkaloid	First trimeric 3-alkyl pyridinium compound from a marine environment. No activity reported	Coast of Blomstrandhalvøya, near Hansneset, Kongsfjorden, Arctic Ocean, 2003	[151]
Phakellia fusca	Axinellida	Compound **1**, **2**, **3**	5-Fluorouracil alkaloid	First report of fluorine containing natural products from a marine source. No activity reported	Yongxiong Island of the Xisha Islands, South China Sea, China 2003	[152]

Table 3. *Cont.*

Agelas clathrodes	Agelasida	Clarhamnoside	Rhamnosylated *R*-Galactosylceramide	The first Rhamnosylated *R*-Galactosylceramide, a glycolipid containing an unusual L-rhamnose unit. No activity reported	Grand Bahamas Island (Sweetings Cay), Bahamas, 2004	[153]
Psammocinia sp.	Dictyoceratida	Psymberin	Cytotoxin (distantly related to the Pederin family)	Several melanoma, breast, and colon cancer cell lines demonstrated high sensitivity ($LC_{50} < 2.5 \times 10^{-9}$ M) to psymberin, and all six leukemia cell lines proved comparably insensitive	Papua New Guinea, 2004	[154]
Callyspongia abnormis	Haploscerida	Callynormine A	Cyclic Peptide	Represents a new class of heterodetic cyclic peptides (designated endiamino peptides). This compound possessing an α-amido-β-aminoacrylamide cyclization functionality	Shimoni reef, Kenya, 2004	[155]
Axinella infundibula	Halichondrida	Axinelloside A	Lipopolysaccharide (Sulfated)	Axinelloside A, a complex polysulfated glycolipid, which strongly inhibited the activity of human telomerase with an IC_{50} value of 0.4 μM	Shikine-jima Island, the Izu Islands, Japan, 2005	[156]
Theonella swinhoei	Lithistida	Plytheonamide A, B	Polypeptide	Showed cytotoxicity against P388 murine leukemia cells with IC_{50} values of 78 and 68 pg/mL, respectively. Linear polypeptides with unprecedented structural features	Hachijo-jima Island, Japan, 2005	[157]
Neopetrosia sp.	Haplosclerida	Neopetrosiamide A, B	Peptide (diastereomeric tricyclic)	Active in inhibiting the amoeboid invasion by human tumor cells	Near Milne Bay, Papua New Guinea, 2005	[158]
Prianos osiros	Haplosclerida	(3*R*,3′*R*,5*S*)-3,3′,5,19′-tetrahydroxy-7′,8′-didehydro-γ,ε-carotene-8-one	Acetylenic carotenoid	Contains an unusual cytotoxic carotenoid	Pohnpei, Micronesia, 2005	[159]

Table 3. *Cont.*

Ircinia sp.	Dictyoceratida	Irciniasulfonic acid B	Fatty acid derivative (taurine conjugated)	Reversed the multi-drug resistance to vincristine in KB/VJ300 cells at the concentration of 100 µM	Tsuzumi Island, Fukuoka Prefecture, Japan, 2006	[160]
Suberites japonicus	Hadromerida	Seragamide A–F	Depsipeptide (actin targeting)	Caused multinuclei formation in cells at 0.01–0.02 µg/mL	Seragaki, Okinawa, Japan, 2006	[161]
Theonella swinhoei	Lithistida	Hurghadolide A	Macrolide	Caused disruption of the actin cytoskeleton at concentrations of 7.3 nM. Active against *Candida albicans* (MIC 31.3 µg/mL)	Red Sea, Egypt, 2006	[89]
Theonella swinhoei	Lithistida	Swinholide I	Macrolide	as above	Red Sea, Egypt, 2006	[89]
Coelocarteria cfr. *singaporensis*	Poecilosclerida	Coelodiol and Coelic acid	Diterpene (ent-isocopalane)	Inhibit the *in vitro* growth of MKN-45 cell line (human gastric adenocarcinoma) at 20 and 40 µg/mL respectively	Bunaken, Marine Park (North Sulawesi), Indonesia, 2006	[162]
Lendenfeldia sp.	Dictyoceratida	(*S*)-2,2′-Dimethoxy-1, 1′-binaphthyl-5,5′,6,6′-tetraol	Naphthalene dimer	Significantly inhibited both hypoxia-induced (IC_{50} values 4.3 µM) and iron chelator (1, 10-phenanthroline)-induced HIF-1 activation in T47D breast tumor cells. This compound inhibited HIF-1 activation at concentrations that were significantly lower than those that suppressed tumor cell viability	Collected at 2 m depth on May 22, 1993 (sample C011337), from a sea grass bed, Indonesia, 2007	[163]
Erylus formosus	Astrophorida	Eryloside F1–F4	Triterpene glycoside	At a concentration of 100 µg/mL were found to activate Ca2 influx into mouse spleenocytes. biosides having aglycons related to penasterol with additional oxidation patterns in their side chains	Puerto Morelos (the Caribbean Sea), Mexico 2007	[164]

Table 3. *Cont.*

Erylus formosus	Astrophorida	Eryloside M–Q	Triterpene glycoside	As above, contain new variants of carbohydrate chains with three, four and six sugar units. Contain 14-carboxy-24-methylenelanost-8(9)-en-3β-ol	Puerto Morelos (the Caribbean Sea), Mexico, 2007	[164]
Cacospongia mycofijiensis	Dictyoceratida	CTP-431	Thiopyrone	Showed only mild cytotoxicity (IC_{50}: 18 μM) against human colon carcinoma HCT-116. This compound has no previous precedent in natural products chemistry. Its structure including absolute configuration as 8*R*,9*R*,10*S*,13*S*	Beqa Lagoon, Fiji, 2008	[165]
Homophymia sp.	Lithistida	Homophymine A	Cyclodepsipeptide	Exhibited cytoprotective activity against HIV-1 infection with a IC_{50} of 75 nM	Coast of New Caledonia, 2008	[166]
Ianthella sp.	Verongida	Petrosterol-3,6-dione and 5α,6α-epoxy-petrosterol	C29 sterol	Showed growth-inhibitory effects with IC_{50} values of 8.4, 19.9, 17.8, 16.2 and 22.1 μM against lung (A549), colon (HT-29), breast (MCF-7), ovary (SK-OV-3), and two types of leukemia (HL-60 and U937) human cancer cell lines	Namyet Island, Khanh Hoa province, Vietnam, 2009	[167]
Topsentia sp.	Halichondrida	Geodisterol-3-*O*-sulfite and 29-demethylgeodisterol-3-*O*-sulfite	Sterol (sulphated)	Reverses efflux pump mediated fluconazole resistance. Also enhances fluconazole activity in a *Saccharomyces cerevisiae* strain overexpressing the *Candida albicans* efflux pump MDR1, as well as in a fluconazole-resistant *Candida albicans* clinical isolate known to overexpress MDR1	Chuuk, Micronesia, 2009	[168]

Table 3. *Cont.*

Spongia (Heterofibria) sp.	Dictyoceratida	Heterofibrin A1–A3 and B1–B3	Fatty acid	Possess a diyne-ene moiety, while the monolactyl and dilactyl moiety featured in selected heterofibrins is unprecedented in the natural products literature. Inhibited lipid droplet formation in A431 fibroblast cells (up to 60% at 10 μM)	Great Australian Bight, Australia, 2010	[169]
Xestospongia sp.	Haplosclerida	Xestosaprol F–M	Xestosaprol (pentacyclic compound)	Showed moderate inhibition of the aspartic protease BACE1 (memapsin-2), which has a central role in the etiology of Alzheimer's disease with the IC_{50} value of 135 ± 11 μM. First examples of a monooxygenated A-ring	Coral reef at Sangalaki, Indonesia, 2010	[170]
Theonella swinhoei	Lithistida	Paltolides A–C	Peptides (Anabaenopeptin like)	Closely related to a group of anabaenopeptins that are submicromolar inhibitors of carboxypeptidase U with greater than 50 fold selectivity over other carboxypeptidases	Uchelbeluu Reef, Palau, 2010	[171]
Neopetrosia proxima	Haplosclerida	Neopetrosiamine A	Alkaloid (tetracyclic bis-piperidine)	Exhibited strong inhibitory activity against MALME-3M melanoma cancer, CCRF-CEM leukemia, and MCF7 breast cancer with IC_{50} values of 1.5, 2.0, and 3.5 μM, respectively. *In vitro* activity *vs.* pathogenic strain of *Mycobacterium tuberculosis* (H37Rv) and *Plasmodium falciparum*	Mona Island, Puerto Rico, 2010	[172]
Iotrochota baculifera	Poecilosclerida	Baculiferins A–O	*O*-sulfated pyrrole alkaloids	Baculiferins C, E–H, and K–N (4, 6–9, 12–15) are potent inhibitors of HIV-1 IIIB virus in both MT4 and MAGI cells. Additionally could bind to the HIV-1 target proteins Vif, APOBEC3G, and gp41	Inner coral reef, Hainan Island, China, 2010	[173]

2.8. Distribution of New Compounds Based on Country/Geographical Area

One important aspect of sponges is their geographical location. A high percentage of bioactive sponge species were reported from different geographical regions [174–176].

Figure 8 shows that the new compounds were mainly isolated from sponges collected from 61 countries. Of these, sponges from Japan had the highest number of compounds (332) followed by Indonesia (235), Korea (211). Sponge samples from Australia and China contributed 187 and 146 compounds, respectively, during the last decade. The Bahamas, Mexico, Palau, Papua New Guinea, Philippines and Vanuatu are the other countries which were the source of more than 50 new compounds found in the last ten years. The obvious question is why do some regions show a high diversity of compounds as well as an abundance of sponges? Studies showed that the highest concentration of toxic or antioxidant sponge metabolites are found in habitats such as coral reefs that are characterized by intense competition and feeding pressure from carnivorous fish [177]. The adaptive significance of the chemical defenses of sponges are that they are highly effective against most species of fish and a group of shell-less gastropods, the nudibranchs that feed on sponges and sequester their chemical armory [177]. An excellent paper on the global diversity of sponges includes global sponge diversity information which was collected from different regional projects and resources and also reviewed was information on invasive sponges that might well have some influence on distribution patterns in the future [83].

Figure 8. Total number of new compounds isolated from different marine sponges and their source locations from 2001 to 2010.

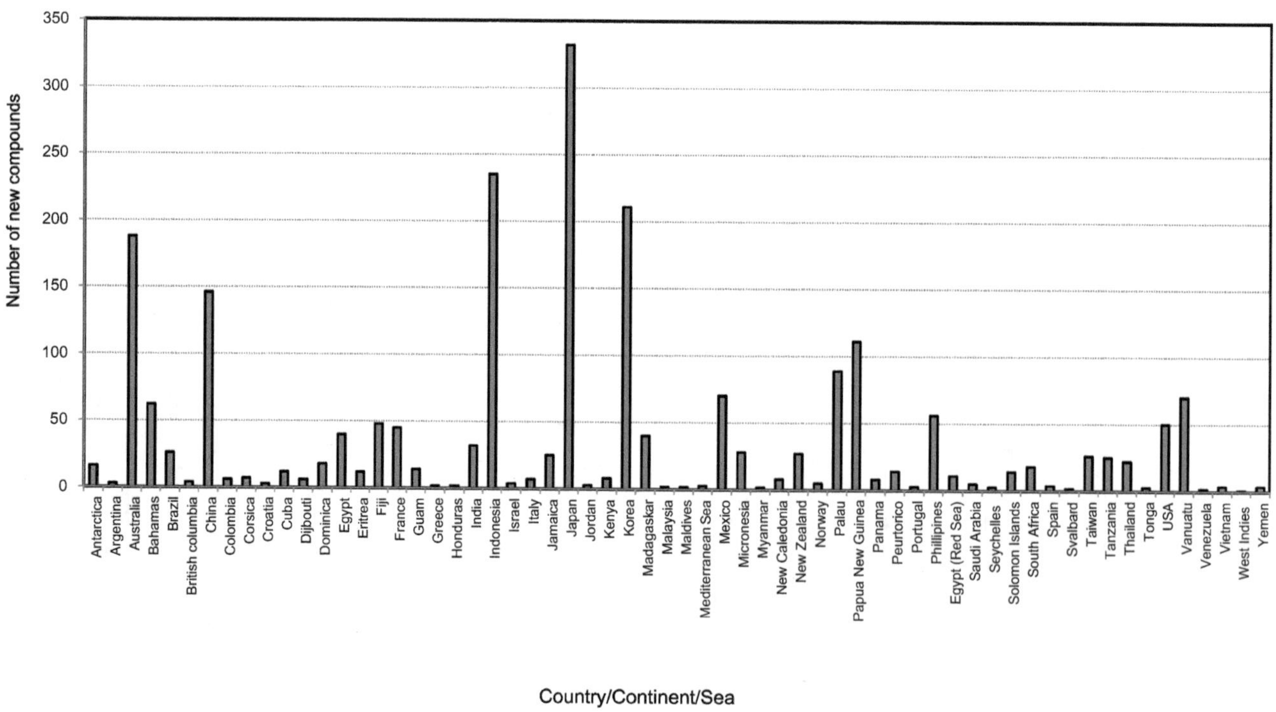

Figure 9 shows the total number of new compounds isolated from the top ten countries. It has been found that each year from 2001 to 2010 at least 20 new compounds were isolated from Japanese sponge samples, with the highest output occurring in 2004 and 2005 with 46 new compounds (Figure S2, Supplementary Information). From 2006 to 2009 at least 20 compounds were found each year in Indonesia. Marine invertebrates, which are plentiful in the Indo-Pacific regions including Indonesia, are rich in secondary metabolites and are becoming targets for the continuing search for bioactive compounds [178].

Figure 9. Total number of new compounds isolated from marine sponges from the top 10 source countries from 2001 to 2010.

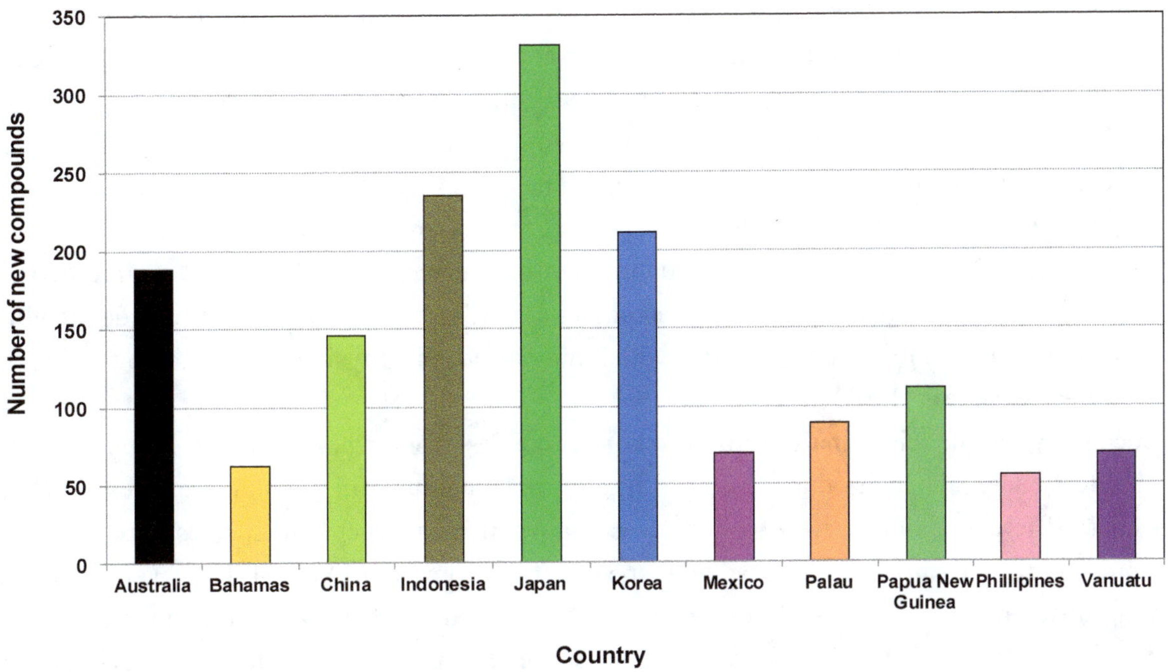

Although in 2001 the highest contributors of new compounds derived from sponges were from Japan, Korea and Australia at 22.9%, 18.9% and 14.3%, respectively, the scenario changed during 2010 where the highest contributions of new compounds came from China, Australia and Indonesia at 23.9%, 21.3%, and 21%, respectively [179]. China started its marine high-tech program since 1996, and after 15 years it emerged as one of the highest contributors to marine sponge natural products discovery in 2010. This achievement demonstrates how significant the government science and technology funding policy can impact the development of marine biotechnology. Australia is one of only 17 recognised megabiodiverse countries primarily based on its highly biodiverse and endemic terrestrial flora and fauna [180,181]. While the full extent of Australian marine biodiversity remains relatively unexplored [182], several marine biodiversity hotspots including centers of endemicity have been recognized, especially in coral reefs [183,184], the temperate coastline [185] and the Great Australian Bight off the coast of South Australia [186]. There have been reports on the high species diversity of sponges in the north west [187–190], in the deep sea off the south west [191,192] of the Bight, and the Great Barrier Reef [189]. Therefore, there is urgent need to explore those potential locations in order to obtain new compounds and drugs for the future.

The Australian Institute of Marine Science (AIMS) conducted a study based on the Australian marine habitat and identified biogeographic bioactivity hotspots that correlated with biodiversity hotspots. AIMS found that high-level phylogeny, and therefore the metabolic machinery available to an organism, is a major basis of bioactivity, while habitat diversity and ecological circumstance are possible drivers in the stimulation of this machinery and bioactive secondary metabolism [193]. Therefore, in near future, knowledge of metabolomics coupled with genomics tool and bioinformatics could be a high level device for the exploration of target specific bioactive compounds from sponges.

In addition, microbes associated with marine sponges could vary with the geographical area, so, if these associated bacteria are responsible for producing compounds, then it is possible that the same sponge species in different geographical locations could produce different secondary metabolites [194,195]. However, a recent study showed the stability of bacterial communities in two temperate sponges exposed to environmental variation, which is consistent with previous research on other temperate sponges. This study used next generation sequencing and revealed how different components of bacterial communities associated with *Ecionemia alata* and *Tethya bergquistae* responded to environmental variation *in situ* [196]. The similarity observed in bacterial communities among specimens occupying different habitats suggests that environmental variation occurring in those habitats does not affect the stability of the community, and hence, most likely does not radically alter the metabolism of these sponges. The study recommends further study to improve the understanding of the role of microbial symbiont communities which may affect the physiology and ecology of sponges on temperate rocky reefs [196].

A recent pyrosequencing analysis of 32 sponge species from eight locations around the world identified few bacterial species that are common to more than a handful of sponge species [177]. They stated that different sponges were found to contain different bacterial species (species-specific community) but share only very few bacterial species (core community) [197]. However, the bacterial species in different sponges are still more closely related to each other than, for example, to seawater bacteria (indicated by Plus-OTUs and sponge-specific clusters), consistent with previous studies suggesting at least partially overlapping communities among different sponges. Sponges therefore contain a uniform, sponge-specific bacterial community although each sponge species contains different bacterial species [197]. Perhaps, this is one of the main reasons for getting new compounds in species level regardless of the genera and order. Although we commented on this, based on our study of sponges of the last decade, further study to prove this hypothesis would be worthwhile.

3. Conclusions

In the decade 2001–2010 marine sponges continued to be the most promising source of marine natural products. Marine microorganisms and phytoplankton grew rapidly to become the third largest source by increasing their contribution from 9% to 39% during the decade. Sponges are a reservoir of marine microorganisms with up to 40%–60% as microbial biomass. In the coming decade sponge associated microorganisms promise to be an outstanding source of the NMNP.

From the current trend of discovery it could be predicted that China, Australia and Indonesia will be the source of more new compounds in the future competing with Japan and Korea. However, although Indonesia is an excellent source of sponges, most of their studies were conducted by scientists in other countries. Certainly, exploration of new compounds from marine sponges is dependent on government funding and policy, industrial interest and investment, research facilities, the expertise of scientists, infrastructure and laboratory facilities, equipment, machinery and institutional support.

Astrophorida, Dictyoceratida, Halichondrida, Haplosclerida and Poecilosclerida were the key orders of sponges studied during the first decade of the 21st century. The examination of the contribution from an individual species revealed that regardless of the order each species contributed on average 3–5 compounds. The high number of new compounds was the result of the high diversity of species from these particular orders. Alkaloids (20%), terpenes (14.7%) or terpenoids (8.3%) and peptides (6.8%) represented the three main chemical classes of compounds discovered from sponges in this period, and together with the other chemical classes showed a range of biological activities. Of all the biological activities investigated cytotoxicity or anticancer activity against different cancer cell lines was most frequently reported at 53.6%. Antibacterial and antifungal activity were two other areas where new compounds showed potential activity at 9.4% and 7.2%, respectively. Anti-Alzheimer's, antibacterial, antituberculosis, anticancer, antifungal, anti-inflammatory, antimalarial, antiviral, anti-HIV were the other activities exhibited by the new compounds from sponges. Because sponge extracts showed potent cytotoxic activity, which is often reported as anticancer activity, it is very important to study the mode of action of these extracts by isolating pure compounds rather than only testing cytotoxicity.

The order Dictyoceratida was found to be the most prolific producer of new compounds among all the sponge orders studied. *Dysidea* sp. and *Ircinia* sp. were found to be the most promising genera because of their capacity for producing new bioactive compounds.

In any event, the discovery of marine natural products from sponges relies particularly on finding new genera and species from the most prolific to the least abundant orders, which is still achievable. In order to overcome the problem of production of sponge-derived compounds, synthesis of bioactive, low molecular weight compounds by cloning biosynthetic gene clusters using recombinant techniques could be applied. Most importantly, sponge derived compounds should be utilized with a combination of innovative technologies which could develop new fields of application that will impact significantly on biotechnology.

Acknowledgments

Financial support from Centre for Marine Bioproducts Development, Flinders University; Flinders University for their IPRS scholarship to Mohammad Mehbub.

Author Contribution

Mohammad Ferdous Mehbub did the coordination for collecting all the information to prepare a comprehensive database for writing this manuscript, proof read the database, wrote a major part of the manuscript and generates the graphs and tables, proof reading the manuscript as well and made necessary correction.

Jie Lei helped to collect the information to develop the database for preparing this manuscript.

Christopher Franco wrote a significant part of the manuscript, decided on the format and figures to be included and did the final proofreading and corrections.

Wei Zhang has the original idea, structure and plan for this manuscript, and planned and refined the graphs and tables with Mohammad Ferdous Mehbub, and wrote the abstract, and wrote a significant portion for the introduction.

References

1. Bergmann, W.; Feeney, R.J. Contributions to the study of marine products. XXXII. The nucleosides of sponges I. *J. Org. Chem.* **1951**, *16*, 981–987.
2. Bergmann, W.; Feeney, R.J. The isolation of a new thymine pentoside from sponges 1. *J. Am. Chem. Soc.* **1950**, *72*, 2809–2810.
3. Carte, B.K. Biomedical potential of marine natural products. *Bioscience* **1996**, *46*, 271–287.
4. Burkholder, P.R.; Pfister, R.M.; Leitz, F.H. Production of a pyrrole antibiotic by a marine bacterium. *Appl. Microbiol.* **1966**, *14*, 649–653.
5. Proksch, P.; Edrada, R.; Ebel, R. Drugs from the seas—Current status and microbiological implications. *Appl. Microbiol. Biotechnol.* **2002**, *59*, 125–134.
6. Weinheimer, A.J.; Spraggins, R.L. The occurrence of two new prostaglandin derivatives (15-*epi*-PGA and its acetate, methyl ester) in the Gorgonian *Plexaura homomalla* chemistry of Coelenterates. XV. *Tetrahedron Lett.* **1969**, *10*, 5185–5188.
7. Ireland, C.M.; Copp, B.R.; Foster, M.P.; McDonald, L.A.; Radisky, D.C.; Swersey, J.C. Biomedical potential of marine natural products. In *Pharmaceutical and Bioactive Natural Products*; Springer: Berlin/Heidelberg, Germany, 1993; pp. 1–43.
8. Gordon, E.M.; Barrett, R.W.; Dower, W.J.; Fodor, S.P.; Gallop, M.A. Applications of combinatorial technologies to drug discovery. 2. Combinatorial organic synthesis, library screening strategies, and future directions. *J. Med. Chem.* **1994**, *37*, 1385–1401.
9. Alonso, D.; Khalil, Z.; Satkunanthan, N.; Livett, B. Drugs from the sea: Conotoxins as drug leads for neuropathic pain and other neurological conditions. *Mini Rev. Med. Chem.* **2003**, *3*, 785–787.
10. Hu, G.P.; Yuan, J.; Sun, L.; She, Z.G.; Wu, J.H.; Lan, X.J.; Zhu, X.; Lin, Y.C.; Chen, S.P. Statistical research on marine natural products based on data obtained between 1985 and 2008. *Mar. Drugs* **2011**, *9*, 514–525.
11. Faulkner, D.J. Marine natural products. *Nat. Prod. Rep.* **2000**, *17*, 7–55.
12. Faulkner, D.J. Marine natural products. *Nat. Prod. Rep.* **2002**, *19*, 1–48.
13. Blunt, J.W.; Copp, B.R.; Munro, M.H.G.; Northcote, P.T.; Prinsep, M.R. Marine natural products. *Nat. Prod. Rep.* **2006**, *23*, 26–78.
14. Blunt, J.W.; Copp, B.R.; Munro, M.H.G.; Northcote, P.T.; Prinsep, M.R. Marine natural products. *Nat. Prod. Rep.* **2010**, *27*, 165–237.
15. Blunt, J.W.; Copp, B.R.; Munro, M.H.G.; Northcote, P.T.; Prinsep, M.R. Marine natural products. *Nat. Prod. Rep.* **2011**, *28*, 196–268.
16. Blunt, J.W.; Copp, B.R.; Keyzers, R.A.; Munro, M.H.G.; Prinsep, M.R. Marine natural products. *Nat. Prod. Rep.* **2012**, *29*, 144–222.

17. Blunt, J.W.; Copp, B.R.; Keyzers, R.A.; Munro, M.H.G.; Prinsep, M.R. Marine natural products. *Nat. Prod. Rep.* **2013**, *30*, 237–323.
18. Blunt, J.W.; Copp, B.R.; Keyzers, R.A.; Munro, M.H.G.; Prinsep, M.R. Marine natural products. *Nat. Prod. Rep.* **2014**, *31*, 160–258.
19. Laport, M.; Santos, O.; Muricy, G. Marine sponges: Potential sources of new antimicrobial drugs. *Curr. Pharma. Biotechnol.* **2009**, *10*, 86–105.
20. Blunt, J.W.; Copp, B.R.; Munro, M.H. G.; Northcote, P.T.; Prinsep, M.R. Marine natural products. *Nat. Prod. Rep.* **2005**, *22*, 15–61.
21. Sagar, S.; Kaur, M.; Minneman, K.P. Antiviral lead compounds from marine sponges. *Mar. Drugs* **2010**, *8*, 2619–2638.
22. Paul, V.J.; Ritson-Williams, R.; Sharp, K. Marine chemical ecology in benthic environments. *Nat. Prod. Rep.* **2011**, *28*, 345–387.
23. Molinski, T.F.; Dalisay, D.S.; Lievens, S.L.; Saludes, J.P. Drug development from marine natural products. *Nat. Rev. Drug Discov.* **2009**, *8*, 69–85.
24. Wijffels, R.H. Potential of sponges and microalgae for marine biotechnology. *Trends Biotechnol.* **2008**, *26*, 26–31.
25. Leal, M.C.; Puga, J.; Serôdio, J.; Gomes, N.C.; Calado, R. Trends in the discovery of new marine natural products from invertebrates over the last two decades—Where and what are we bioprospecting? *PLoS One* **2012**, *7*, e30580.
26. Blunt, J.W.; Copp, B.R.; Munro, M.H. G.; Northcote, P.T.; Prinsep, M.R. Marine natural products. *Nat. Prod. Rep.* **2003**, *20*, 1–48.
27. Blunt, J.W.; Copp, B.R.; Munro, M.H. G.; Northcote, P.T.; Prinsep, M.R. Marine natural products. *Nat. Prod. Rep.* **2004**, *21*, 1–49.
28. Blunt, J.W.; Copp, B.R.; Hu, W.-P.; Munro, M.H.G.; Northcote, P.T.; Prinsep, M.R. Marine natural products. *Nat. Prod. Rep.* **2008**, *25*, 35–94.
29. Blunt, J.W.; Copp, B.R.; Hu, W.-P.; Munro, M.H.G.; Northcote, P.T.; Prinsep, M.R. Marine natural products. *Nat. Prod. Rep.* **2009**, *26*, 170–244.
30. Blunt, J.W.; Copp, B.R.; Hu, W.-P.; Munro, M.H. G.; Northcote, P.T.; Prinsep, M.R. Marine natural products. *Nat. Prod. Rep.* **2007**, *24*, 31–86.
31. Hooper, J.N. *Sponguide. Guide to Sponge Collection and Identification*; Queensland Museum: Queensland, Australia, 2000.
32. Paul, V.J.; Puglisi, M.P. Chemical mediation of interactions among marine organisms. *Nat. Prod. Rep.* **2004**, *21*, 189–209.
33. Paul, V.J.; Puglisi, M.P.; Ritson-Williams, R. Marine chemical ecology. *Nat. Prod. Rep.* **2006**, *23*, 153–180.
34. McClintock, J.B.; Baker, B.J. *Marine Chemical Ecology*; CRC Press: Boca Raton, FL, USA, 2010; pp. 1–624.
35. Lee, Y.K.; Lee, J.-H.; Lee, H.K. Microbial symbiosis in marine sponges. *J. Microbiol.* **2001**, *39*, 254–264.
36. Jensen, P.R.; Fenical, W. Strategies for the discovery of secondary metabolites from marine bacteria: Ecological perspectives. *Ann. Rev. Microbiol.* **1994**, *48*, 559–584.

37. Bernan, V.; Greenstein, M.; Maiese, W. Marine microorganisms as a source of new natural products. *Adv. Appl. Microbiol.* **1997**, *43*, 57–90.
38. Haygood, M.G.; Schmidt, E.W.; Davidson, S.K.; Faulkner, D.J. Microbial symbionts of marine invertebrates: Opportunities for microbial biotechnology. *J. Mol. Microbiol. Biotechnol.* **1999**, *1*, 33–43.
39. Osinga, R.; Armstrong, E.; Burgess, J.G.; Hoffmann, F.; Reitner, J.; Schumann-Kindel, G. Sponge—Microbe associations and their importance for sponge bioprocess engineering. *Hydrobiologia* **2001**, *461*, 55–62.
40. Fusetani, N. Biofouling and antifouling. *Nat. Prod. Rep.* **2004**, *21*, 94–104.
41. Li, C.-W.; Chen, J.-Y.; Hua, T.-E. Precambrian sponges with cellular structures. *Science* **1998**, *279*, 879–882.
42. Hentschel, U.; Hopke, J.; Horn, M.; Friedrich, A.B.; Wagner, M.; Hacker, J.; Moore, B.S. Molecular evidence for a uniform microbial community in sponges from different oceans. *Appl. Environ. Microbiol.* **2002**, *68*, 4431–4440.
43. Selvin, J.; Shanmugha Priya, S.; Seghal Kiran, G.; Thangavelu, T.; Sapna Bai, N. Sponge-associated marine bacteria as indicators of heavy metal pollution. *Microbiol. Res.* **2009**, *164*, 352–363.
44. Wilkinson, C.; Fay, P. Nitrogen fixation in coral reef sponges with symbiotic cyanobacteria. *Nature* **1979**, *279*, 527–529.
45. Diaz, M.C.; Rutzler, K. Sponges: An essential component of Caribbean coral reefs. *Bull. Mar. Sci.* **2001**, *69*, 535–546.
46. Thoms, C.; Horn, M.; Wagner, M.; Hentschel, U.; Proksch, P. Monitoring microbial diversity and natural product profiles of the sponge *Aplysina cavernicola* following transplantation. *Mar. Biol.* **2003**, *142*, 685–692.
47. Thoms, C.; Schupp, P. Biotechnological potential of marine sponges and their associated bacteria as producers of new pharmaceuticals (Part II). *J. Int. Biotechnol. Law* **2005**, *2*, 257–264.
48. Proksch, P. Defensive roles for secondary metabolites from marine sponges and sponge-feeding nudibranchs. *Toxicon* **1994**, *32*, 639–655.
49. Pawlik, J.R.; McFall, G.; Zea, S. Does the odor from sponges of the genus *Ircinia* protect them from fish predators? *J. Chem. Ecol.* **2002**, *28*, 1103–1115.
50. Mahon, A.R.; Amsler, C.D.; McClintock, J.B.; Amsler, M.O.; Baker, B.J. Tissue-specific palatability and chemical defenses against macropredators and pathogens in the common articulate brachiopod *Liothyrella uva* from the Antarct. *Penins. J. Exp. Mar. Biol. Ecol.* **2003**, *290*, 197–210.
51. Paul, V.; Cruz-Rivera, E.; Thacker, R. Chemical mediation of macroalgal-herbivore interactions: Ecological and evolutionary perspectives. In *Marine Chemical Ecology*; CRC Press: Boca Raton, FL, USA, 2001; pp. 227–265.
52. Unson, M.D.; Holland, N.D.; Faulkner, D.J. A brominated secondary metabolite synthesized by the cyanobacterial symbiont of a marine sponge and accumulation of the crystalline metabolite in the sponge tissue. *Mar. Biol.* **1994**, *119*, 1–11.
53. Vacelet, J.; Vacelet, E.; Gaino, E.; Gallissian, M. Bacterial attack of spongin skeleton during the 1986–1990 Mediterranean sponge disease. In *Sponges in Time and Space*; Balkema: Rotterdam, The Netherlands, 1994; pp. 355–362.

54. Boehm, M.; Hentschel, U.; Friedrich, A.; Fieseler, L.; Steffen, R.; Gamulin, V.; Mueller, I.; Müller, W. Molecular response of the sponge *Suberites domuncula* to bacterial infection. *Mar. Biol.* **2001**, *139*, 1037–1045.
55. Maldonado, M.; Sánchez-Tocino, L.; Navarro, C. Recurrent disease outbreaks in corneous demosponges of the genus *Ircinia*: Epidemic incidence and defense mechanisms. *Mar. Biol.* **2010**, *157*, 1577–1590.
56. Selvin, J.; Ninawe, A.; Seghal Kiran, G.; Lipton, A. Sponge-microbial interactions: Ecological implications and bioprospecting avenues. *Crit. Rev. Microbiol.* **2010**, *36*, 82–90.
57. Thakur, N.L.; Müller, W.E. Biotechnological potential of marine sponges. *Curr. Sci.* **2004**, *86*, 1506–1512.
58. Tilvi, S.; Rodrigues, C.; Naik, C.; Parameswaran, P.; Wahidhulla, S. New bromotyrosine alkaloids from the marine sponge *Psammaplysilla purpurea*. *Tetrahedron* **2004**, *60*, 10207–10215.
59. Abad, M.; Bedoya, L.; Bermejo, P. Marine compounds and their antimicrobial activities. In *Science Against Microbial Pathogens: Communicating Current Research and Technological Advances*; Formatex Research Center: Badajoz, Spain, 2011.
60. Van Soest, R.; Boury-Esnault, N.; Hooper, J.; Rützler, K.; de Voogd, N.; Alvarez de Glasby, B.; Hajdu, E.; Pisera, A.; Manconi, R.; Schoenberg, C. World Porifera Database. The World Register of Marine Species (WoRMS). Available online: http://www.marinespecies.org/porifera (accessed on 25 October 2012).
61. Appeltans, W.; Decock, W.; Vanhoorne, B.; Hernandez, F.; Bouchet, P.; Boxshall, G.; Fauchald, K.; Gordon, D.P.; Poore, G.C.B.; van Soest, R.; *et al.* The World Register of Marine Species. In Proceedings of the future of the 21st century ocean: Marine Sciences and European Research Infrastructures, Brest, France, 28 June–1 July 2011; p. 30.
62. Ausubel, J.; Trew Christ, D.; Waggoner, P. First Census of Marine Life 2010: Highlights of A Decade of Discovery. Available online: http://www.coml.org/pressreleases/census2010/PDF/Highlights-2010-Report-Low-Res.pdf (accessed on 11 January 2014).
63. Valerie, P.; Raphael, R.-W. Marine chemical ecology. *Nat. Prod. Rep.* **2008**, *25*, 662–695.
64. Skropeta, D. Deep-sea natural products. *Nat. Prod. Rep.* **2008**, *25*, 1131–1166.
65. Faulkner, D.J. Marine natural products. *Nat. Prod. Rep.* **1998**, *15*, 113–158.
66. Crews, P.; Gerwick, W.; Schmitz, F.; France, D.; Bair, K.; Wright, A.; Hallock, Y. Molecular approaches to discover marine natural product anticancer leads-an update from a drug discovery group collaboration. *Pharm. Biol.* **2003**, *41*, 39–52.
67. Sipkema, D.; Franssen, M.C.; Osinga, R.; Tramper, J.; Wijffels, R.H. Marine sponges as pharmacy. *Mar. Biotechnol.* **2005**, *7*, 142–162.
68. Bhanot, A.; Sharma, R.; Noolvi, M.N. Natural sources as potential anti-cancer agents: A review. *Int. J. Phytomed.* **2011**, *3*, 9–26.
69. Newman, D.J.; Cragg, G.M. Marine natural products and related compounds in clinical and advanced preclinical trials. *J. Nat. Prod.* **2004**, *67*, 1216–1238.
70. Thakur, A.N.; Thakur, N.L.; Indap, M.M.; Pandit, R.A.; Datar, V.V.; Müller, W.E. Antiangiogenic, antimicrobial, and cytotoxic potential of sponge-associated bacteria. *Mar. Biotechnol.* **2005**, *7*, 245–252.

71. Cragg, G.M.; Newman, D.J.; Snader, K.M. Natural products in drug discovery and development. *J. Nat. Prod.* **1997**, *60*, 52–60.
72. Petit, K.; Biard, J.-F. Marine natural products and related compounds as anticancer agents: An overview of their clinical status. *Anti-Cancer Agents Med. Chem.* **2013**, *13*, 603–631.
73. Newman, D.J.; Cragg, G.M. Marine-sourced anti-cancer and cancer pain control agents in clinical and late preclinical development. *Mar. Drugs* **2014**, *12*, 255–278.
74. Kinghorn, A.D.; Chin, Y.-W.; Swanson, S.M. Discovery of natural product anticancer agents from biodiverse organisms. *Curr. Opin. Drug Discov. Dev.* **2009**, *12*, 189–196.
75. Duckworth, A. Farming sponges to supply bioactive metabolites and bath sponges: A review. *Mar. Biotechnol.* **2009**, *11*, 669–679.
76. Müller, W.E.G.; Böhm, M.; Batel, R.; de Rosa, S.; Tommonaro, G.; Müller, I.M.; Schröder, H.C. Application of cell culture for the production of bioactive compounds from sponges: Synthesis of Avarol by primmorphs from *Dysidea avara*. *J. Nat. Prod.* **2000**, *63*, 1077–1081.
77. Zhang, X.; Cao, X.; Zhang, W.; Yu, X.; Jin, M. Primmorphs from archaeocytes-dominant cell population of the sponge *Hymeniacidon perleve*: Improved cell proliferation and spiculogenesis. *Biotechnol. Bioeng.* **2003**, *84*, 583–590.
78. Müller, W.E.; Grebenjuk, V.A.; le Pennec, G.; Schröder, H.-C.; Brümmer, F.; Hentschel, U.; Müller, I.M.; Breter, H.-J. Sustainable production of bioactive compounds by sponges—Cell culture and gene cluster approach: A review. *Mar. Biotechnol.* **2004**, *6*, 105–117.
79. Koopmans, M.; Martens, D.; Wijffels, R.H. Towards commercial production of sponge medicines. *Mar. Drugs* **2009**, *7*, 787–802.
80. Wilson, M.C.; Mori, T.; Rückert, C.; Uria, A.R.; Helf, M.J.; Takada, K.; Gernert, C.; Steffens, U.A.; Heycke, N.; Schmitt, S. An environmental bacterial taxon with a large and distinct metabolic repertoire. *Nature* **2014**, *506*, 58–62.
81. Jaspars, M.; Challis, G. Microbiology: A talented genus. *Nature* **2014**, *506*, 38–39.
82. Piel, J. Metabolites from symbiotic bacteria. *Nat. Prod. Rep.* **2004**, *21*, 519–538.
83. Van Soest, R.W.; Boury-Esnault, N.; Vacelet, J.; Dohrmann, M.; Erpenbeck, D.; de Voogd, N.J.; Santodomingo, N.; Vanhoorne, B.; Kelly, M.; Hooper, J.N. Global diversity of sponges (Porifera). *PLoS One* **2012**, *7*, e35105.
84. Kalaitzis, J.A. 2005. Chemical Investigations of Australian Marine Sponges. Available online: https://www120.secure.griffith.edu.au/rch/file/786c6d3d-925a-dea0-3864-2db686e40950/1/02 Chapter1.pdf (accessed on 29 April 2014).
85. Thomas, T.R.A.; Kavlekar, D.P.; LokaBharathi, P.A. Marine drugs from sponge-microbe association—A review. *Mar. Drugs* **2010**, *8*, 1417–1468.
86. Matsunaga, S.; Fusetani, N. Nonribosomal peptides from marine sponges. *Curr. Org. Chem.* **2003**, *7*, 945–966.
87. Matsunaga, S.; Fusetani, N.; Kato, Y.; Hirota, H. Aurantosides A and B: Cytotoxic tetramic acid glycosides from the marine sponge *Theonella* sp. *J. Am. Chem. Soc.* **1991**, *113*, 9690–9692.
88. Carmely, S.; Kashman, Y. Structure of swinholide-a, a new macrolide from the marine sponge *Theonella swinhoei*. *Tetrahedron Lett.* **1985**, *26*, 511–514.
89. Youssef, D.T.; Mooberry, S.L. Hurghadolide A and Swinholide I, potent actin-microfilament disrupters from the Red Sea sponge *Theonella swinhoei*. *J. Nat. Prod.* **2006**, *69*, 154–157.

90. Webster, N.S.; Blackall, L.L. What do we really know about sponge-microbial symbioses? *Int. Soc. Micobial. Ecol. J.* **2009**, *3*, 1–3.
91. Webster, N.S.; Taylor, M.W. Marine sponges and their microbial symbionts: Love and other relationships. *Environ. Microbiol.* **2012**, *14*, 335–346.
92. Mohamed, N.M.; Rao, V.; Hamann, M.T.; Kelly, M.; Hill, R.T. Monitoring bacterial diversity of the marine sponge *Ircinia strobilina* upon transfer into aquaculture. *Appl. Environ. Microbiol.* **2008**, *74*, 4133–4143.
93. Wilkinson, C.R. Net primary productivity in coral reef sponges. *Science* **1983**, *219*, 410–412.
94. Corredor, J.E.; Wilkinson, C.R.; Vicente, V.P.; Morell, J.M.; Otero, E. Nitrate release by Caribbean reef sponges. *Limnol. Oceanogr.* **1988**, *33*, 114–120.
95. Diaz, M.C.; Ward, B.B. Sponge-mediated nitrification in tropical benthic communities. *Mar. Ecol. Prog. Ser.* **1997**, *156*, 97–107.
96. Shieh, W.Y.; Lin, Y.M. Association of heterotrophic nitrogen-fixing bacteria with a marine sponge of *Halichondria* sp. *Bull. Mar. Sci.* **1994**, *54*, 557–564.
97. Wilkinson, C.C.; Summons, R.R.; Evans, E. Nitrogen fixation in symbiotic marine sponges: Ecological significance and difficulties in detection. *Mem. Qld. Mus.* **1999**, *44*, 667–673.
98. Hoffmann, F.; Larsen, O.; Thiel, V.; Rapp, H.T.; Pape, T.; Michaelis, W.; Reitner, J. An anaerobic world in sponges. *Geomicrobiol. J.* **2005**, *22*, 1–10.
99. Hentschel, U.; Fieseler, L.; Wehrl, M.; Gernert, C.; Steinert, M.; Hacker, J.; Horn, M. Microbial diversity of marine sponges. In *Sponges (Porifera)*; Springer: Berlin/Heidelberg, Germany, 2003; pp. 59–88.
100. Kennedy, J.; Baker, P.; Piper, C.; Cotter, P.D.; Walsh, M.; Mooij, M.J.; Bourke, M.B.; Rea, M.C.; O'Connor, P.M.; Ross, R.P. Isolation and analysis of bacteria with antimicrobial activities from the marine sponge *Haliclona simulans* collected from Irish waters. *Mar. Biotechnol.* **2009**, *11*, 384–396.
101. Rahman, H.; Austin, B.; Mitchell, W.J.; Morris, P.C.; Jamieson, D.J.; Adams, D.R.; Spragg, A.M.; Schweizer, M. Novel anti-infective compounds from marine bacteria. *Mar. Drugs* **2010**, *8*, 498–518.
102. Lu, X.; Cao, X.; Liu, X.; Jiao, B. Marine microbes-derived anti-bacterial agents. *Mini Rev. Med. Chem.* **2010**, *10*, 1077–1090.
103. Dudler, R.; Eberl, L. Interactions between bacteria and eukaryotes via small molecules. *Curr. Opin. Biotechnol.* **2006**, *17*, 268–273.
104. Schmidt, E.W. Trading molecules and tracking targets in symbiotic interactions. *Nat. Chem. Biol.* **2008**, *4*, 466–473.
105. Hu, J.-F.; Hamann, M.T.; Hill, R.; Kelly, M. The manzamine alkaloids. *Alkaloids Chem. Biol.* **2003**, *60*, 207–285.
106. Jares-Erijman, E.A.; Sakai, R.; Rinehart, K.L. Crambescidins: New antiviral and cytotoxic compounds from the sponge *Crambe crambe*. *J. Org. Chem.* **1991**, *56*, 5712–5715.
107. Berlinck, R.; Braekman, J.C.; Daloze, D.; Bruno, I.; Riccio, R.; Ferri, S.; Spampinato, S.; Speroni, E. Polycyclic guanidine alkaloids from the marine sponge *Crambe crambe* and Ca++ channel blocker activity of crambescidin 816. *J. Nat. Prod.* **1993**, *56*, 1007–1015.

108. Croué, J.; West, N.J.; Escande, M.-L.; Intertaglia, L.; Lebaron, P.; Suzuki, M.T. A single betaproteobacterium dominates the microbial community of the crambescidine-containing sponge *Crambe crambe*. *Sci. Rep.* **2013**, *3*, 1–8.
109. Sladic, D.; Gasic, M.J. Reactivity and biological activity of the marine sesquiterpene hydroquinone avarol and related compounds from sponges of the order Dictyoceratida. *Molecules* **2006**, *11*, 1–33.
110. Assmann, M.; Lichte, E.; Pawlik, J.R.; Köck, M. Chemical defenses of the Caribbean sponges *Agelas wiedenmayeri* and *Agelas conifera*. *Mar. Ecol. Prog. Ser.* **2000**, *207*, 255–262.
111. Neeman, I.; Fishelson, L.; Kashman, Y. Isolation of a new toxin from the sponge *Latrunculia magnifica* in the Gulf of Aquaba (Red Sea). *Mar. Biol.* **1975**, *30*, 293–296.
112. Keyzers, R.A.; Davies-Coleman, M.T. Anti-inflammatory metabolites from marine sponges. *Chem. Soc. Rev.* **2005**, *34*, 355–365.
113. Taylor, M.W.; Radax, R.; Steger, D.; Wagner, M. Sponge-associated microorganisms: Evolution, ecology, and biotechnological potential. *Microbiol. Mol. Biol. Rev.* **2007**, *71*, 295–347.
114. Dumdei, E.J.; Blunt, J.W.; Munro, M.H.G.; Battershill, C.N.; Page, M.J. The whys and whats of sponge chemistry: Why chemists extract sponges and what problems does this cause? In *Sponge Sciences. Multidiscipinary Perspectives*; Watanabe, Y., Fusetani, N., Eds.; Springer-Verlag: Tokyo, Japan, 1998; pp. 353–364.
115. Sacristán-Soriano, O.; Banaigs, B.; Casamayor, E.O.; Becerro, M.A. Exploring the links between natural products and bacterial assemblages in the sponge *Aplysina aerophoba*. *Appl. Environ. Microbiol.* **2011**, *77*, 862–870.
116. Rinehart, K.L. Secondary metabolites from marine organisms. *Ciba Found. Symp.* **1992**, *171*, 236–249.
117. Jenssen, H.; Hamill, P.; Hancock, R.E. Peptide antimicrobial agents. *Clin. Microbiol. Rev.* **2006**, *19*, 491–511.
118. Pasupuleti, M.; Schmidtchen, A.; Malmsten, M. Antimicrobial peptides: Key components of the innate immune system. *Crit. Rev. Biotechnol.* **2012**, *32*, 143–171.
119. Mohammed, R.; Peng, J.; Kelly, M.; Hamann, M.T. Cyclic heptapeptides from the Jamaican sponge *Stylissa caribica*. *J. Nat. Prod.* **2006**, *69*, 1739–1744.
120. Matsunaga, S.; Fusetani, N.; Konosu, S. Bioactive marine metabolites, IV. Isolation and the amino acid composition of discodermin A, an antimicrobial peptide, from the marine sponge *Discodermia kiiensis*. *J. Nat. Prod.* **1985**, *48*, 236–241.
121. Otero-González, A.J.; Magalhães, B.S.; Garcia-Villarino, M.; López-Abarrategui, C.; Sousa, D.A.; Dias, S.C.; Franco, O.L. Antimicrobial peptides from marine invertebrates as a new frontier for microbial infection control. *FASEB J.* **2010**, *24*, 1320–1334.
122. Pushpanathan, M.; Gunasekaran, P.; Rajendhran, J. Antimicrobial peptides: Versatile biological properties. *Int. J. Pept.* **2013**, *2013*, doi:10.1155/2013/675391.
123. Cooper, E.L. Comparative immunology. *Integr. Comp. Biol.* **2003**, *43*, 278–280.
124. Brogden, K.A. Antimicrobial peptides: Pore formers or metabolic inhibitors in bacteria? *Nat. Rev. Microbiol.* **2005**, *3*, 238–250.
125. Langenheim, J.H. Higher plant terpenoids: A phytocentric overview of their ecological roles. *J. Chem. Ecol.* **1994**, *20*, 1223–1280.

126. Schmitz, F.J.; Bowden, B.F.; Toth, S.I. Antitumor and cytotoxic compounds from marine organisms. In *Pharmaceutical and Bioactive Natural Products*; Springer: Berlin/Heidelberg, Germany, 1993; pp. 197–308.
127. Zheng, L.; Yan, X.; Han, X.; Chen, H.; Lin, W.; Lee, F.S.; Wang, X. Identification of norharman as the cytotoxic compound produced by the sponge (*Hymeniacidon perleve*)-associated marine bacterium *Pseudoalteromonas* piscicida and its apoptotic effect on cancer cells. *Biotechnol. Appl. Biochem.* **2006**, *44*, 135–142.
128. Folmer, F.; Jaspars, M.; Dicato, M.; Diederich, M. Marine cytotoxins: Callers for the various dances of death. *Gastroenterol. Hepatol. Bed Bench* **2009**, *2*, S34–S50.
129. Bao, B.; Sun, Q.; Yao, X.; Hong, J.; Lee, C.-O.; Cho, H.Y.; Jung, J.H. Bisindole alkaloids of the topsentin and hamacanthin classes from a marine sponge *Spongosorites* sp. *J. Nat. Prod.* **2007**, *70*, 2–8.
130. Luo, X.; Li, F.; Hong, J.; Lee, C.-O.; Sim, C.J.; Im, K.S.; Jung, J.H. Cytotoxic oxylipins from a marine sponge *Topsentia* sp. *J. Nat. Prod.* **2006**, *69*, 567–571.
131. Mansoor, T.A.; Lee, Y.M.; Hong, J.; Lee, C.-O.; Im, K.S.; Jung, J.H. 5,6:8,9-Diepoxy and other cytotoxic sterols from the marine sponge *Homaxinella* sp. *J. Nat. Prod.* **2006**, *69*, 131–134.
132. Nagle, D.G.; Zhou, Y.-D.; Mora, F.D.; Mohammed, K.A.; Kim, Y.-P. Mechanism targeted discovery of antitumor marine natural products. *Curr. Med. Chem.* **2004**, *11*, 1725.
133. Aoki, S.; Cho, S.-H.; Ono, M.; Kuwano, T.; Nakao, S.; Kuwano, M.; Nakagawa, S.; Gao, J.-Q.; Mayumi, T.; Shibuya, M. Bastadin 6, a spongean brominated tyrosine derivative, inhibits tumor angiogenesis by inducing selective apoptosis to endothelial cells. *Anti-Cancer Drugs* **2006**, *17*, 269–278.
134. Singh, R.; Sharma, M.; Joshi, P.; Rawat, D.S. Clinical status of anti-cancer agents derived from marine sources. *Anti-Cancer Agents Med. Chem.* **2008**, *8*, 603–617.
135. Gong, H.; Zuliani, P.; Komuravelli, A.; Faeder, J.R.; Clarke, E.M. Analysis and verification of the HMGB1 signaling pathway. *BMC Bioinform.* **2010**, *11*, S10.
136. Essack, M.; Bajic, V.B.; Archer, J.A. Recently confirmed apoptosis-inducing lead compounds isolated from marine sponge of potential relevance in cancer treatment. *Mar. Drugs* **2011**, *9*, 1580–1606.
137. Valeriote, F.A.; Tenney, K.; Pietraszkiewicz, H.; Edelstein, M.; Johnson, T.A.; Amagata, T.; Crews, P. Discovery and development of anticancer agents from marine sponges: Perspectives based on a chemistry-experimental therapeutics collaborative program. *J. Exp. Ther. Oncol.* **2012**, *10*, 119–134.
138. Ford, P.W.; Gustafson, K.R.; McKee, T.C.; Shigematsu, N.; Maurizi, L.K.; Pannell, L.K.; Williams, D.E.; Dilip de Silva, E.; Lassota, P.; Allen, T.M. Papuamides AD, HIV-inhibitory and cytotoxic depsipeptides from the sponges *Theonella mirabilis* and *Theonella swinhoei* collected in Papua New Guinea. *J. Am. Chem. Soc.* **1999**, *121*, 5899–5909.
139. Qureshi, A.; Faulkner, D.J. Haplosamates A and B: New steroidal sulfamate esters from two haplosclerid sponges. *Tetrahedron* **1999**, *55*, 8323–8330.
140. Müller, W.E.; Sobel, C.; Diehl-Seifert, B.; Maidhof, A.; Schröder, H.C. Influence of the antileukemic and anti-human immunodeficiency virus agent avarol on selected immune responses *in vitro* and *in vivo*. *Biochem. Pharmacol.* **1987**, *36*, 1489–1494.

141. Zhou, X.; Liu, J.; Yang, B.; Lin, X.; Yang, X.-W.; Liu, Y. Marine natural products with anti-HIV activities in the last decade. *Curr. Med. Chem.* **2013**, *20*, 953–973.
142. Aoki, S.; Watanabe, Y.; Sanagawa, M.; Setiawan, A.; Kotoku, N.; Kobayashi, M. Cortistatins A, B, C, and D, anti-angiogenic steroidal alkaloids, from the marine sponge *Corticium simplex*. *J. Am. Chem. Soc.* **2006**, *128*, 3148–3149.
143. Aoki, S.; Watanabe, Y.; Tanabe, D.; Arai, M.; Suna, H.; Miyamoto, K.; Tsujibo, H.; Tsujikawa, K.; Yamamoto, H.; Kobayashi, M. Structure-activity relationship and biological property of cortistatins, anti-angiogenic spongean steroidal alkaloids. *Bioorg. Med. Chem.* **2007**, *15*, 6758–6762.
144. Mousseau, G.; Clementz, M.A.; Bakeman, W.N.; Nagarsheth, N.; Cameron, M.; Shi, J.; Baran, P.; Fromentin, R.; Chomont, N.; Valente, S.T. An analog of the natural steroidal alkaloid cortistatin a potently suppresses tat-dependent hiv transcription. *Cell Host Microbe* **2012**, *12*, 97–108.
145. Andavan, G.S.B.; Lemmens-Gruber, R. Cyclodepsipeptides from marine sponges: Natural agents for drug research. *Mar. Drugs* **2010**, *8*, 810–834.
146. Shin, J.; Rho, J.R.; Seo, Y.; Lee, H.S.; Cho, K.W.; Sim, C.J. Sarcotragins A and B, new sesterterpenoid alkaloids from the sponge *Sarcotragus* sp. *Tetrahedron Lett.* **2001**, *42*, 3005–3007.
147. Santafé, G.; Paz, V.; Rodríguez, J.; Jiménez, C. Novel cytotoxic oxygenated C29 Sterols from the Colombian marine sponge *Polymastia tenax*. *J. Nat. Prod.* **2002**, *65*, 1161–1164.
148. Pattenden, G.; Wickramasinghe, W.A.; Bandaranayake, W.M. Benzylthiocrellidone, a novel thioether with strong UV A and B absorption from the Great Barrier Reef sponge Crella spinulata, (Poecilosclerida: Crellidae). *Trends Comp. Biochem. Physiol.* **2002**, *9*, 205–216.
149. Costantino, V.; Fattorusso, E.; Imperatore, C.; Mangoni, A. Ectyoceramide, the first natural hexofuranosylceramide from the marine sponge *Ectyoplasia ferox*. *Eur. J. Org. Chem.* **2003**, *2003*, 1433–1437.
150. Nakao, Y.; Kuo, J.; Yoshida, W.Y.; Kelly, M.; Scheuer, P.J. More Kapakahines from the marine Sponge *Cribrochalina olemda*. *Org. Lett.* **2003**, *5*, 1387–1390.
151. Volk, C.A.; Kŏck, M. Viscosamine: The first naturally occurring trimeric 3-alkyl pyridinium alkaloid. *Org. Lett.* **2003**, *5*, 3567–3569.
152. Xu, X.-H.; Yao, G.-M.; Li, Y.-M.; Lu, J.-H.; Lin, C.-J.; Wang, X.; Kong, C.-H. 5-Fluorouracil derivatives from the sponge *Phakellia fusca*. *J. Nat. Prod.* **2003**, *66*, 285–288.
153. Costantino, V.; Fattorusso, E.; Imperatore, C.; Mangoni, A. Glycolipids from sponges. 13.1 Clarhamnoside, the first rhamnosylated α-galactosylceramide from *Agelas clathrodes*. Improving spectral strategies for glycoconjugate structure determination. *J. Org. Chem.* **2004**, *69*, 1174–1179.
154. Cichewicz, R.H.; Valeriote, F.A.; Crews, P. Psymberin, a potent sponge-derived cytotoxin from *Psammocinia* distantly related to the pederin family. *Org. Lett.* **2004**, *6*, 1951–1954.
155. Berer, N.; Rudi, A.; Goldberg, I.; Benayahu, Y.; Kashman, Y. Callynormine A, a new marine cyclic peptide of a novel class. *Org. Lett.* **2004**, *6*, 2543–2545.
156. Warabi, K.; Hamada, T.; Nakao, Y.; Matsunaga, S.; Hirota, H.; van Soest, R.W.M.; Fusetani, N. Axinelloside A, an unprecedented highly sulfated lipopolysaccharide inhibiting telomerase, from the marine sponge, *Axinella infundibula*. *J. Am. Chem. Soc.* **2005**, *127*, 13262–13270.
157. Hamada, T.; Matsunaga, S.; Yano, G.; Fusetani, N. Polytheonamides A and B, highly cytotoxic, linear polypeptides with unprecedented structural features, from the marine sponge, *Theonella swinhoei*. *J. Am. Chem. Soc.* **2005**, *127*, 110–118.

158. Williams, D.E.; Austin, P.; Diaz-Marrero, A.R.; Soest, R.V.; Matainaho, T.; Roskelley, C.D.; Roberge, M.; Andersen, R.J. Neopetrosiamides, peptides from the marine sponge *Neopetrosia* sp. that inhibit amoeboid invasion by human tumor cells. *Org. Lett.* **2005**, *7*, 4173–4176.
159. Rogers, E.W.; Molinski, T.F. A cytotoxic carotenoid from the marine sponge *Prianos osiros*. *J. Nat. Prod.* **2005**, *68*, 450–452.
160. Emura, C.; Higuchi, R.; Miyamoto, T. Irciniasulfonic acid B, a novel taurine conjugated fatty acid derivative from a Japanese marine sponge, *Ircinia* sp. *Tetrahedron* **2006**, *62*, 5682–5685.
161. Tanaka, C.; Tanaka, J.; Bolland, R.F.; Marriott, G.; Higa, T. Seragamides A–F, new actin-targeting depsipeptides from the sponge *Suberites japonicus* Thiele. *Tetrahedron* **2006**, *62*, 3536–3542.
162. Fattorusso, E.; Romano, A.; Taglialatela-Scafati, O.; Bavestrello, G.; Bonelli, P.; Calcinai, B. Coelodiol and coeloic acid, ent-isocopalane diterpenes from the Indonesian sponge *Coelocarteria* cfr. *singaporensis*. *Tetrahedron Lett.* **2006**, *47*, 2197–2200.
163. Dai, J.; Liu, Y.; Zhou, Y.-D.; Nagle, D.G. Cytotoxic metabolites from an Indonesian sponge *Lendenfeldia* sp. *J. Nat. Prod.* **2007**, *70*, 1824–1826.
164. Antonov, A.S.; Kalinovsky, A.I.; Stonik, V.A.; Afiyatullov, S.S.; Aminin, D.L.; Dmitrenok, P.S.; Mollo, E.; Cimino, G. Isolation and structures of Erylosides from the Carribean sponge *Erylus formosus*. *J. Nat. Prod.* **2007**, *70*, 169–178.
165. Johnson, T.A.; Amagata, T.; Oliver, A.G.; Tenney, K.; Valeriote, F.A.; Crews, P. The unexpected isolation of CTP-431, a novel thiopyrone from the sponge *Cacospongia mycofijiensis*. *J. Org. Chem.* **2008**, *73*, 7255–7259.
166. Zampella, A.; Sepe, V.; Luciano, P.; Bellotta, F.; Monti, M.C.; D'Auria, M.V.; Jepsen, T.; Petek, S.; Adeline, M.-T.R.S.; Laprévôte, O. Homophymine A, an anti-HIV cyclodepsipeptide from the sponge *Homophymia* sp. *J. Org. Chem.* **2008**, *73*, 5319–5327.
167. Tung, N.H.; Minh, C.V.; Ha, T.T.; Kiem, P.V.; Huong, H.T.; Dat, N.T.; Nhiem, N.X.; Tai, B.H.; Hyun, J.-H.; Kang, H.-K.; *et al.* C29 sterols with a cyclopropane ring at C-25 and 26 from the Vietnamese marine sponge *Ianthella* sp. and their anticancer properties. *Bioorg. Med. Chem. Lett.* **2009**, *19*, 4584–4588.
168. DiGirolamo, J.A.; Li, X.-C.; Jacob, M.R.; Clark, A.M.; Ferreira, D. Reversal of fluconazole resistance by sulfated sterols from the marine sponge *Topsentia* sp. *J. Nat. Prod.* **2009**, *72*, 1524–1528.
169. Salim, A.A.; Rae, J.; Fontaine, F.; Conte, M.M.; Khalil, Z.; Martin, S.; Parton, R.G.; Capon, R.J. Heterofibrins: Inhibitors of lipid droplet formation from a deep-water southern Australian marine sponge, *Spongia* (Heterofibria) sp. *Org. Biomol. Chem.* **2010**, *8*, 3188–3194.
170. Dai, J.; Sorribas, A.; Yoshida, W.Y.; Kelly, M.; Williams, P.G. Xestosaprols from the Indonesian marine sponge *Xestospongia* sp. *J. Nat. Prod.* **2010**, *73*, 1188–1191.
171. Plaza, A.; Keffer, J.L.; Lloyd, J.R.; Colin, P.L.; Bewley, C.A. Paltolides A–C, Anabaenopeptin-type peptides from the Palau sponge *Theonella swinhoei*. *J. Nat. Prod.* **2010**, *73*, 485–488.
172. Wei, X.; Nieves, K.; Rodríguez, A.D. Neopetrosiamine A, biologically active bis-piperidine alkaloid from the Caribbean sea sponge *Neopetrosia proxima*. *Bioorg. Med. Chem. Lett.* **2010**, *20*, 5905–5908.

173. Fan, G.; Li, Z.; Shen, S.; Zeng, Y.; Yang, Y.; Xu, M.; Bruhn, T.; Bruhn, H.; Morschhäuser, J.; Bringmann, G.; Lin, W. Baculiferins A–O, *O*-sulfated pyrrole alkaloids with anti-HIV-1 activity, from the Chinese marine sponge *Iotrochota baculifera*. *Bioorg. Med. Chem.* **2010**, *18*, 5466–5474.
174. Schmitt, S.; Hentschel, U.; Taylor, M.W. Deep sequencing reveals diversity and community structure of complex microbiota in five Mediterranean sponges. *Hydrobiologia* **2012**, *687*, 341–351.
175. Erwin, P.M.; López-Legentil, S.; González-Pech, R.; Turon, X. A specific mix of generalists: Bacterial symbionts in Mediterranean *Ircinia* spp. *FEMS Microbiol. Ecol.* **2012**, *79*, 619–637.
176. Thiel, V.; Leininger, S.; Schmaljohann, R.; Brümmer, F.; Imhoff, J.F. Sponge-specific bacterial associations of the Mediterranean sponge *Chondrilla nucula* (Demospongiae, Tetractinomorpha). *Microb. Ecol.* **2007**, *54*, 101–111.
177. Perdicaris, S.; Vlachogianni, T.; Valavanidis, A. Bioactive natural substances from marine sponges: New developments and prospects for future pharmaceuticals. *Nat. Prod. Chem. Res.* **2013**, *1*, 1–8.
178. Sabdono, A. Microbial symbionts in marine sponges: Marine natural product factory. *J. Coast. Dev.* **2011**, *11*, 57–61.
179. Laird, S.; Monagle, C.; Johnston, S. Queensland biodiscovery collaboration: The Griffith University AstraZeneca Partnership for natural product discovery: An access and benefit sharing case study. In *An Access and Benefit Sharing Case Study*; United Nations University, Institute of Advanced Studies: Yokohama, Japan, 2008.
180. Mittermeier, R.A.; Goettsch Mittermeier, C. *Megadiversity: Earth's Biologically Wealthiest Nations. Megadiversity: Most Biological Rich Countries of the World*; Cemex: San Pedro Garza García, Mexico, 1997.
181. Tolley, M. *UNEP-WCMC World Conservation Monitoring Centre Website*; UNEP: Cambridge, UK, 2011.
182. Butler, A.J.; Rees, T.; Beesley, P.; Bax, N.J. Marine biodiversity in the Australian region. *PLoS One* **2010**, *5*, e11831.
183. Hughes, T.P.; Bellwood, D.R.; Connolly, S.R. Biodiversity hotspots, centres of endemicity, and the conservation of coral reefs. *Ecol. Lett.* **2002**, *5*, 775–784.
184. Roberts, C.M.; McClean, C.J.; Veron, J.E.; Hawkins, J.P.; Allen, G.R.; McAllister, D.E.; Mittermeier, C.G.; Schueler, F.W.; Spalding, M.; Wells, F. Marine biodiversity hotspots and conservation priorities for tropical reefs. *Science* **2002**, *295*, 1280–1284.
185. Wernberg, T.; Russell, B.D.; Moore, P.J.; Ling, S.D.; Smale, D.A.; Campbell, A.; Coleman, M.A.; Steinberg, P.D.; Kendrick, G.A.; Connell, S.D. Impacts of climate change in a global hotspot for temperate marine biodiversity and ocean warming. *J. Exp. Mar. Biol. Ecol.* **2011**, *400*, 7–16.
186. Ward, T.M.; Sorokin, S.J.; Currie, D.R.; Rogers, P.J.; McLeay, L.J. Epifaunal assemblages of the eastern Great Australian Bight: Effectiveness of a benthic protection zone in representing regional biodiversity. *Cont. Shelf Res.* **2006**, *26*, 25–40.
187. Fromont, J.; Vanderklift, M.A.; Kendrick, G.A. *Marine Sponges of the Dampier Archipelago, Western Australia: Patterns of Species Distributions, Abundance and Diversity*; Springer: Berlin/Heidelberg, Germany, 2006; Volume 15, pp. 3731–3750.
188. Hooper, J.; Kennedy, J.A. Small-scale patterns of sponge biodiversity (Porifera) from the Sunshine Coast reefs, eastern Australia. *Invertebr. Syst.* **2002**, *16*, 637–653.

189. Hooper, J.N.; Kennedy, J.A.; Quinn, R.J. Biodiversity "hotspots", patterns of richness and endemism, and taxonomic affinities of tropical Australian sponges (Porifera). *Biodivers. Conserv.* **2002**, *11*, 851–885.
190. Heyward, A.A.; Fromont, J.J.; Schoenberg, C.C.; Colquhoun, J.J.; Radford, B.B.; Gomez, O.O. The sponge gardens of Ningaloo Reef, Western Australia. *Open Mar. Biol.* **2010**, *4*, 3–11.
191. Fromont, J.; Althaus, F.; McEnnulty, F.R.; Williams, A.; Salotti, M.; Gomez, O.; Gowlett-Holmes, K. Living on the edge: The sponge fauna of Australia's southwestern and northwestern deep continental margin. *Hydrobiologia* **2012**, *687*, 127–142.
192. Williams, A.; Althaus, F.; Dunstan, P.K.; Poore, G.C.; Bax, N.J.; Kloser, R.J.; McEnnulty, F.R. Scales of habitat heterogeneity and megabenthos biodiversity on an extensive Australian continental margin (100–1100 m depths). *Mar. Ecol.* **2010**, *31*, 222–236.
193. Evans-Illidge, E.A.; Logan, M.; Doyle, J.; Fromont, J.; Battershill, C.N.; Ericson, G.; Wolff, C.W.; Muirhead, A.; Kearns, P.; Abdo, D. Phylogeny drives large scale patterns in australian marine bioactivity and provides a new chemical ecology rationale for future biodiscovery. *PLoS One* **2013**, *8*, e73800.
194. Dayton, P.K.; Robilliard, G.A.; Paine, R.T.; Dayton, L.B. Biological accommodation in the benthic community at McMurdo Sound, Antarctica. *Ecol. Monogr.* **1974**, *44*, 105–128.
195. Haefner, B. Drugs from the deep: Marine natural products as drug candidates. *Drug Discov. Today* **2003**, *8*, 536–544.
196. Cárdenas, C.A.; Bell, J.J.; Davy, S.K.; Hoggard, M.; Taylor, M.W. Influence of environmental variation on symbiotic bacterial communities of two temperate sponges. *FEMS Microbiol. Ecol.* **2014**, *88*, 516–527.
197. Schmitt, S.; Tsai, P.; Bell, J.; Fromont, J.; Ilan, M.; Lindquist, N.; Perez, T.; Rodrigo, A.; Schupp, P.J.; Vacelet, J. Assessing the complex sponge microbiota: Core, variable and species-specific bacterial communities in marine sponges. *Int. Soc. Micob. Ecol. J.* **2012**, *6*, 564–576.

2

Analysis of the Biomass Composition of the Demosponge *Amphimedon queenslandica* on Heron Island Reef, Australia

Jabin R. Watson [1], Timothy C. R. Brennan [2], Bernard M. Degnan [1], Sandie M. Degnan [1] and Jens O. Krömer [3,*]

[1] School of Biological Science, University of Queensland, Brisbane, Queensland 4072, Australia; E-Mails: jabin.watson@uq.edu.au (J.R.W.); b.degnan@uq.edu.au (B.M.D.); s.degnan@uq.edu.au (S.M.D.)

[2] Systems and Synthetic Biology Group, Australian Institute for Bioengineering and Nanotechnology, University of Queensland, Brisbane, Queensland 4072, Australia; E-Mail: t.brennan3@uq.edu.au

[3] Centre for Microbial Electrosynthesis (CEMES), Advanced Water Management Centre, University of Queensland, Brisbane, Queensland 4072, Australia

* Author to whom correspondence should be addressed; E-Mail: j.kromer@uq.edu.au

---

Abstract: Marine sponges are a potential source of important pharmaceutical drugs, the commercialisation of which is restricted by the difficulties of obtaining a sufficient and regular supply of biomass. One way to optimize commercial cell lines for production is the in-depth characterization and target identification through genome scale metabolic modeling and flux analysis. By applying these tools to a sponge, we hope to gain insights into how biomass is formed. We chose *Amphimedon queenslandica* as it has an assembled and annotated genome, a prerequisite for genome scale modeling. The first stepping stone on the way to metabolic flux analysis in a sponge holobiont, is the characterization of its biomass composition. In this study we quantified the macromolecular composition and investigated the variation between and within sponges of a single population. We found lipids and protein to be the most abundant macromolecules, while carbohydrates were the most variable. We also analysed the composition and abundance of the fatty acids and amino acids, the important building blocks required to synthesise the abundant macromolecule types, lipids, and protein. These data complement the extensive genomic information available for *A. queenslandica* and lay the basis for genome scale modelling and flux analysis.

Keywords: Porifera; biochemical composition; *Amphimedon queenslandica*; amino acid analysis; FAME analysis; spongin

---

1. Introduction

Marine sponges and their associated bacteria are prolific producers of secondary metabolites that are an important source of bioactive compounds with pharmacological potential. More than 5000 compounds have been identified to date, including 355 new compounds reported in 2012 alone [1]. However, despite this enormous potential, approval for use as a therapeutic compound is scarce [2]. The lack of commercial development of sponge-derived compounds is attributed to both a biomass supply problem—no reliable method to culture either whole sponges or cells exists—and a lack of understanding of sponge-bacterial interactions. Together, these constraints limit the applicability of metabolic engineering approaches to microbial isolates or sponge cells for over-production of secondary metabolites. To move forward in the development of sponge-derived drugs requires extending the research activities beyond the current major focus on identifying secondary compounds and, to a limited extent, characterizing the biosynthetic pathways that produce these secondary metabolites [3]. We need to understand how the sponge holobiont (the animal and its resident microbes) is able to grow; understanding how the central metabolic pathways contribute to biomass production is crucial to overcoming the biomass supply problem.

The metabolic processes underlying sponge growth can be comprehensively investigated by the application of genome scale modeling and flux analyses [4]. An essential step of such analyses is characterization of the biochemical composition of the sponge holobiont, with a focus on the components that contribute significantly to overall biomass production [5]. Characterizing the biochemical composition allows the formation of a biomass equation, essential for genome scale modeling and flux analysis as it constrains hundreds of fluxes through the system [6]. The focus on secondary metabolic compounds has left the products of central metabolism comparatively overlooked, including analysis of the major cellular macromolecules—protein, lipid, carbohydrate, DNA, and RNA—and their constituent building blocks—amino acids, fatty acids, sugars and nucleotides. Although protein, lipid, and carbohydrate content has been quantitated in a number of marine sponges [7–12], a complete analysis of the main biomass components has not been reported.

In addition to quantifying the components that contribute to biomass formation, information about how variable the composition is across the population and within an individual sponge is important for guiding the design of experiments associated with flux analysis. Although sponges appear to be morphologically simple and homogenous, it has been shown that production of secondary metabolites can be controlled and restricted to specific areas of an individual [13]. This apparent heterogeneity may be related to physiological differences in the sponge itself or in the associated microbial community. Marine sponges typically host large and diverse communities of microbes [14], often in a symbiotic relationship [15]. These microbes have been shown to be the source of certain secondary metabolites [16], and also to interact with central metabolic pathways in the host sponge [17]. Thus the distribution of particular bacterial species and communities within an individual sponge has the potential to influence local metabolic activity. Nonetheless, subsampling the main biomass components within an individual sponge typically has not been performed, as most biochemical analyses performed in sponges have been restricted to a single sample from an individual or small number of individuals within a population [7–12]. Analysing multiple samples within individuals will give insights into the variation in the macromolecule composition within an individual sponge.

We report here the first analysis of the products of central metabolism of the marine demosponge, *Amphimedon queenslandica* from the Southern Great Barrier Reef. We have quantified the protein, lipid, carbohydrate, DNA and RNA content of this sponge. These analyses are supplemented with a detailed analysis of amino acid and fatty acid composition. By analysing multiple areas within multiple individuals within one *A. queenslandica* population we reveal marked variation within and between sponges for most of these macromolecular classes. These results, along with the extensive genomic resources available for this demosponge [18], provide a strong foundation for future detailed understanding of the metabolic pathways that contribute to biomass production.

2. Results

2.1. Displacement to Dry Weight Composition Conversion Factor

Metabolic flux analysis requires the ability to quantify the biomass involved in the turn-over of the substrate. In order to relate the amount of live sponge to the dry weight composition, we established a correlation between displacement volume of a sponge piece and its dry weight. The displacement volume and dry weight from 97 samples were used to estimate a conversion factor of 0.091 (std. dev. = 0.012) grams of dry sponge weight (gDW) per mL of displacement.

Figure 1. Mean skeleton and macromolecular composition of *A. queenslandica* calculated from all biomass samples summarised in Table 1. The box represents the upper and lower quartile, split by the mean value. The upper and lower whiskers denote the minimum and maximum values. Units are grams per gram of dry weight.

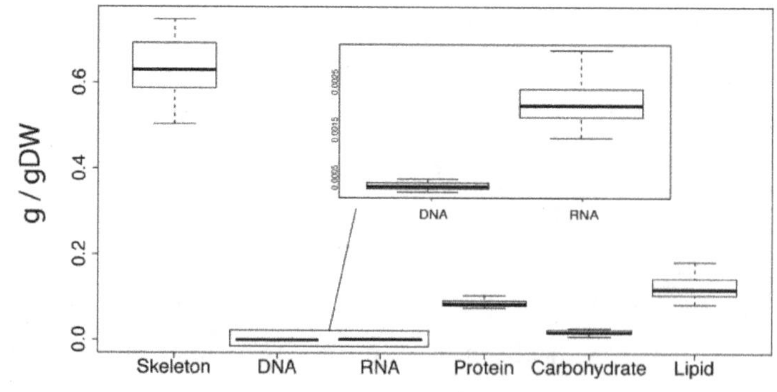

2.2. Overall Macromolecular Composition

For each biochemical analysis, we analysed five biopsies from several parts of four *A. queenslandica* individuals, all of which were collected from a single wild population in Shark Bay on Heron Island Reef, Australia (for detailed methods see Section 4). This sampling regime allowed us to document biochemical differences both within and between individual sponges. Analysis of mean values of the macro-components shows the skeleton (0.6342 g/gDW) to be the dominant component of the dry weight of *A. queenslandica* (Figure 1 and Table 1). The most abundant macromolecule type was lipid with 0.1251 g/gDW, followed by protein (0.0881 g/gDW), carbohydrate (0.0197 g/gDW), RNA (0.0021 g/gDW) and DNA (0.0003 g/gDW). The coefficient of variation revealed that carbohydrates were the most variable, followed by DNA, RNA, and lipids (Table 1). Protein and skeleton had the least overall variability (Table 1).

Table 1. Mean skeletal and macromolecular composition of *A. queenslandica* calculated from all biomass samples.

Macro-Component	Mean (g/gDW)	Standard Deviation	Standard Error	Coefficient of Variation (%)	*n*
Skeleton	0.6343	0.0647	0.0152	9.81	18
Lipid	0.1252	0.0251	0.0058	20.09	19
Protein	0.0881	0.0082	0.0018	9.31	20
Carbohydrate	0.0197	0.0055	0.0012	27.53	20
RNA	0.0021	0.0005	0.0001	22.24	20
DNA	0.0003	0.0001	0.00002	25.78	19

2.3. The Source of Variation in the Overall Composition

We used a linear mixed effects model to investigate how much of the variation observed in the mean values (Table 1, Coefficient of variation column) was attributable to the between-individual variation. Over half of the variation seen in the skeleton (58%) and lipids (54%) was due to differences between individuals. The significance of this relationship was tested using a x^2 test. The variation of skeleton (p = 0.009) and lipids (p = 0.016) between individuals had significant effects on the overall variation of both macromolecule types (Table 2). Figure 2 shows the variation seen within each individual for the respective macro-components.

Table 2. The source and degree of variation.

Macro-Component	Percentage of Overall Variation Caused by between Individual Variation (%)	x^2	*P*-value
Skeleton	58	6.691	0.009
Lipid	54	5.763	0.016
Protein	24	1.393	0.237
Carbohydrate	15	0.597	0.439
RNA	1	2.309	0.128
DNA	32	3.53×10^{-8}	0.999

2.4. Fatty Acid and Sterol Analysis

The lipid extractions were analysed using a fatty acid methyl ester (FAME) analysis using gas chromatography-mass spectrometry (GC-MS). We were interested in the main components that contribute to forming sponge lipids. Compounds were considered essential if they were found in most of the 23 replicate pieces of sponge. Eleven fatty acid (FA) compounds were identified and confirmed with standards. Palmitic acid was the most abundant FA, contributing 98.14 μmol/gDW. Octadecanoic acid (62.21 μmol/gDW) and an unknown FA (UU4-FA, 71.749 in Table 3) were other significant FAs. Three additional compounds matched compounds in the NIST database but were not confirmed with a standard (Table 3), this included cholesterol. Based on the fragmentation patterns, eleven compounds were classified as unknown FAs and one as an unknown sterol, highlighting the diversity and complexity of the lipid metabolism in *A. queenslandica.* If the total area under all peaks in the chromatogram is assumed to represent the total lipid fraction, then the main FA and sterols reported here cover 82.5% of the lipid fraction.

Figure 2. Variation of the composition of macro-components within and between individuals of *A. queenslandica.* Four individual sponges were sampled for each macromolecule. Each box and whisker plot represents samples collected from an individual sponge. The box represents the upper and lower quartile, split by the mean value. The upper and lower whiskers denote the minimum and maximum values. All of the components apart from DNA showed marked variation between and within individual sponges.

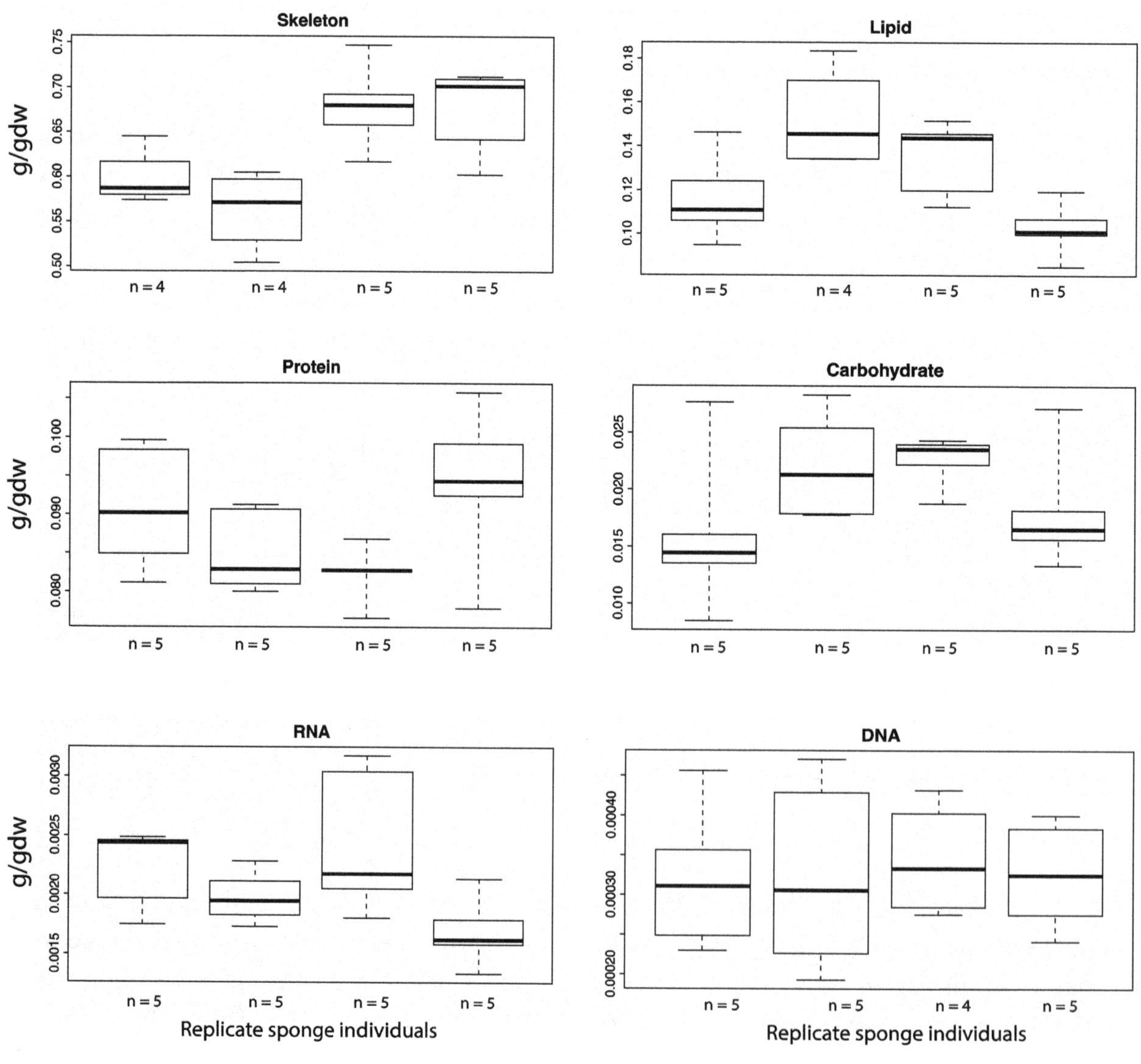

Table 3. Main compounds that contribute to the lipid biomass.

Compound	Lipid Number	µmol/gDW	Standard Deviation
Fatty Acids			
Docosanoic acid FAME	C22:0	9.769	4.299
Eicosanoic acid FAME	C20:1	1.947	0.579
Erucic acid FAME	C22:1ω9	1.721	0.668
Heptadecanoic acid FAME	C17:0	0.922	0.234
Myristic acid FAME	C14:0	3.957	1.455
Nervonic acid FAME	C24:1ω9	10.090	3.451
Octadecanoic acid FAME	C18:0	62.210	19.095
Palmitic acid FAME	C16:0	98.140	22.426
Pentadecanoic acid FAME	C15:0	7.065	2.468
Tetracosanoic acid FAME	C24:0	6.769	3.546
Tricosanoic acid FAME	C23:0	1.147	0.323
11-Eicosenoic acid FAME *	C20:1ω9	5.342	1.737
U6-FA **		1.005	0.580
U7-FA **		5.973	3.845
U8-FA **		2.279	0.722
U10-FA **		3.288	3.014
UU2-FA **		7.732	2.805
UU3-FA **		5.669	1.897
UU4-FA **		71.749	17.820
UU7-FA **		0.690	0.604
UU8-FA **		4.215	2.155
UU9-FA **		29.843	10.551
UU11-FA **		4.460	1.846
UU1 ***		1.728	1.766
UU6 ***		9.617	4.268
Sterols			
Cholesterol *		4.706	2.477
Brassicasterol *		7.581	4.806
UU13-sterol **		6.264	3.375

* Compounds that did not match a compound in the standard mix but matched a compound in a database; ** interpreted from the fragmentation pattern; *** compounds that are neither FA nor sterols.

2.5. *Amino Acid Analysis*

We analysed the amino acid composition of 20 samples of complete sponge biomass (Table 4). Samples of biomass were lyophilized and the hydrolysis incubation time was optimized before quantifying the amino acids with high performance liquid chromatography (HPLC). We found glycine to be the most abundant amino acid with 4.582 mmol/gDW, followed by alanine (1.538 mmol/gDW). Due to the oxidation of asparagine and glutamine to aspartate and glutamate respectively during acid hydrolysis, a 50:50 abundance ratio was assumed between asparagine:aspartate and glutamine:glutamate. Tryptophan (0.089 mmol/gDW), tyrosine (0.157 mmol/gDW), and histidine (0.110 mmol/gDW) were the least abundant amino acids. Tryptophan and methionine are known to be degraded during the hydrolysis. While methionine was not detected tryptophan was measurable but *in vivo* concentrations may be higher than what was observed. Neither cysteine, nor cystine was measured with the methods used and thus not reported.

Table 4. Mean amino acid composition of complete sponge biomass.

Amino Acid	**mmol/gDW**	**Standard Deviation (mmol/gDW)**	**Amino Acid Contribution (mol/mol)**
ALA	1.538	0.565	0.097
ARG	0.694	0.258	0.044
ASP	0.841	0.312	0.053
ASN	0.841	0.312	0.053
GLN	0.767	0.287	0.048
GLU	0.767	0.287	0.048
GLY	4.582	1.702	0.289
HIS	0.110	0.042	0.007
ILE	0.402	0.149	0.025
LEU	0.597	0.222	0.038
LYS	0.688	0.268	0.043
PHE	0.369	0.137	0.023
PRO	1.050	0.387	0.066
SER	0.806	0.318	0.051
THR	0.826	0.304	0.052
TRP *	0.089	0.053	0.006
TYR	0.157	0.060	0.010
VAL	0.753	0.281	0.047

* Tryptophan is largely destroyed during the hydrolysis, so concentrations *in vivo* may be higher than observed.

The skeleton is the most abundant component in *A. queenslandica* and is comprised of inorganic silica spicules and a collagen based extracellular matrix. Sponges produce a specific collagen, called spongin [19]. As the skeleton contributes significantly to the overall sponge composition, we analysed the amino acid composition of the skeleton (Table 5). The hydrolysis incubation time was optimized before the analysis was performed on 20 samples. As above, a 50:50 ratio between aspartate and asparagine and between glutamine and glutamate was assumed. The most abundant amino acids in the skeleton were glycine with 4.248 mmol/gDW and alanine with 1.257 mmol/gDW. The least abundant amino acids were tyrosine (0.066 mmol/gDW) and histidine (0.034 mmol/gDW).

A comparison of the relative amino acid composition of the skeleton and the cellular component revealed that glycine was the most abundant amino acid for both components (Figure 3). The skeleton had relatively more glycine, proline, and alanine than the cellular fraction. The cellular fraction had a higher relative abundance of isoleucine, leucine, and valine than the skeleton. Histidine and tyrosine were the least abundant of the amino acids measured for both the skeleton and cellular components.

Table 5. Mean amino acid composition of the skeleton.

Amino Acid	mmol/gDW	Standard Deviation (mmol/gDW)	Amino Acid Contribution (mol/mol)
ALA	1.257	0.537	0.103
ARG	0.545	0.226	0.045
ASP	0.594	0.250	0.049
ASN	0.594	0.250	0.049
GLN	0.573	0.243	0.047
GLU	0.573	0.243	0.047
GLY	4.248	1.807	0.349
HIS	0.034	0.016	0.003
ILE	0.176	0.080	0.014
LEU	0.290	0.121	0.024
LYS	0.442	0.173	0.036
PHE	0.239	0.102	0.020
PRO	0.916	0.351	0.075
SER	0.534	0.211	0.044
THR	0.559	0.226	0.046
TRP	0.100	0.041	0.008
TYR	0.066	0.028	0.005
VAL	0.440	0.177	0.036

Figure 3. A comparison of the relative amino acid composition of the cellular biomass and the skeleton. The amino acid composition of the cellular components was calculated by subtracting the quantified amino acid values of the skeleton (Table 5) from the values for the complete sponge biomass (Table 4). The skeletal and estimated cellular amino acid values were then normalized by converting to percentages.

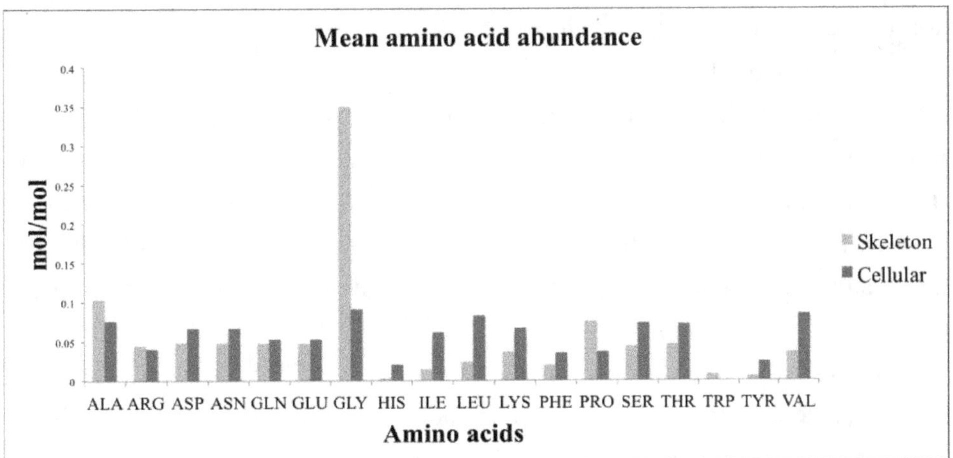

2.6. Nucleotide Composition

The molar contribution of nucleotides to the DNA and RNA composition was estimated using available genome and transcriptome data. The genomic GC content was estimated to be 31.14% using the publicly available genome [18]. The RNA GC content was estimated to be 39.76%, using an adult transcriptome [20]. These percentages were then converted into micromoles per gram of dry weight, using the molecular weight of a polymerized nucleotide (Tables 6 and 7).

Table 6. Deoxyribonucleotide composition.

	MW (g/mol)	mol/ mol DNA	μmol/gDW
dAMP	313	0.3443	0.37
dCMP	289	0.1557	0.17
dGMP	329	0.1557	0.17
dTMP	304	0.3443	0.37

Table 7. Ribonucleotide composition.

	MW (g/mol)	mol/ mol DNA	μmol/gDW
AMP	313	0.3012	2.03
CMP	289	0.1988	1.34
GMP	329	0.1988	1.34
UMP	308	0.3012	2.03

3. Discussion

3.1. Macromolecule Composition

To gain insights into how sponges produce biomass, the complex metabolic processes must be investigated as a complete functioning system. The systems biology approach of genome scale metabolic modeling requires the biochemical composition of the organism to be known. The biochemical composition identifies the products of central metabolism that contribute to growth. With the products of central metabolism known, the synthesis pathways can be identified and the fluxes of metabolites modeled. As a first step towards genome scale modeling, we analysed the macromolecular abundance and composition of the marine sponge *A. queenslandica* and investigated the variation seen across the population and within individuals.

The skeleton was the dominant component of the biomass, contributing 63% of the dry weight. Lipids and protein were the most abundant macromolecules, while carbohydrates were the most variable (Table 1). Variation in the amount of skeleton and lipids between individual sponges had significant effects on the overall variability of the respective macro-components (Table 2). The apparent variation in the abundance of macromolecules within an individual (Figure 2) was not significant and had limited effect on the overall population variability. This implies that not only do the central metabolic pathways have an overall response to the individual's microenvironment, but localised stimulus could also cause compositional differences across an individual. This highlights the need for compositional and metabolic studies to sample appropriately to ensure accurate representation.

Overall, the macromolecular composition of *A. queenslandica* in the current study fits within the ranges reported for protein, lipids, and carbohydrates for 71 tropical marine sponge species [7]. *A. queenslandica* notably contained more lipids than protein (Table 1), a characteristic that has been reported previously for two marine demosponges, *Aplysinia archeri* and *Plaktortis halichondroides* [7]. The high lipid content in *A. queenslandica* could indicate the accumulation of energy storage products in response to increased food availability and, thus, variably depending on food abundance. Alternatively, Elvin [8] found that, for *Haliclona permollis*, an increase in lipid content was associated with oogenesis and the formation of mesenchymal amoebocytes. Although *A. queenslandica* is reproductively active year round, larval production peaks in the warmer months [21]. It is not clear if oogenesis is the reason for the high lipid content as the sampling time of this study corresponds with the end of summer when reproductive output is starting to decline. In addition, *A. queenslandica* concentrates its oocytes and embryos in localised brood chambers [21], unlike many sponges, suggesting that these levels should vary markedly within an individual sponge.

In *A. queenslandica*, the cellular biomass accounts for less than half of the dry weight. We defined the skeleton as the solid, inorganic spicules and the flexible, organic collagen. The skeleton, being a product of cellular metabolism, is a considerable sink of metabolic resources, accounting for 64% of the total dry weight (Table 1). The amount of skeleton in *A. queenslandica* is within the ranges of both Antarctic [12] and tropical [7] sponges. The high skeletal requirement by all sponges, greater than 50% dry weight, is likely due to the lack of defined tissues, including connective tissues, that are required to support complex body plans. Without the significant support and framework that the skeleton provides, sponges would not be able to achieve the sizes observed across the phylum [22].

Although sponges do not have defined tissues, the distribution of cells in *A. queenslandica* does not appear to be even if DNA is used as a proxy for cell number. This could be in part due to the large size range of sponge cells, as smaller cells are able to occupy less space yet all cells contain the same amount of DNA. Cells typically contain more RNA than DNA [23], and this is also the case for *A. queenslandica*, in which RNA was more abundant than DNA.

3.2. The Variation in Macromolecule Composition

Metabolic networks are dynamic, plastic, and redundant [23]. These characteristics have evolved to give organisms the flexibility to rapidly adapt to variations in environment or external stimulus [23]. As the aim of this study is to gather baseline data for use in metabolic modeling, we sampled different individuals and took sub-samples within each individual to investigate the source of potential variation. This is important information that will help guide future work. There was marked variation both between and within individual sponges in the abundance of macromolecules (Figure 2).

The skeleton was the most abundant component, and the variation in between-individual abundance was shown to have a significant effect on the overall population variation. The skeleton is the structural support for the cellular biomass and it guides the irregular canal network through which water is pumped to supply food and oxygen [14]. The variability seen in the amount of skeleton is likely a result of *A. queenslandica*'s encrusting-lobate growth form [24], combined with the irregular substrate and different structural requirements of specific internal features, such as canals and brood chambers.

It was previously mentioned that the high lipid content could be due to the reproductive state of the population. The reproductive state would also explain the variability in the lipid content within an individual. *A. queenslandica* forms localised brood chambers where embryos are matured before release [25] and not all of the biopsies would have contained brood chambers.

The variation in macromolecule composition may underlie the metabolic responses of the sponge to an unpredictable and competitive environment. Thus, at the scale of within an individual, differences in microhabitat such as food abundance, predatory pressures and conflicts for space can influence the macromolecular composition. This is potentially exacerbated by the nature of the biochemical assays and the error associated with correcting for salt content. Our results reveal significant effects of between-individual variation and a lesser effect of within-individual variation. This indicates that future work on the composition or metabolic processes of *A. queenslandica* should maintain a minimum of four representative individuals, supplemented with at least some subsampling within each individual. Care should still be taken with experimental design as this dataset is representative of a single metabolic state, and assumptions relating these results to different seasonal or ontogenetic stages should be made with caution. These results highlight the importance of biomass composition analysis, rather than the use of literature values.

3.2. Fatty Acid and Sterol Composition

The lipidome consists of a mixture of fatty acids and sterols, of which sponges are renowned for their diversity and often unique compounds [26–29]. Despite this diversity among FA structures in sponges, there are no clear or reliable taxonomic trends [26,28]; this can in part be attributed to the microbial communities associated with sponges [28]. We did not want to focus on identifying each compound in the lipid fraction as this is not necessary for modeling the sponge holobiont and performing flux balance analysis. As the precursor molecule for all lipid synthesis is acetyl coenzyme A, flux balance analysis may be performed if the contribution of lipids towards the overall biomass is known. We identified what compounds matched our databases as their inclusion will increase the models accuracy. Unidentified compounds were classified as FAs, sterols or neither based on their fragmentation pattern.

The two most abundant FAs in *A. queenslandica* were palmitic acid and octadecanoic acid. Both of these FAs are common in nature and have been reported in sponges previously [30–34]. Palmitic acid is a precursor with chain elongation, desaturation and addition of side chain modifications possible [35]. Apart from pentadecanoic acid (15:0) and heptadecanoic acid (17:0), the remaining identified FAs are long chain FA. FA with carbon chain length 20 (eocosanoic acid), 22 (docosanoic acid) and 24 (tetracosanoic acid), along with desaturated FAs of the same length (11-eicosanoic acid, erucic acid and nervonic acid, respectively) were identified as contributing to biomass formation. All of these FAs have been reported in sponges previously (see references above). Eleven out of twenty-three FAs considered important for biomass production were unknown. Sponges, and their associated bacteria, have been shown to produce extensive modifications including methylation [28] and hydroxylation [33] that would prevent the matching of novel compounds to commercial FA databases. This diversity and high number of unknown compounds highlights the complexity of the *A. queenslandica* lipidome and the potential for novel compound discovery.

The FAME analysis also identified the sterol fraction of the extracted lipids. The three sterols that fit our criteria of contributing to the overall biomass were brassicasterol, cholesterol, and an unknown sterol (UU13). Surprisingly, the most abundant of these was the brassicasterol, a plant steroid hormone [35] that is synthesized by terrestrial and marine plants, including macro and unicellular algae. Brassicasterol has been previously reported in a range of filter feeding and grazing marine invertebrates including sponges [36], gorgonian corals (Cnidaria; Anthozoa; Octocorallia; Gorgonacea) [37] and sea urchins (Echinodermata; Echinozoa; Echinoidea) [38]. In these other classes of animals, brassicasterol is a minor sterol and its presence could be a result of dietary uptake. This is likely the case for the *A. queenslandica* holobiont, as neither animals nor bacteria have been shown to synthesize brassicasterol [36–38]. Nevertheless, the high abundance implies an important biological function that this study is unable to discern. While brassicasterol is an unusual, dominant sterol for an animal, cholesterol is considered an important component of all cellular membranes in animals [23]. The concentration of cholesterol in a membrane performs two important functions. The overall fluidity of a membrane is regulated by the amount of cholesterol. Increases in cholesterol content reduce the fluidity of the membrane and permability of neutral ions [39]. Membrane bound cholesterol can also act as a promoter or inhibitor of membrane bound protein reactions [40]. Additionally, cholesterol functions as an important substrate molecule for bioactive compounds and other sterols, including brassicasterol [35]. Without identifying the second most abundant sterol, UU13, we are unable to infer its biological role, or synthesis pathway.

3.3. Amino Acid Composition

Analysing the amino acid composition identifies the relative amounts of precursor molecules that are required to synthesis the protein component of the biomass. When complemented with information from genome scale modeling of the synthetic pathways that are present in an organism, these data allow for the identification of essential amino acids that cannot be produced by the organism [5]. Subsequent flux analysis can identify bottlenecks where the high requirement of a compound for biomass synthesis slows overall growth. The protein component of *A. queenslandica* comprises of two distinct products of metabolism. The overall sponge biomass consists of soluble, cellular derived protein, and an insoluble, extracellular protein matrix. This extracellular matrix is dominated by a sponge specific collagen called spongin [19], which is the main component of the sponge skeleton.

We defined the skeleton of *A. queenslandica* as the inorganic silica spicules, together with the organic collagen matrix. Collagen is an important structural component of metazoans. We found that the spongin in *A. queenslandica* contained 34.9% glycine (Table 6). This is close to the values reported for *Chondria reniformis* (30.5%) ([41], *Scypha graminae* (31.5%–32.3%) [42] and *Ircinia* sp. (32.3%) [43]. Alanine (10.3%) and proline (7.5%) were the next two most abundant amino acids (Table 6) and are very close in abundance to the values of these other sponges. Although the amino acid composition of sponge collagens is well documented [41–43], the amino acid composition of the complete sponge biomass has not been reported.

The complete amino acid composition may give insights into the many primary cell culture attempts that so far have had limited success [44]. Many studies have utilised mammalian cell culture media, with and without serum [44]. A comparison of the amino acid composition of the complete *A. queenslandica* biomass (Table 4) with a hybridoma cell line [45] revealed nine amino acids that are at least twice as abundant in *A. queenslandica*. Proline and glycine stand out at 3.3 and 8.5 times more abundant in *A. queenslandica* than the hybridoma line. If we use DMEM F12 medium as an example of mammalian cell culture media, both proline and glycine are minor contributors to the amino acid fraction [46] and could present significant bottlenecks for the production of biomass. In addition, due to the different amino acid bias in serum compared to sponge biomass, in the serum experiments excess waste amino acids could lead to growth limitations and may point to necessary changes in the approach to sponge cell culture.

We analysed the amino acid composition of the complete sponge biomass. From this we were able to estimate the average amino acid composition of the cellular fraction by subtracting the average composition of the skeleton. Values were then normalized for comparison by converting to a percentage. This showed that the cellular composition was distinct from the skeleton. The skeleton proportionally contained more glycine, alanine and proline than the cellular biomass. These three amino acids are key in giving collagen its triple helix shape and physical properties [23]. Amino acids that are essential in higher animals—in particular isoleucine, leucine, and valine [23]—were proportionally higher in abundance in the cellular biomass than in the skeleton. Whether these are also essential amino acids for *A. queenslandica* is yet to be shown.

4. Experimental Section

4.1. Sample Collection and Preparation

A. queenslandica were collected from Shark Bay, Heron Island, Australia (23°26′37.92″ S, 151°55′8.81″ E). Individual sponges and their coral rubble substrate were collected with a hammer and chisel [25] and transported submerged to the Heron Island Research Station. Sponges were held in a flow through aquaria system and processed within 8 h of collection. Biopsy samples were collected from a total of 29 sponges. The samples were cut with a sterile scalpel and contained a representative cross section of sponge biomass that consistently included the pinacoderm layer through to biomass just above the substratum. A thin layer of biomass was removed from the bottom of the sponge to avoid contamination by the substrata. Pieces of sponge biomass that had foreign material incorporated were avoided. To investigate the variation in composition within individual sponges, biopsies were collected from each sponge in multiples of 5. This resulted in 20 sponge samples per analysis, comprising 5 biopsy samples each of 4 individual adult sponges. Samples were snap frozen in liquid nitrogen then stored at −80 °C and transported at on dry ice. Lyophilisation was performed overnight. Dry weights were recorded on an analytical balance and corrected for salt content by multiplying the amount of water lost through lyophilisation by the specific gravity of seawater. We assumed that the salt content of the cells was the same as the seawater. Throughout the study, the weight of each reagent in each reaction was recorded. The specific gravity of each reagent was then used to ensure that accurate calculations were made.

4.2. Skeleton Dry Weight

Cellular material was removed from the *A. queenslandica* biopsies by re-hydrating with a 2% SDS solution made with MilliQ water and incubating for 2 h at room temperature on a shaker table. The biopsies were then washed thoroughly for 5 min with running RO water. The SDS incubation and washing times were optimized by inspecting with a Nikon Eclipse Ti-E inverted microscope (Nikon,

Tokyo, Japan) to ensure all cellular material was removed. The pieces were then lyophilised overnight and dry weights recorded. These samples were stored at −80 °C.

4.3. Nucleic Acids

DNA was quantified by Hoechst fluorescence assay. The lyophilised samples from 4.1 were re-hydrated with 3.0 mL of TE buffer (pH 8.0) with 0.01 g/mL lysozyme and incubated at 37 °C for 1 h on a shaker table set to 100 rpm. Re-hydrating the lyophilised biopsies caused complete lysis of the *A. queenslandica* cells, but not the bacterial fraction. Lysozyme was added to ensure lysis of the bacterial fraction. After incubation, sterile tweezers were used to remove the sponge skeleton from the tube. The piece of skeleton was squeezed and rolled against the inside wall to ensure that the lysis buffer was retained. Samples were then vortexed and centrifuged for 5 min at 4000 g to pellet cellular debris. Aliquots of the supernatant were used in the Hoechst fluorescence assay. Calf thymus DNA was used to generate a standard curve [47]. All samples and standards were measured in triplicate.

RNA content was quantified by digesting the lyophilised samples from 4.1 with 3.0 mL of 0.3 M KOH for 1 h at 37 °C [48]. The reaction was then acidified with the addition of 1.0 mL of 3 M perchloric acid. [49]. Following centrifugation to remove DNA and protein, the RNA-containing supernatant was quantified in triplicate by UV absorbance at 260 nm [48].

The nucleotide composition of both the DNA and RNA was calculated using the GC content of the genome and an adult transcriptome. The calculations assumed that the DNA and RNA is all sponge derived as the bacterial genomic and transcriptomic sequence data, in addition to bacterial diversity and abundance data is not yet available. The GC content of the *A. queenslandica* genome and adult transcriptome are 31.14% [18] and 39.76% [20], respectively. The percentage contribution of each nucleotide was calculated, assuming an even guanine-cytosine and thymine-adenine/thymine-uracil ratios. Next the average molecular weight of one nucleotide was calculated. This was done by multiplying the abundance of each nucleotide by its molecular weight. These values were then added together to give the average molecular weight. This was multiplied by the mean total amount of DNA or RNA (g/gDW) to give the number of mols of average nucleotide per gram of dry weight. The amount of each nucleotide (mols) was then calculated using this value and the guanine-cytosine to thymine-adenine/thymine-uracil ratios. Values were converted to micromoles for reporting.

4.4. Protein

Sponge biopsies from 4.1 were digested with 3.0 mL of 1.0 M NaOH incubated at 95 °C for 5 min. The reaction was cooled in iced water and then centrifuged at 4000× *g* for 5 min to pellet suspended skeleton solids. The supernatant was assayed in triplicate using the Coomassie Plus™ (Bradford) Assay Kit (Thermo Scientific, Rockford, MI, USA). Bovine serum albumin was used as the standard [50].

4.5. Carbohydrate

Carbohydrate content was quantified using the phenol-sulfuric acid method [51]. The sponge biopsies from 4.1 were rehydrated with 1.0 mL of MilliQ water. 5.0 mL of concentrated sulfuric acid was added and the reaction let proceed for 10 min. The reaction was rapidly cooled by placing the tube in iced water for 1 min. [52]. 50 μL of phenol was added and incubated for 30 min in a water bath set to 25 °C. The reaction was diluted 1:50 using concentrated sulfuric acid and the absorbance read at 488 nm. All measurements were taken in triplicate and glucose was used as the standard [52].

4.6. Lipid

Biopsy samples from 4.1 were rehydrated with 1.0 mL MilliQ water and extracted overnight with 3:2 hexane:isopropanol [53]. 5.0 mL of 0.47 M $NaSO_4$ was added, mixed and then separated by centrifugation at 4000× *g* for 5 min. The hydrophobic upper hexane phase contains the lipid fraction and was transferred to a clean glass tube. The initial solution was washed with 7:2 hexane:isopropanol to ensure that all of the lipids were recovered [53]. This second hexane phase was again recovered with centrifugation and combined with the first extract. The combined hexane extracts were evaporated under a constant flow of nitrogen to prevent oxidation. The remaining solid lipids were quantified gravimetrically. The lipid solid was then stored under nitrogen at −80 °C for fatty acid analysis.

4.7. Fatty Acids and Sterols

A fatty acid methyl ester (FAME) analysis was performed on the extracted lipids as previously described [54]. The lipid solid was dissolved in hexane and 5.1 μg of nondecanoic acid (Sigma-Aldrich, St. Louis, MO, USA) was used as an internal standard (ISTD) to a final concentration of 3.6 μg/mL. Saponification, methylation and GC/MS analysis of the lipid mass fraction followed the protocol described in [54] The software program AMDIS (version 2.64, National Institute of Standards and Technology (NIST), Gaithersburg, MD, USA) was used for the analysis of the fatty acid and sterols. A minimum match factor of 80 and signal threshold of 700 was set for identifying compounds. A FAME library was built using the target ions and retention times of thirty-two known fatty acid standards (Sigma-Aldrich, St. Louis, MO, USA). An alkane ($C_{7\text{-}30}$) standard was used as a retention time index. To define compounds important for biomass formation, we used a cut-off of 0.05% of the largest peak in the MS analysis, which means that peaks below those signal intensities would not be reported. For unknown compounds (*i.e.*, that were not identified using the standard library), their fragmentation patterns were compared to the NIST database library (version 2.0, NIST, Gaithersburg, MD, USA). For a range of peaks both the target library as well as the NIST library did not provide accurate hits. Based on manual interpretation of fragmentation patterns, unknowns that were found consistently in the majority (>75%) of the samples were classified into sterols or fatty acids. From the quantified lipid fraction (g/gDW) the estimated weight of glycerol and phosphate groups was subtracted assuming a glycerolipid to phospholipid ratio of 20:80. It was then assumed that the reminder of the lipid fraction would generate a signal in GC-MS and that the MS response and ionization efficiency for all FA and sterol ions was comparable (due to the unknown character of many FA's this cannot be tested, but individual errors for individual compounds will not impact dramatically on the later modelling). All peak areas were normalized to the total signal generated for that sample. Using the molecular weight of the knowns and an average molecular weight (calculated from all identified FA's and sterols) for the unknowns the individual concentrations were calculated in mol/gDW. A total of 23 separate fractions of biomass were analyzed and the standard deviation was calculated.

4.8. Amino Acids

Amino acid analysis was performed by reversed phase high performance liquid chromatography. 20 samples containing cellular and skeleton components were analysed in addition to 20 samples of skeleton, prepared as above. 5 mL of 6 M HCl was added to each sample and incubated at 105 °C. Hydrolysis times were first optimized and found to be 54 h for the complete biomass samples and 48 h for skeleton samples. Optimization monitored the concentration increase of all amino acids over the time course and 54/48 h was found to be a time at which most amino acids reached the peak concentration. Longer times may favor some compounds but will lead to degradation of others. This compromise was necessary since otherwise individual experiments need to be conducted for all 20 amino acids. Samples were allowed to cool, mixed, and then a 500 µL aliquot was taken and filtered with a 0.22 µm Ultrafree-MC GV centrifugal filter (Merck Millipore, Darmstadt, Germany). The HCl was evaporated by heating to 40 °C under a constant flow of nitrogen. The solid pellet was stored at −80 °C. Derivatisation was performed in a high-performance autosampler (Agilent HiP-ALS SL, G1367C, Santa Clara, CA, USA). 0.5 µL of sample containing 250 µM of internal standards, sarcosine and 2-aminobutyric acid, was added into 2.5 µL of borate buffer (0.4 N, pH 10.2, Agilent PN: 5061-3339, Santa Clara, CA, USA), mixed and incubated for 20 s at 4 °C. 1 µL of OPA reagent (10 mg *o*-pthalaldehyde/mL in 3-mercaptopropionic acid, Agilent PN: 5061-3335, Santa Clara, CA, USA) was then added to initially derivatise primary amino acids. The reaction was mixed and incubated for 20 s at 4 °C. Then 0.4 µL of FMOC reagent (2.5 mg 9-fluorenylmethyl chloroformate/mL in acetonitrile, Agilent PN:5061-3337, Santa Clara, CA, USA) was added, mixed and incubated for 20 s at 4 °C to derivatise the secondary amines proline and sarcosine. 45.6 µL of Buffer A (40 mM Na_2HPO_4, 0.02% NaN_3, pH 7.8) was added to lower the pH of the reaction prior to injecting the 50 µL reaction onto an Agilent Zorbax Extend C-18 column (3.5 µm, 4.6 × 150 mm, Agilent PN: 763953-902, Santa Clara, CA, USA) with a guard column (SecurityGuard Gemini C18, Phenomenex PN: AJO-7597, Santa Clara, CA, USA). Column temperature was kept at 39 °C in a thermostatted column compartment (Agilent TCC, G1316B, Santa Clara, CA, USA). Chromatography was performed using an Agilent 1200-SL HPLC system, equipped with an active seal wash and a degasser (Agilent Degasser, G1379B, Santa Clara, CA, USA). The HPLC gradient was 2%–45% buffer B (45% acetonitrile, 45% methanol and 10% water) from 0 to 18 min, 50%–60% buffer B from 18.1 to 20 min. 100% buffer B from 20.1 to 24 min, and 2% buffer B from 24.1 to 27 min—using a binary pump (Agilent Bin Pump SL, G1312B, Santa Clara, CA, USA). Flow rate was 2 mL/min. Derivatised amino acids were monitored using a fluorescence detector (Agilent FLD, G1321A, Santa Clara, CA, USA). OPA-derivatised amino acids were detected at 340_{ex} and 450_{em} nm from 1 to 18 min, and FMOC-derivatised amino acids at 266_{ex} and 305_{em} nm from 18 to 27 min. Chromatograms were integrated using ChemStation (Rev B.03.02, Agilent, Santa Clara, CA, USA).

4.9. Statistical Analysis

To analyse how the variation between and within individual sponges contributed to the overall variation for each macromolecule separately, the data were fit to a linear mixed effects model using the *nlme* package [55] in R [56].

The amount of macromolecule (g/gDW) was the response variable and Sponge ID (individual) was the random effect. The among-individual variance was divided by the total variance and expressed as the percentage of the overall variability that was due to variation between individuals. To address whether this between-individual variation was significant, we used a likelihood ratio x^2 test to compare two models: one with the random effect and one without the random effect.

5. Conclusions

As a prerequisite for a systems scale investigation of the metabolic processes responsible for sponge biomass production, we analysed the composition and abundance of the macro-components of *A. queenslandica.* We found the composition to be within the reported ranges for other tropical marine sponges and that there is significant variation in the composition at the population level and notable variation within the biomass of an individual sponge. The skeleton was the most abundant component of the overall biomass, while lipids and protein were the most abundant macromolecules. Further analysis of the fatty acid and amino acids was undertaken. The FAME analysis revealed a diverse and complex fatty acid composition. The composition of the sterols revealed that brassicasterol and an unknown sterol had a higher mean abundance than cholesterol. There was a distinct difference in amino acid composition between the soluble, cellular fraction and the insoluble skeletal component. This highlights the different protein products of metabolism that share the same pool of precursor molecules. This complete analysis of the biochemical composition of *A. queenslandica* provides baseline information that complements the extensive genomic resources available for *A. queenslandica*, a key species for understanding sponge biochemistry and metabolism.

Acknowledgments

This work was funded by Australian Research Council grants to SMD and BMD in addition to an Australian Postgraduate Award Scholarship to JW. JK was financially supported by the Australian Research Council (No. DE120101549). The authors thank Manuel Plan from Metabolomics Australia for his help with amino acid analysis.

Author Contributions

JK, SD, and BD initiated the project. JW conducted the fieldwork and performed the macromolecular assays. JW analysed the macromolecular data. TB performed the amino acid and FAME analyses. JW, TB and JK analysed the amino acid and FAME data. JW drafted the initial manuscript. JK, SD and BD edited the manuscript. All authors read and approved the final manuscript.

References

1. Blunt, J.W.; Copp, B.R.; Keyzers, R.A.; Munro, M.H.G.; Prinsep, M.R. Marine natural products. *Natl. Prod. Rep.* **2014**, *31*, 160–258.
2. Huyck, T.K.; Gradishar, W.; Kirkpatrick, P. 4 Eribulin mesylate. *Nat. Rev. Drug Dis.* **2011**, *10*, 173–174.

3. Genta-Jouve, G.; Thomas, O.P. *Sponge Chemical Diversity*, 1st ed.; Elsevier Ltd.: Toronto, ON, Canada, 2012; Volume 62, pp. 183–230.
4. Oberhardt, M.A.; Palsson, B.Ø.; Papin, J.A. Applications of genome-scale metabolic reconstructions. *Mol. Sys. Biol.* **2009**, *5*, 1–15.
5. Thiele, I.; Palsson, B.Ø. A protocol for generating a high-quality genome-scale metabolic reconstruction. *Nat. Protoc.* **2010**, *5*, 93–121.
6. Feist, A.M.; Palsson, B.Ø. The biomass objective function. *Curr. Opin. Microbiol.* **2010**, *13*, 344–349.
7. Chanas, B.; Pawlik, J.R. Defenses of Caribbean sponges against predatory reef fish. II. Spicules, tissue toughness, and nutritional quality. *Mar. Ecol. Prog. Ser.* **1995**, *127*, 195–211.
8. Elvin, D.W. The relationship of seasonal changes in the biochemical components to the reproductive behavior of the intertidal sponge, *Haliclona permollis*. *Biol. Bull.* **1979**, *156*, 47–61.
9. Pallela, R.; Koigoora, S.; Gunda, V.G.; Sunkara, M.S.; Janapala, V.R. Comparative morphometry, biochemical and elemental composition of three marine sponges (Petrosiidae) from Gulf of Mannar, India. *Chem. Speciat. Bioavailab.* **2011**, *23*, 16–23.
10. Koigoora, S.; Pallela, R.; Nelakurti, C.R.; Sunkara, M.S.; Janapala, V.R. Morphometric, elemental and biochemical characterisation of three *Haliclona* species (Demospongiae) from the Gulf of Mannar, India. *Chem. Speciat. Bioavailab.* **2013**, *25*, 15–23.
11. Stone, A.R. Seasonal variation in the gross biochemical composition of *Hymeniacidon perleve* (Montagu). *J. Exp. Mar. Biol. Ecol.* **1970**, *5*, 265–271.
12. McClintock, J.B. Investigation of the relationship between invertebrate predation and biochemical composition, energy content, spicule armament and toxicity of benthic sponges at McMurdo Sound, Antarctica. *Mar. Biol.* **1987**, *94*, 479–487.
13. Freeman, C.J.; Gleason, D.F. Chemical defenses, nutritional quality, and structural components in three sponge species: *Ircinia felix*, *I. campana*, and *Aplysina fulva*. *Mar. Biol.* **2010**, *157*, 1083–1093.
14. Hentschel, U.; Piel, J.; Degnan, S.M.; Taylor, M.W. Genomic insights into the marine sponge microbiome. *Nat. Rev. Microbiol.* **2012**, *10*, 641–654.
15. Thomas, T.R.A.; Kavlekar, D.P.; LokaBharathi, P.A. Marine Drugs from Sponge-Microbe Association—A Review. *Mar. Drugs* **2010**, *8*, 1417–1468.
16. Wilson, M.C.; Mori, T.; Rückert, C.; Uria, A.R.; Helf, M.J.; Takada, K.; Gernert, C.; Steffens, U.A.E.; Heycke, N.; Schmitt, S.; *et al.* An environmental bacterial taxon with a large and distinct metabolic repertoire. *Nature* **2014**, *506*, 58–62.
17. Radax, R.; Hoffmann, F.; Rapp, H.T.; Leininger, S.; Schleper, C. Ammonia-oxidizing archaea as main drivers of nitrification in cold-water sponges. *Environ. Microbiol.* **2011**, *14*, 909–923.
18. Srivastava, M.; Simakov, O.; Chapman, J.; Fahey, B.; Gauthier, M.E.A.; Mitros, T.; Richards, G.S.; Conaco, C.; Dacre, M.; Hellsten, U.; *et al.* The *Amphimedon queenslandica* genome and the evolution of animal complexity. *Nature* **2010**, *466*, 720–726.
19. Gross, J.; Sokal, Z.; Rougvie, M. Structural and chemical studies on the connective tissue of marine sponges. *J. Histochem. Cytochem.* **1956**, *4*, 227–246.

20. Fernandez-Valverde, S.L.; Degnan, B.M. *Amphimedon queenslandica* gene reannotation. 2013, Unpublished work.
21. Leys, S.P.; Degnan, B.M. Cytological basis of photoresponsive behavior in a sponge larva. *Biol. Bull.* **2001**, *201*, 323–338.
22. Hooper, J.N.A.; Soest, R.W.M. *Systema Porifera. A Guide to the Classification of Sponges*; Springer US: Boston, MA, USA, 2002.
23. Lehninger, A.; Nelson, D.L.; Cox, M.M. *Lehninger Principles of Biochemistry & eBook*, 5th ed.; W.H. Freeman: New York, NY, USA, 2008; pp. 385–399.
24. Hooper, J.N.; van Soest, R. A new species of *Amphimedon* (Porifera, Demospongiae, Haplosclerida, Niphatidae) from the Capricorn-Bunker Group of Islands, Great Barrier Reef, Australia: Target species for the "sponge genome project". *Zootaxa* **2006**, *1314*, 31–39.
25. Leys, S.P.; Larroux, C.; Gauthier, M.; Adamska, M.; Fahey, B.; Richards, G.S.; Degnan, S.M.; Degnan, B.M. Isolation of *Amphimedon* Developmental Material. *Cold Spring Harbor Protoc.* **2008**, doi:10.1101/pdg.prot5095.
26. Rodríguez, W.; Osorno, O.; Ramos, F.A.; Duque, C.; Zea, S. Biochemical Systematics and Ecology. *Biochem. Syst. Ecol.* **2010**, *38*, 774–783.
27. Lawson, M.P.; Bergquist, P.R.; Cambie, R.C. Fatty acid composition and the classification of the Porifera. *Biochem. Syst. Ecol.* **1984**, *12*, 375–393.
28. Rod'kina, S.A. Fatty acids and other lipids of marine sponges. *Russ. J. Mar. Biol.* **2005**, *31*, S49–S60.
29. Erpenbeck, D.; Soest, R.W.M. Status and Perspective of Sponge Chemosystematics. *Mar. Biotechnol.* **2006**, *9*, 2–19.
30. Barnathan, G.; Korpnrobst, J.M.; Doumenq, P.; Mirallès, J. New unsaturated long-chain fatty acids in the phospholipids from the axinellida sponges *Trikentrion loeve* and *Pseudaxinella* cf. *lunaecharta*. *Lipids* **1996**, *31*, 193–200.
31. Carballeira, N.M.; Alicea, J. The first naturally occurring α-methoxylated branched-chain fatty acids from the phospholipids of *Amphimedon complanata*. *Lipids* **2001**, *36*, 83–87.
32. Christie, W.W.; Brechany, E.Y.; Stefanov, K.; Popov, S. The fatty acids of the sponge *Dysidea fragilis* from the black sea. *Lipids* **1992**, *27*, 640–644.
33. Carballeira, N.M.; Lopez, M.R. On the isolation of 2-hydroxydocosanoic and 2-hydroxytricosanoic acids from the marine sponge *Amphimedon compressa*. *Lipids* **1989**, *24*, 89–91.
34. Nechev, J.; Christie, W.W.; Robaina, R.; de Diego, F.; Popov, S.; Stefanov, K. Chemical composition of the sponge *Hymeniacidon sanguinea* from the Canary Islands. *Comp. Biochem. Physiol. Part A* **2004**, *137*, 365–374.
35. Michal, G.; Schomburg, D. *Biochemical Pathways*; John Wiley & Sons: New York, NY, USA, 2013; pp. 107–122.
36. Erdman, T.R.; Thomson, R.H. Sterols from the sponges *Cliona celata* Grant and *Hymeniacidon perleve* Montagu. *Tetrahedron* **1972**, *28*, 5163–5173.
37. Block, J.H. Marine sterols from some gorgonians. *Steroids* **1974**, *23*, 421–424.
38. Voogt, P.A. Sterols of some echinoids. *Arch. Physiol. Biochem.* **1972**, *80*, 883–891.

39. Haines, T.J. Do sterols reduce proton and sodium leaks through lipid bilayers? *Prog. Lipid Res.* **2001**, *40*, 299–324.
40. Yeagle, P.L. Modulation of membrane function by cholesterol. *Biochimie* **1991**, *73*, 1303–1310.
41. Imhoff, J.M.; Garrone, R. Solubilization and characterization of *Chondrosia reniformis* sponge collagen. *Connect. Tissue Res.* **1983**, *11*, 193–197.
42. Peiz, K.A.; Gross, J. The amino acid composition and morphology of some invertebrate and vertebrate collagens. *Biochim. Biophys. Acta* **1959**, *34*, 24–39.
43. Junqua, S.; Robert, L.; Garrone, R.; Pavans de Ceccatty, M.; Vacelet, J. Biochemical and morphological studies on collagens of horny sponges. *Ircinia* filaments compared to spongines. *Connect. Tissue Res.* **1974**, *2*, 193–203.
44. Schippers, K.J.; Sipkema, D.; Osinga, R.; Smidt, H.; Pomponi, S.A.; Martens, D.E.; Wijffels, R.H. Cultivation of sponges, sponge cells and symbionts: Achievements and future prospects. *Adv. Mar. Biol.* **2012**, *62*, 273–337.
45. Quek, L.; Dietmar, S.; Krömer, J.O.; Nielsen, L.K. Metabolic flux analysis in mammalian cell culture. *Metab. Eng.* **2010**, *12*, 161–171.
46. Freshney, R.I. *Culture of Animal Cells*; Wiley: New Jersey, NJ, USA, 2010; pp. 102–104.
47. Green, M.R.; Sambrook, J. *Molecular Cloning*; Cold Spring Harbor Laboratory: New York, NY, USA, 2012; pp. 68–70.
48. Benthin, S.; Nielsen, J.; Villadsen, J. A simple and reliable method for the determination of cellular RNA content. *Biotechnol. Tech.* **1990**, *5*, 39–42.
49. Fleck, A.; Munro, H.N. The precision of ultraviolet absorption measurements in the Schmidt-Thannhauser procedure for nucleic acid estimation. *Biochim. Biophys. Acta* **1962**, *55*, 571–583.
50. Bradford, M.M. A rapid and sensitive method for the quantitation of microgram quantities of protein utilizing the principle of protein-dye binding. *Anal. Biochem.* **1976**, *72*, 248–254.
51. DuBois, M.; Gilles, K.A.; Hamilton, J.K.; Rebers, P.A.; Smith, F. Colorimetric Method for Determination of Sugars and Related Substances. *Anal. Chem.* **1956**, *28*, 350–356.
52. Rao, P.P.; Pattabiraman, T.N.T. Reevaluation of the phenol-sulfuric acid reaction for the estimation of hexoses and pentoses. *Anal. Biochem.* **1989**, *181*, 18–22.
53. Hara, A.; Radin, N.S. Lipid extraction of tissues with a low-toxicity solvent. *Anal. Biochem.* **1978**, *90*, 420–426.
54. Brennan, T.C.R.; Krömer, J.O.; Nielsen, L.K. Physiological and Transcriptional Responses of *Saccharomyces. cerevisiae* to D-Limonene Show Changes to the Cell Wall but Not to the Plasma Membrane. *Appl. Environ. Microbiol.* **2013**, *79*, 3590–3600.
55. Pinheiro, J.; Bates, D.; DebRoy, S.; Sarkar, D. R Development Core Team. nlme: Linear and Nonlinear Mixed Effects Models. R Package Version 2013. Available online: http://CRAN.R-project.org/package=nlme (accessed on 26 October 2013).
56. R Core Team. R: A language and environment for statistical computing. R Foundation for Statistical Computing: Vienna, Austria, 2013. Available online: http://www.R-project.org/ (accessed on 26 October 2013).

3

Molecular Response to Toxic Diatom-Derived Aldehydes in the Sea Urchin *Paracentrotus lividus*

Stefano Varrella [1], Giovanna Romano [1], Adrianna Ianora [1], Matt G. Bentley [2], Nadia Ruocco [1] and Maria Costantini [1,*]

[1] Stazione Zoologica Anton Dohrn, Villa Comunale, Naples 80121, Italy;
E-Mails: stefano.varrella@szn.it (S.V.); giovanna.romano@szn.it (G.R.);
adrianna.ianora@szn.it (A.I.); nadia.ruocco @szn.it (N.R.)

[2] Dove Marine Laboratory, School of Marine Science and Technology, Newcastle University, Newcastle upon Tyne, UK; E-Mail: matt.bentley@newcastle.ac.uk

* Author to whom correspondence should be addressed; E-Mail: maria.costantini@szn.it

---

Abstract: Diatoms are dominant photosynthetic organisms in the world's oceans and represent a major food source for zooplankton and benthic filter-feeders. However, their beneficial role in sustaining marine food webs has been challenged after the discovery that they produce secondary metabolites, such as polyunsaturated aldehydes (PUAs), which negatively affect the reproductive success of many invertebrates. Here, we report the effects of two common diatom PUAs, heptadienal and octadienal, which have never been tested before at the molecular level, using the sea urchin, *Paracentrotus lividus*, as a model organism. We show that both PUAs are able to induce teratogenesis (*i.e.*, malformations), as already reported for decadienal, the better-studied PUA of this group. Moreover, post-recovery experiments show that embryos can recover after treatment with all three PUAs, indicating that negative effects depend both on PUA concentrations and the exposure time of the embryos to these metabolites. We also identify the time range during which PUAs exert the greatest effect on sea urchin embryogenesis. Finally, we report the expression levels of thirty one genes (having a key role in a broad range of functional responses, such as stress, development, differentiation, skeletogenesis and detoxification processes) in order to identify the common targets affected by PUAs and their correlation with morphological abnormalities. This study opens new perspectives for understanding how marine organisms afford protection from environmental toxicants through an integrated network of genes.

Keywords: aldehydes; molecular targets; sea urchin; teratogenesis

---

Abbreviations

FSW, filtered sea water; hpf, hours post fertilization; µL, microliter; µM, micromolar; min, minutes; mM, millimolar; nm, nanometer.

1. Introduction

Diatoms are a highly productive class of microalgae, widespread in both marine and freshwater habitats, that are widely fed upon by both planktonic and benthic invertebrates. However, while diatoms may provide a source of energy for larval growth, they often reduce fecundity and/or hatching success or cause malformations (teratogenesis) during growth, due to the production of secondary metabolites, such as polyunsaturated aldehydes (PUAs) and other products deriving from the oxidation of polyunsaturated fatty acids (PUFAs), collectively termed oxylipins (reviewed by [1] and [2]). This biological model is new and has no other equivalent in marine plant-herbivore systems, since most of the known negative plant-animal interactions are generally related to repellent or poisoning processes, but never to reproductive failure.

Oxylipins, and PUAs in particular, have important biological and biochemical properties, including the disruption of gametogenesis, gamete functionality, fertilization, embryonic mitosis and larval fitness and competence [1]. The dominant bioactive PUAs released by diatoms are C_{10} 2-*trans*-4-*trans*-decadienal, 2-*trans*-4-*cis*-7-*cis*-decatrienal and 2-*trans*-4-*trans*-7-*cis*-decatrienal [3], but also C_8 2-*trans*-4-*cis*-7-octatrienal, 2-*trans*-4-*trans*-7-octatrienal, 2-*trans*-4-*cis*-7-octadienal, 2-*trans*-4-*trans*-7-octadienal, C_7 2-*trans*-4-*cis*-7-heptadienal and 2-*trans*-4-*trans*-7-heptadienal [4,5]. The first two PUAs were isolated from the freshwater diatom, *Melosira varians*, by Wendel and Jüttner [6], but the biological activity of these molecules was not known at the time. Miralto *et al.* [3] showed, for the first time, that they arrested the embryonic development of copepod and sea urchin embryos in a dose-dependent manner and also had anti-proliferative and apoptotic effects on human carcinoma cells. Successive studies have shown that sea urchin gametes incubated with decadienal have impaired fertilization success, due to the inhibition of sperm motility [7,8]. At concentrations higher than the dose required to arrest cell cleavage progression, decadienal induced apoptotic events in *Paracentrotus lividus* embryos by inducing caspase-3-like protease activity [9].

Hansen *et al.* [10] studied the effects of decadienal on the sea urchin, *Sphaerechinus granularis*, and showed that this PUA inhibited cyclin B/Cdk1 kinase activity, DNA replication and tubulin polymerization, leading to arrest of the cell cycle. Romano *et al.* [11] showed that PUAs compromised embryonic and larval development of sea urchins even at low doses and that the most deleterious of the PUAs tested were the longer chain aldehydes, such as decadienal.

The first molecular studies on the effects of PUAs on the sea urchin, *P. lividus*, were reported very recently [12,13]. In particular, newly fertilized sea urchin eggs were exposed to low concentrations of decadienal, and the expression levels of seventeen genes, implicated in a broad range of functional responses, were followed by real-time qPCR. At low decadienal concentrations, the sea urchin, *P. lividus*, activated different classes of genes to defend itself against this toxic aldehyde, ranging from canonical stress genes to developmental and skeletogenic genes [13]. The authors suggested that this orchestrated defence system against decadienal represents part of the chemical defensome of *P. lividus*, affording protection from environmental toxicants.

Since information on the molecular effects of PUAs are scant and mostly related to the effects induced by decadienal, the aim of the present study was to explore the effects of two other ecologically important aldehydes, heptadienal and octadienal, which have never been tested before on *P. lividus* embryos from the molecular point of view, and to compare these effects with those induced by decadienal. The sea urchin, *P. lividus*, is considered a good model system to study the ecotoxicological response of marine invertebrates to environmental pollutants for several reasons: its ecological relevance, benthic and relatively sedimentary lifestyle, rapid response and high sensitivity to many types of contaminants, transparent embryos that grow rapidly in the laboratory and its long reproductive period. For these reasons, we decided to adopt it for our study and treated sea urchin embryos with increasing concentrations of heptadienal and octadienal to analyse morphological changes induced by exposure to these natural products and to define their mechanism of action and possible teratogenic activity. We also followed, by real-time qPCR, thirty one genes to identify potential target genes. Fourteen of these genes, having a key role in a broad range of functional responses, such as development, differentiation and detoxification processes, are now compared to those investigated in our previous study on the effects of decadienal on *P. lividus* development [13].

2. Results

2.1. Determination of Teratogenic Effects and PUA Dose-Dependent Concentrations

As reported in previous studies [11,12], decadienal induced teratogenesis at low concentrations, with an increase in the number of abnormal plutei in the sea urchin, *P. lividus*. Marrone *et al.* [13] also reported a dose-dependent effect of this PUA in *P. lividus*, with severe malformations in plutei, such as asymmetrical arms and spicules and reduced arm length and shortening of the apex. Romano *et al.* [12] showed that a decadienal concentration of 1.6 μM (0.25 μg/mL) was ideal to study morphogenetic changes in embryo development and gene expression levels in *P. lividus*, with the production of about 35% abnormal plutei.

The present study explores further the effects of decadienal at the molecular level and compares these effects with those induced by two other PUAs, heptadienal and octadienal, which have never been tested before on *P. lividus* embryo development. Treatment with the three PUAs induced similar malformations in the plutei (see Figure 1), with effects on the arms, spicules and apex. Such plutei had a poorly-formed apex (Figure 1B) because of the spicules that appeared either parallel or disjoined at the tip (Figure 1C,D) or crossed at the apex (Figure 1E), with arms that appeared longer and broader or crooked and asymmetrical (Figure 1F) or completely degenerated (Figure 1G), compared to controls

(Figure 1A). At times, the entire body plan of the plutei was strongly compromised and malformed (Figure 1H). For a more detailed overview of the abnormal plutei produced by PUAs, see Supplementary Figure S1.

Figure 1. Examples of malformations induced in *Paracentrotus lividus* plutei at 48 hours post fertilization (hpf) after incubation with the three polyunsaturated aldehydes (PUAs), heptadienal, octadienal and decadienal (**B–H**) in comparison with (**A**), the control; embryos are in sea water without aldehydes.

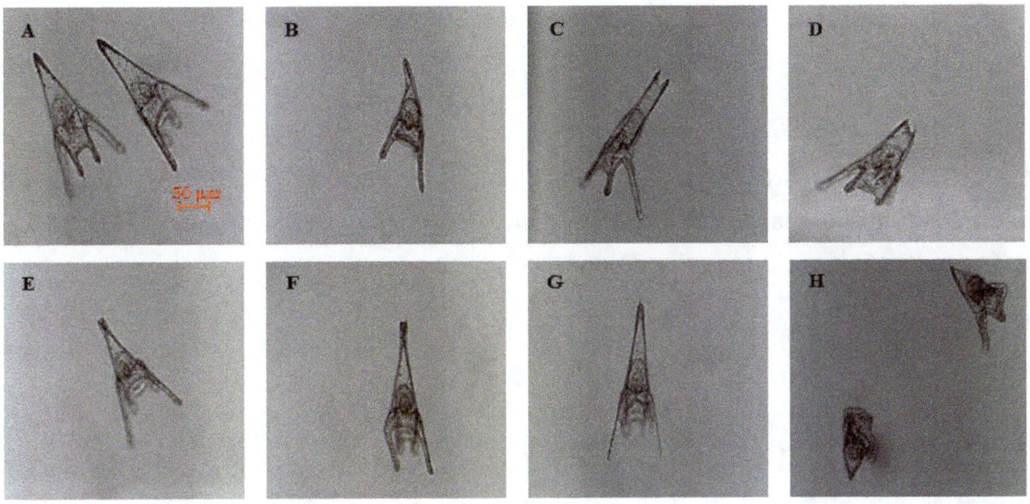

The percentage of abnormal plutei was calculated at the different concentrations tested (Figure 2; see also the Experimental Section for more details).

The effect was dose-dependent for the three PUAs tested, even if the range of concentrations inducing teratogenesis differed (from 0.5 to 2.5 μM for decadienal, from 1.0 to 6.0 μM for heptadienal and from 2.0 to 9.0 μM for octadienal). Decadienal was the strongest of the three, because of the very narrow range (2.0 μM) that affected embryonic development; heptadienal and octadienal required higher ranges of concentrations to reach the same effects as decadienal. Therefore, we chose as teratogenic concentrations 3.0 μM for heptadienal and 4.5 μM for octadienal, in order to obtain the same percentage of abnormal plutei (about 35%), as in the case of decadienal. Controls were also performed in filtered sea water (FSW) and in FSW plus methanol, and we found that methanol had no interference with the embryo development. In fact, the percentage of abnormal plutei was the same for embryos in FSW, as well as in FSW plus methanol.

2.2. Post-Recovery Experiments

Post-recovery experiments were performed in order to investigate if sea urchin embryos were able to recover after exposure to these PUAs. Eggs were incubated with three different concentrations of PUAs and then fertilized. The lowest concentrations represented the concentration inducing teratogenesis for each PUA (decadienal 1.6 μM, heptadienal 3.0 μM, octadienal 4.5 μM) and were also used to study the gene stress response (see below); the other concentrations (decadienal two and 2.5 μM; heptadienal five and 6 μM; octadienal six and 8 μM) were chosen to have comparable percentages of abnormal plutei (about 60% and 75%). Embryos were washed at different times after fertilization, 40 min, two, five, nine and 24 hours post fertilization (hpf), corresponding to two-cell, eight-cell, early blastula, swimming blastula and prism, respectively. After washing, embryos were grown to the pluteus stage to calculate the number of abnormal embryos. The results indicate that embryos were able to recover at all concentrations tested and at all times after fertilization (Figure 3), with the exception of the highest concentrations at 24 hpf. Moreover, the results suggest that octadienal could have a different mechanism of action compared to decadienal and heptadienal, because embryos were less able to recover at all times after fertilization at the highest concentrations. According to these results, we conclude that the post-recovery effects depend both on PUA concentrations and the exposure time of embryos to these metabolites, with the exception of octadienal at the highest concentration.

Figure 2. Percentage of abnormal plutei (%) produced when *Paracentrotus lividus* newly-fertilized eggs are exposed to different concentrations (in μM) of heptadienal, octadienal and decadienal for 48 h post fertilization (*** with a *p*-value <0.001, Student's *t*-test, GraphPad Software Inc., San Diego, CA, USA).

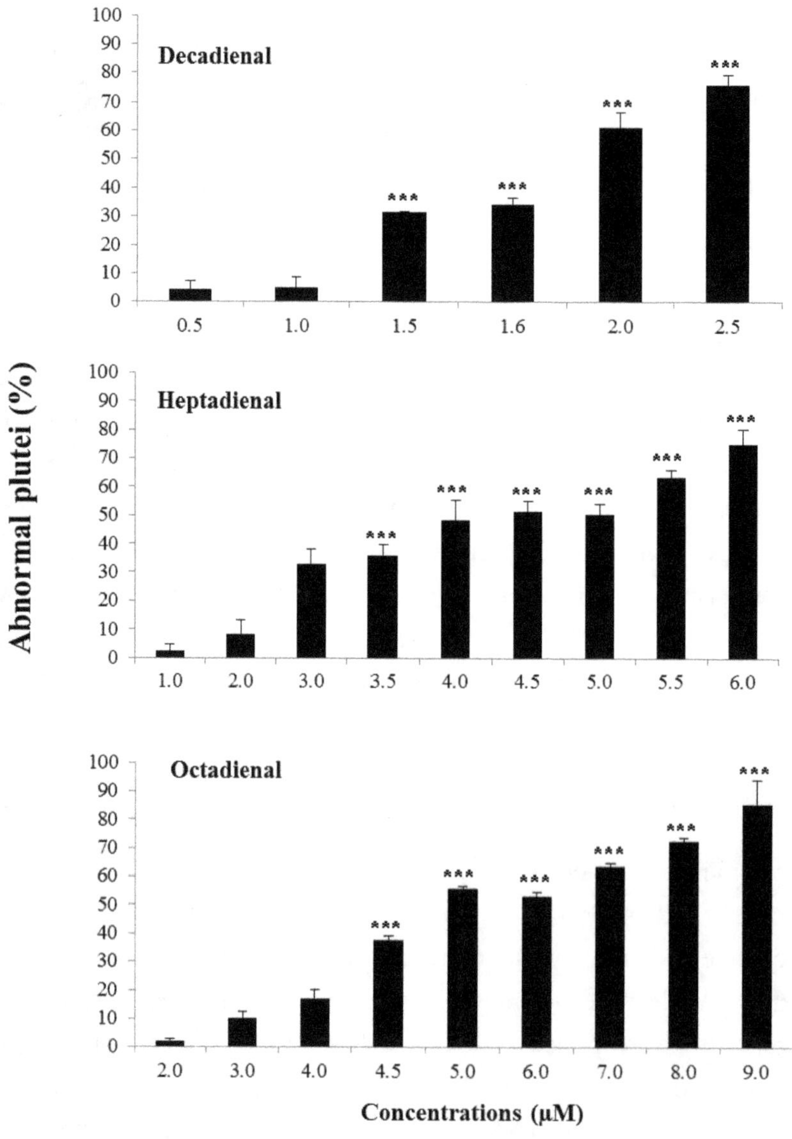

Figure 3. Percentage of abnormal *Paracentrotus lividus* plutei (%) produced after exposure to different concentrations of heptadienal, octadienal and decadienal and at different development times after fertilization. The percentages of abnormal plutei without washing are also reported.

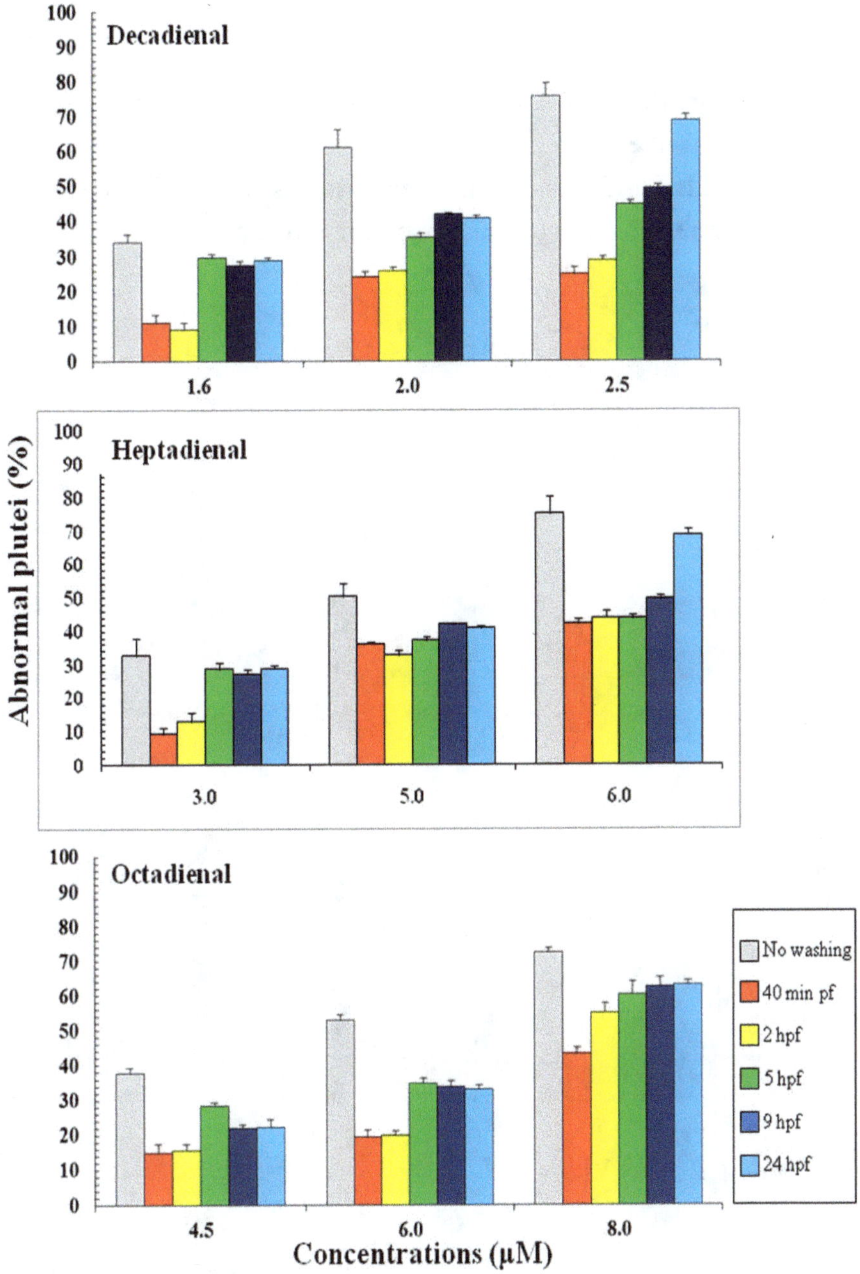

2.3. *Addition of PUAs at Different Developmental Stages*

To define the stage at which PUAs affect embryo development, aldehydes were added: 10 min before fertilization (bf), 10 min pf, 40 min pf, 2, 3, 5 and 8 hpf. These stages represented key stages during sea urchin embryogenesis. In fact, 40 min pf corresponds to the two-cell stage; from this stage until 5 hpf, only mitotic cell division occurs; at 5 hpf (corresponding to early blastula stage) the differentiation of blastomeres begins, and cell fates in specific embryonic territories are defined; at

8 hpf, the embryo envelope starts to be digested, so the embryos comes into direct contact with the external environment for the first time. The results show that the addition of PUAs affect embryonic development in the same way, whether they are added 10 min before fertilization (bf) and/or 10 min pf, resulting in the same percentage of abnormal plutei (Figure 4). The addition of PUAs in later developmental stages does not seem to affect the embryonic development of *P. lividus*. In fact, the percentage of abnormal plutei remains very low (about 10%–20%) for each concentration tested.

Figure 4. The percentage of abnormal *Paracentrotus lividus* plutei (%) produced after the exposure of newly fertilized eggs to different concentrations of heptadienal, octadienal and decadienal and examined at different developmental stages: 10 min before fertilization (bf), 10 min and 40 min post fertilization (pf) and two, three, five and 8 hpf.

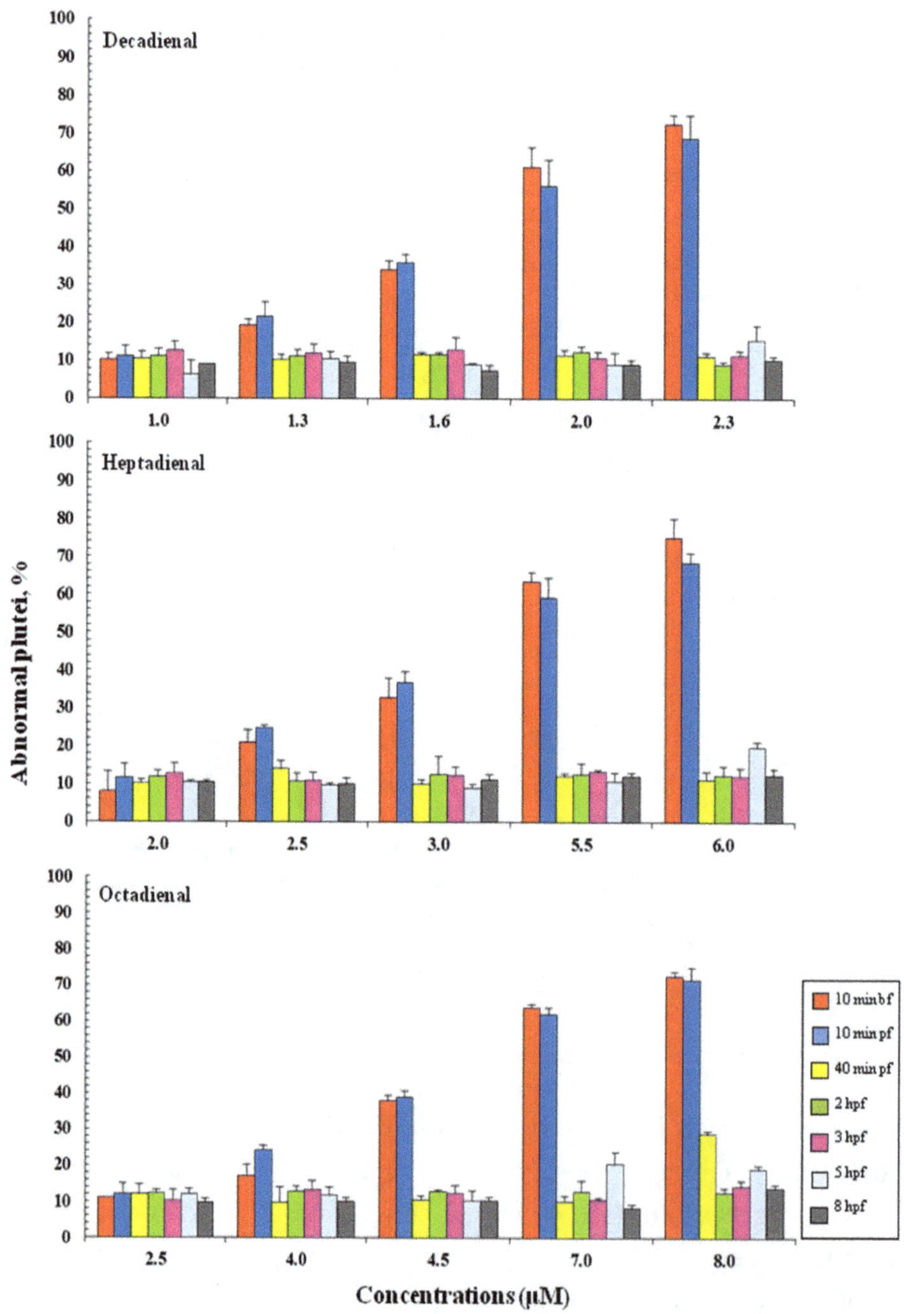

2.4. Gene Stress Response to PUAs

To better understand the morphological effects at the molecular level, *P. lividus* embryos were allowed to develop in the presence of PUAs at concentrations inducing teratogenesis (1.6 μM for decadienal, 3.0 μM for heptadienal and 4.5 μM for octadienal), and embryos were collected at different development times after fertilization, corresponding to the stages of early blastula (5 hpf), swimming blastula (9 hpf), prism (24 hpf) and pluteus (48 hpf). The expression levels of thirty one genes (see the Experimental Section for more details) were followed by real-time qPCR to identify potential gene targets. These genes are implicated in a broad range of functional responses, such as stress, development, differentiation, skeletogenesis and detoxification processes (see Table 1).

Table 1. The function for the genes analysed in the present study.

Gene Name	Acronym	Function	Reference
Pl-p19	*p19*	small acidic proteins, involved in the formation of the	[14]
Pl-p16	*p16*	biomineralised skeleton of sea urchin embryos and adults	
ALG-2 interacting protein X/1	*Alix*	protein involved in endocytic membrane trafficking, filamentous (F)-actin remodelling and cytokinesis	[15]
Blimp	*Blimp*	zinc finger transcription factor, which plays a central role in both early and late endomesoderm specification	[16]
Wnt 5	*Wnt5*	initiates the specification of the sea urchin posterior ectoderm	[17]
Wnt 6	*Wnt6*	activates endoderm in the sea urchin gene regulatory network	[18]
Wnt 8	*Wnt8*	endomesodermal specification, embryo patterning, early primary mesenchyme cells-gene regulatory network regulator	[19]
Multi drug resistance protein 1	*MDR1*	ATP-binding cassette protein	[20]
Metallothionein 4	*MT4*		
Metallothionein 5	*MT5*	proteins capable of binding to heavy metals, involved in	[21]
Metallothionein 6	*MT6*	the transport of heavy metals and cellular detoxification	
Metallothionein 7	*MT7*		
Metallothionein 8	*MT8*		
Catalase	*Cat*	antioxidant defensive protein	

The control gene for real-time qPCR was ubiquitin, the expression of which remained constant in all sea urchin developmental stages. The histograms reported in Figures 5–8 show the relative expression ratios of the analysed genes with respect to control embryos in sea water without PUAs. Only expression values greater than a two-fold difference with respect to the controls were considered significant.

Figure 5. The histograms show the differences in the expression levels of thirty one genes followed by real-time qPCR. *Paracentrotus lividus* embryos were grown in the presence of decadienal, heptadienal and octadienal at teratogenic concentrations (1.6, 3.0 and 4.5 μM, respectively) and collected at 5 hpf. Data are reported as a fold difference (mean ± SD), compared to the control embryos in sea water without aldehydes. Fold differences greater than ±2 (see the dotted horizontal guide lines at the values of +2 and −2) were considered significant.

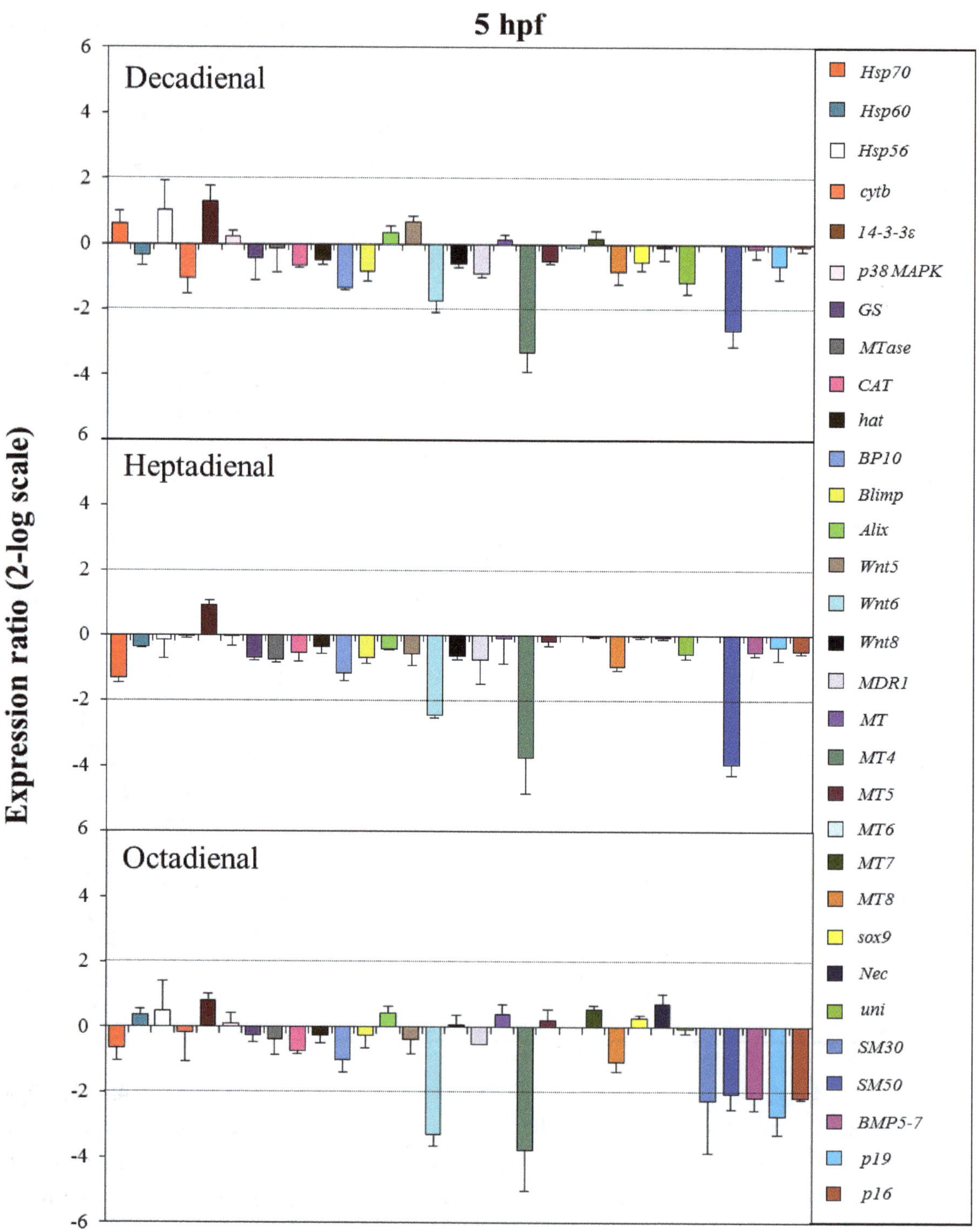

Figure 6. The histograms show the differences in the expression levels of thirty one genes followed by real-time qPCR. *Paracentrotus lividus* embryos were grown in the presence of decadienal, heptadienal and octadienal at teratogenic concentrations (1.6, 3.0 and 4.5 μM, respectively) and collected at 9 hpf. For more details, see also the legend to Figure 5.

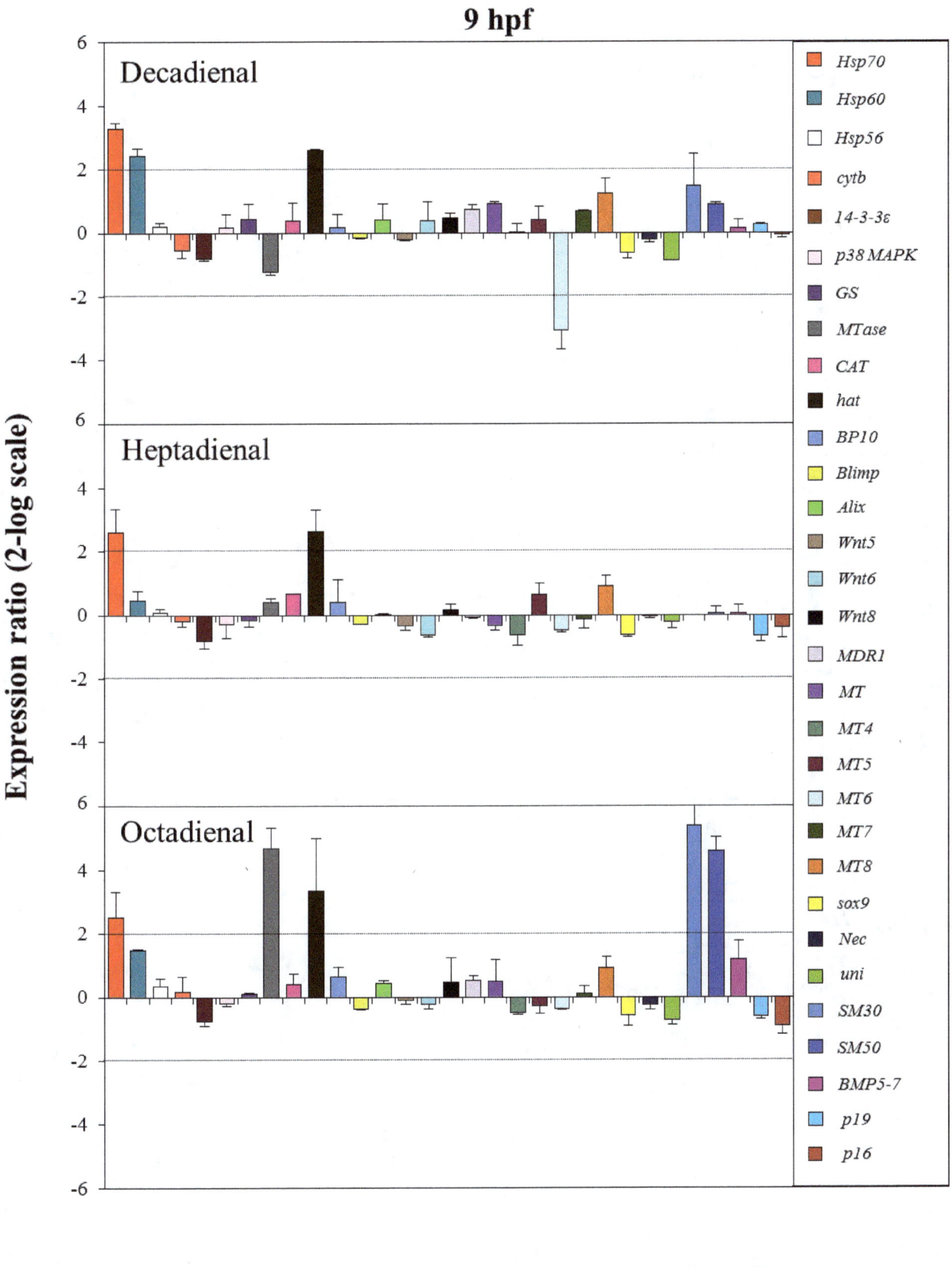

Figure 7. The histograms show the differences in the expression levels of thirty one genes followed by real-time qPCR. *Paracentrotus lividus* embryos were grown in the presence of decadienal, heptadienal and octadienal at teratogenic concentrations (1.6, 3.0 and 4.5 μM, respectively) and collected at 24 hpf. For more details, see also the legend to Figure 5.

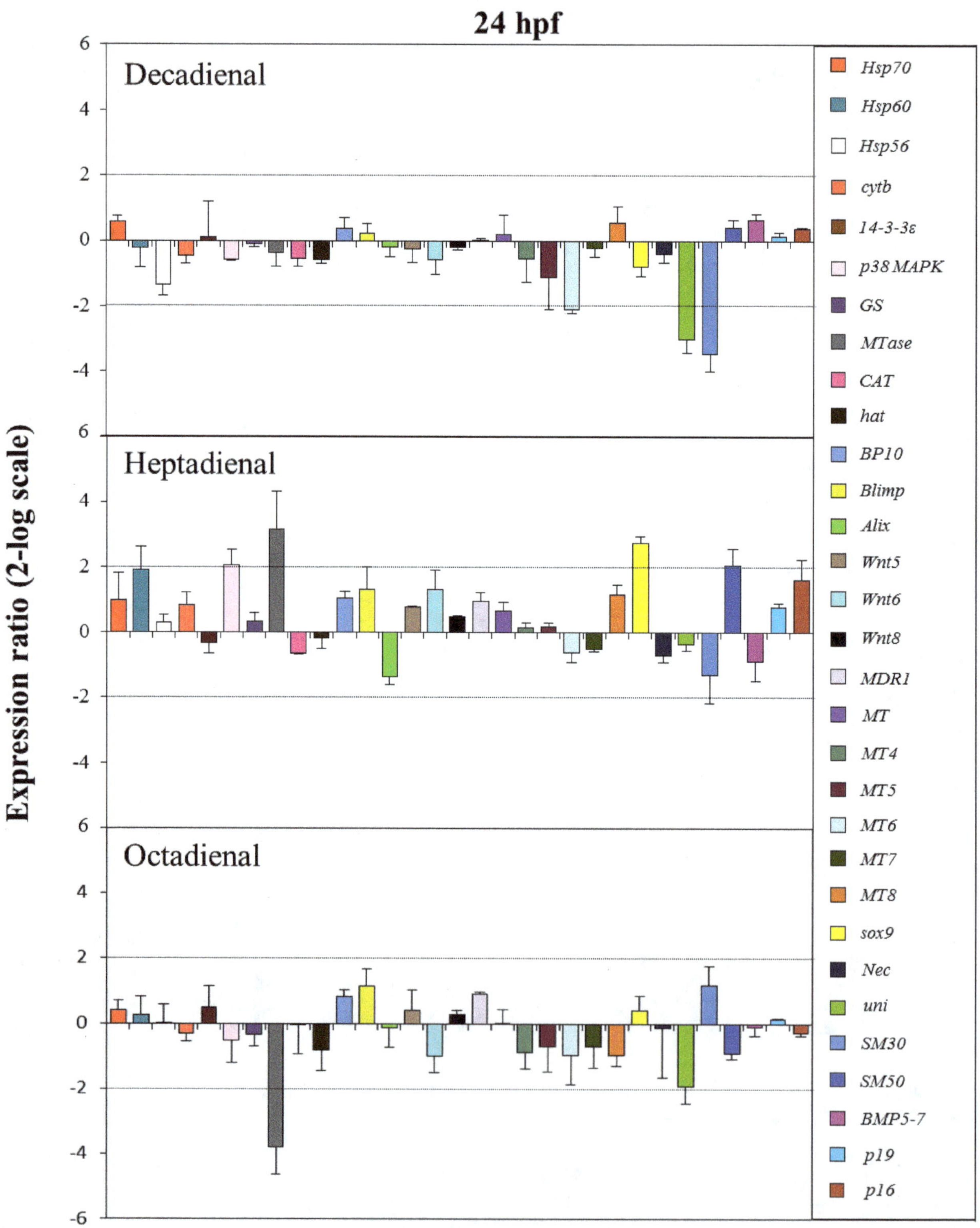

Figure 8. The histograms show the differences in the expression levels of thirty one genes followed by real-time qPCR. *Paracentrotus lividus* embryos were grown in the presence of decadienal, heptadienal and octadienal at teratogenic concentrations (1.6, 3.0 and 4.5 µM, respectively) and collected at 48 hpf. For more details, see also the legend to Figure 5.

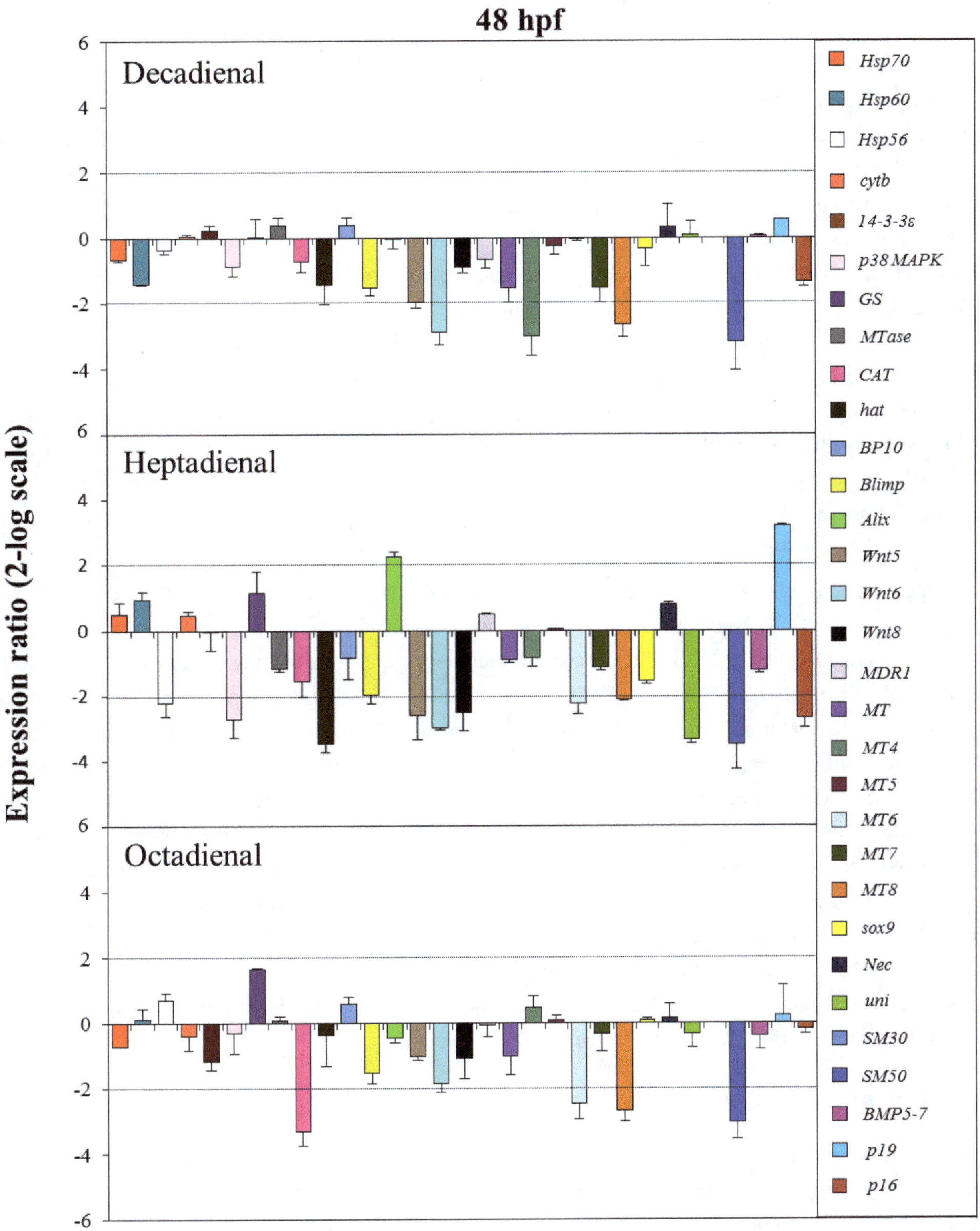

At the early blastula stage (5 hpf; see Figure 5), one metallothionein and one skeletogenic gene, *MT4* and *SM50*, were targeted by all three PUAs, showing a 2.6-, 4.0- and 2.1-fold decrease in expression levels with respect to the control, respectively. The expression level of *Wnt6*, a gene implicated in the developmental and differentiation processes, was significantly reduced by heptadienal (2.4-fold) and octadienal (3.3-fold) compared to the control. Moreover, at this stage, only octadienal affected the expression levels of four skeletogenic genes, which were downregulated: *SM30*, *Bmp5-7*, *p19* and *p16*, showing a 2.2-, 2.2-, 2.7- and 2.2-fold decrease with respect to the control, respectively.

At the swimming blastula stage (9 hpf; see Figure 6), two genes were upregulated: the stress gene *hsp70* (3.2- for decadienal, 2.6- for heptadienal and 2.5-fold for octadienal) and the protease *hat* (2.6-fold for decadienal and heptadienal; 3.5-fold for octadienal). Moreover, treatment with decadienal showed a 2.4-fold increase in the expression level of the stress gene, *hsp60*, and a three-fold decrease for metallothionein *MT6*. At this stage, octadienal activated one stress gene and two skeletogenic genes: *MTase*, *SM30* and *SM50*, with a 4.7-, 5.4- and 4.6-fold increase with respect to the control.

At the prism stage (24 hpf; see Figure 7), heptadienal and octadienal differentially affected the expression levels of the stress gene, *MTase*. Whereas heptadienal upregulated this gene with a 3.2-fold increase, octadienal downregulated with a 3.8-decrease in expression levels. At this developmental stage, the genes targeted by heptadienal were the stress gene, *p38 MAPK*, and the skeletogenic gene, *SM50*, both of which showed a 2.1-fold increase in expression levels with respect to the control, and the developmental gene, *sox9*, with a 2.7-fold increase. Decadienal affected the expression levels of the metallothionein, *MT6*, and the two skeletogenic genes, *uni* and *SM30*, with a 2.1-, 3- and 3.5-fold downregulation, respectively.

At the pluteus stage (48 hpf; see Figure 8), all three PUAs targeted the metallothionein, *MT8*, and the skeletogenic *SM50* genes. In particular, the expression levels of these genes were all downregulated: *MT8* had a 2.7-fold decrease for decadienal and octadienal and a 2.1-fold decrease for heptadienal; *SM50* a 3.2-, 3.5- and 3.1-fold decrease for decadienal, heptadienal and octadienal, respectively. The metallothionein *MT6* gene was targeted by both heptadienal and octadienal, with a 2.2- and 2.5-fold downregulation of their expression levels, respectively. Decadienal and heptadienal induced a 2.9-fold downregulation of *Wnt6*. Moreover, decadienal also affected the expression level of the metallothionein *MT4* gene, causing a 3-fold decrease with respect to the control. Heptadienal upregulated the expression level of two genes, the developmental gene, *Alix* (2.2-fold), and the skeletogenic gene, *p19* (3.2-fold), and downregulated several genes: two stress genes, *hsp56* and *p38 MAPK* (2.2- and 2.7-fold decrease, respectively); three developmental genes, *hat, Wnt5* and *Wnt 8* (3.5, 2.6- and 2.5-fold decrease, respectively); and two skeletogenic genes, *uni* and *p16* (3.4 and 2.7-fold decrease, respectively). Octadienal was the only PUA that downregulated the *CAT* gene (3.3-fold decrease) implicated in detoxification processes.

3. Discussion

Our results greatly expand on previous investigations on the stress response to the toxic PUA, decadienal, during sea urchin development [12,13]. Here, we probe more deeply into the effects induced by two ecologically important, but relatively unknown, PUAs, heptadienal and octadienal, which have never been tested before on *P. lividus* embryos from the molecular point of view, in comparison with the better-known PUA, decadienal. Since mainly heptadienal and octadienal are released when diatom cells are wounded during grazing [5,22,23] or lysed from senescent cells during bloom periods [24], it should be interesting to determine the direct effects of these pure molecules. Until now, very few studies have tested the effects of pure PUAs on copepods [25–27] or reported their effects at the molecular level in copepods [28,29] or the sea urchin, *P. lividus* [12,13]. Sea urchins have been widely used as a sensible bioindicator of biochemical, morphological and physiological changes related to environmental stressors, such as pesticides, essential and heavy metals, ionizing radiations, ocean warming and acidification, metal nanoparticles and natural toxins [30–39].

Figure 9. Synopsis of the patterns of up- and down-regulation of different classes of genes in the sea urchin, *Paracentrotus lividus*, in the presence of the PUAs decadienal, heptadienal and octadienal.

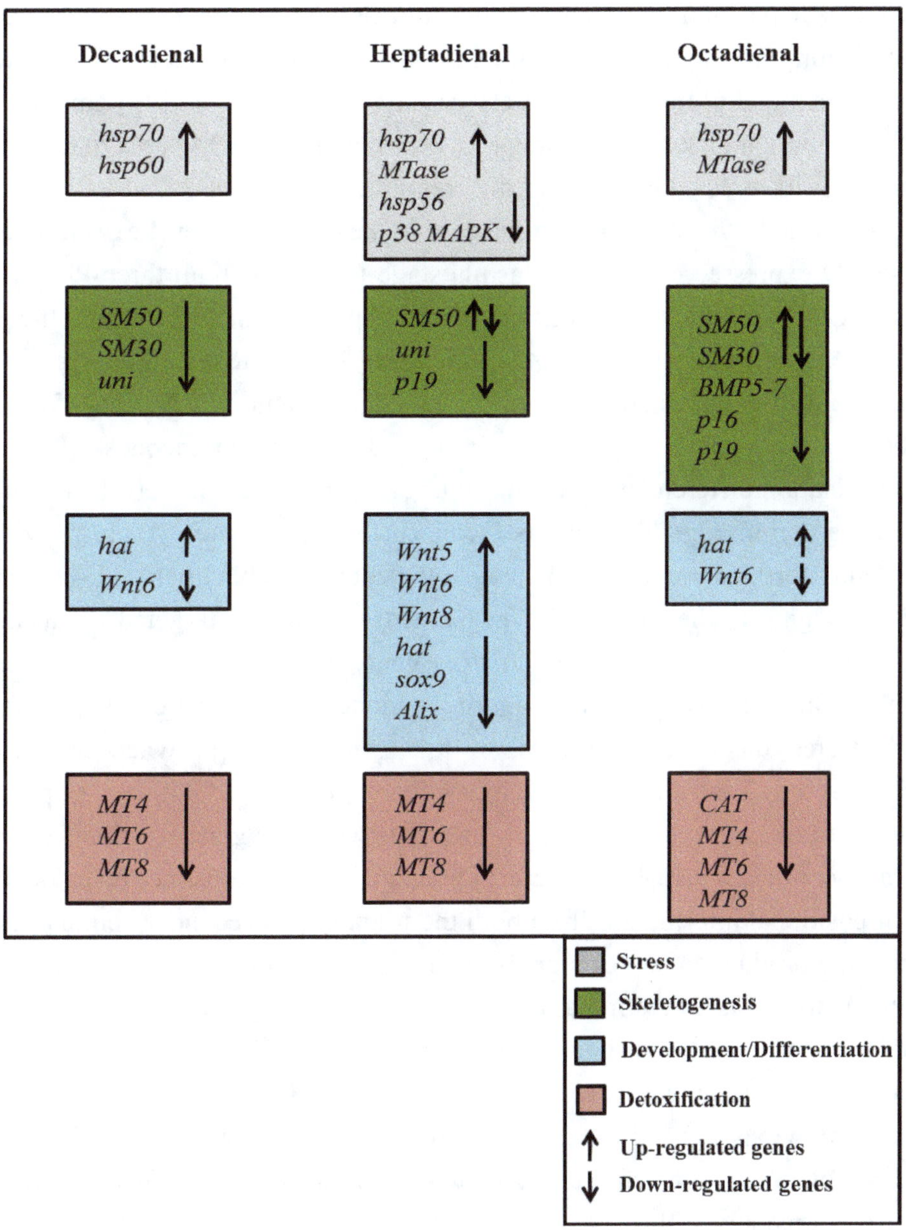

An important outcome of this study was the finding that the expression levels of a great number of genes in the sea urchin, *P. lividus*, appears to be modulated by the PUAs, decadienal, heptadienal and octadienal. A synopsis showing the patterns of up- and down-regulation of different classes of genes is shown in Figure 9. These three aldehydes have very few common targets, but specifically affect different classes of genes and at different times of development.

Of the stress genes, the canonical stress gene, *hsp70*, was affected by the three aldehydes at the same development time (at the swimming blastula stage), confirming that embryos were subjected to stress, resulting in the activation of this gene as a first defence system [12]. The other two heat shock proteins, *hsp60* and *hsp56*, were also targeted by PUAs; these genes are reported as a part of a protection system against different stressors enhancing cell survival and normal cellular homeostasis [12,28,40–42].

The possible role of DNA methylation as a molecular marker in response to stress [43] was confirmed by the increase in the expression level of *MTase* after treatment with heptadienal at the prism stage and octadienal at the swimming blastula and prism stages. The expression level of *p38 MAPK*, participating in a signalling cascade in response to different stimuli [33], was affected by heptadienal at the prism and pluteus stages. Very recently, Pinsino *et al.* [44] emphasized the role of p38 MAPK in the regulation of sea urchin embryonic development after exposure to manganese.

Of the genes involved in development and differentiation processes, PUAs affected the expression levels of *hat*, *sox9*, *alix* and *Wnt*. We found an increase in the expression level of *hat*, an early embryonic messenger transiently expressed during the blastula stage [45,46] with all three PUAs. The *sox9* gene was upregulated by heptadienal at the prism stage, suggesting an effect on left-right asymmetry processes [47]. *Alix*, a multifunctional protein involved in different cellular processes, including endocytic membrane trafficking, filamentous-actin remodelling and cytokinesis [15], was upregulated only by heptadienal at the pluteus stage. In sea urchins, this transcript encodes for a maternal protein involved in determination/differentiation events that is expressed from fertilization to the two-cell embryo stage. In sea urchin eggs, the gene is localized throughout the cytoplasm with a punctuated pattern, and soon after fertilization, it accumulates in the cytosol and in microvilli-like protrusions.

Of the genes belonging to the canonical Wnt pathway, *Wnt6* was targeted by all three PUAs that induced a decrease in the expression levels of this gene. This represents a very interesting result, supporting the essential role of Wnt6 to in triggering endoderm specification [18]. In fact, heptadienal and octadienal affected the expression level of this gene at 5 hpf, when the differentiation of blastomeres occurs and cell fates in specific embryonic territories are defined. Heptadienal was also able to determine a decrease in the expression levels of two other genes of the Wnt pathway: *Wnt5*, which acts as a short-range inter-blastomere signal, activating cells in the ectoderm, integrating different types of positional information from both the primary and secondary embryonic axes in order to correctly locate the production site of signals needed for skeleton formation to take place [17]; and *Wnt8* is associated with cell fate determination through canonical signalling pathways and is important for the morphogenetic movement of primary mesenchyme cells [19].

All genes involved in skeletogenesis were downregulated by PUAs, except for the *Nec* gene. *SM30*, *SM50* and *Uni* were previously found to be target genes for decadienal [13]. Octadienal downregulated the expression level of *BMP5-7*, a gene family that is reported as being positive regulators of oral and aboral ectoderm specification [17]. *P19* was upregulated by heptadienal, but was downregulated by octadienal together with *p16*; these are two small acidic proteins involved in the formation of the biomineralised skeleton of sea urchin embryos and adults [14].

We also analysed eight genes involved in detoxification processes. In particular, of six metallothioneins that were analysed in this work, only the expression levels of three genes appeared to be perturbed by PUAs. Ragusa *et al.* [21] reported that two metallothionein genes (*MT7* and *MT8*) appeared to be constitutively expressed and upregulated upon cadmium treatment, whereas other genes (*MT4*, *MT5* and *MT6*) were not transcribed in control embryos and were specifically activated in response to cadmium treatment. Our results show the down-expression of the induced metallothioneins, *MT4* and *MT6*, and the constitutive one, *MT8*, at a later stage of development (48 hpf). The induction of other metallothioneins may require exposure to higher concentrations of PUAs or may depend on the nature of the stress agent, considering that metallothioneins seem to respond very specifically to heavy metal exposure. The gene, *MDR1*, belongs to ATP-binding cassette transporters, which are activated by sub-lethal doses of specific contaminants (such as oxybenzone, mercuric chloride and tributyltin) during embryonic development (from the zygote to the blastula stage) of sea urchins [20]. Moreover, sea urchin embryos utilize it in cell signalling and lysosomal and mainly mitochondrial homeostasis [48]. The expression level of this gene was not affected by PUA exposure, probably due to the low levels of exposure. In fact, we demonstrated the loss of mitochondrial functionality only at higher concentrations of decadienal in a previous study [12]. The down-expression of the *CAT* gene only in the case of octadienal suggests that a specific detoxification system could be activated in sea urchins after exposure to this PUA.

These molecular results are in accordance with our morphological results that revealed that the majority of malformations affected the skeleton and the plan of the development and differentiation of sea urchin embryos, as reported in Figure 1. In fact, several genes belonging to the skeletogenic, developmental and differentiation classes were affected by PUAs. Even if, according to our morphological analyses, octadienal seems to be the weakest of the three aldehydes, our study on gene expression levels reveals that this aldehyde affects a very early stage of embryonic development. In fact, octadienal differentially affected the expression levels of five skeletogenic genes at the early blastula stage compared to the other aldehydes. These data are also in accordance with our post-recovery experiments, in which we hypothesized that octadienal could have a different mechanism of action, because embryos cannot recover in the presence of this aldehyde at the highest concentration. This different behaviour could be correlated with the downregulation of skeletogenic genes at an early stage of embryo development, which does not give embryos the chance to recover. The strongest molecular effects seem to occur after treatment with heptadienal, but at a later stage of development. In fact, at 48 hpf (pluteus stage), heptadienal affected the expression levels of thirteen genes. Taken together, these results suggest that although treatment with the three aldehydes did not induce any visible differences at the morphological level, they affected different physiological processes.

The genes identified in this work as targets for PUAs may represent possible biomarkers to detect exposure to pollutants that may include microbial products, heavy metals and phytotoxins. More generally, the new genes analysed in this work can be considered as an additional part of the stress surveillance system or chemical defensome of the sea urchin, as proposed in Marrone *et al.* [13], affording protection from environmental xenobiotics.

4. Experimental Section

4.1. Ethics Statement

Paracentrotus lividus (Lamarck) sea urchins were collected from a location that is not privately-owned or protected in any way, according to Italian legislation of the Marina Mercantile (Decreto del Presidente della Repubblica DPR 1639/68, 09/19/1980 confirmed on 01/10/2000). The field studies did not involve endangered or protected species. All animal procedures were in compliance with the guidelines of the European Union (Directive 609/86).

4.2. Gamete Collection, Embryo Culture, Exposure to Aldehydes and Morphological Analysis

Adult sea urchins of the species, *P. lividus*, were collected during the breeding season by scuba-diving in the Gulf of Naples, transported in an insulated box to the laboratory within 1 h after collection and maintained in tanks with circulating sea water until testing. Sea urchins were injected with 2 M KCl through the peribuccal membrane to obtain the emission of gametes. Eggs were washed with filtered sea water (FSW) and kept in FSW until use. Concentrated sperm was collected, dried and kept undiluted at +4 °C until use.

Before fertilization, eggs were incubated at room temperature for 10 min in the presence of different concentrations of the three PUAs: 2-*trans*,4-*trans*-decadienal at 1.0, 1.3, 1.6, 2.0, 2.3 μM (similar to the concentrations tested in reference [13]); 2-*trans*,4-*trans*-heptadienal at 1.0, 2.0, 3.0, 3.5, 4.0, 4.5, 5.0, 5.5, 6.0 μM; 2-*trans*,4-*trans*-octadienal (Sigma-Aldrich, Milan, Italy) at 2.0, 3.0, 4.0, 4.5, 5.0, 6.0, 7.0, 8.0, 9.0 μM; and the controls were in FSW without PUAs.

Eggs were fertilized utilising sperm-to-egg ratios of 100:1 for both controls and treated embryos. Fertilized eggs were kept at 20 °C in a controlled temperature chamber on a 12 h:12 h light:dark cycle. PUAs were diluted in methanol, considering a methanol to FSW ratio of 10 μL:1 mL, so as to avoid interference with embryo development. Controls were also performed in FSW and in FSW in the presence of methanol.

Experiments were conducted in triplicate using three egg groups collected from three different females. After 48 h of incubation, morphological malformations were determined for at least 200 plutei using a light microscope (Zeiss Axiovert 135TV, Carl Zeiss, Jena, Germany) in comparison to control embryos in FSW without PUAs.

4.3. Post-Recovery Experiments

The procedure for the treatments with the three PUAs was the same as reported above. For each PUA, three concentrations were tested according to the percentage of abnormal plutei recorded during previous tests. The lowest concentrations were those inducing teratogenesis: 1.6 μM for decadienal, 3.0 μM for heptadienal and 4.5 μM for octadienal, all three of which induced the production of about 35% abnormal plutei that were also used for molecular experiments (see below). We then tested the concentrations that induced the production of about 60% and 75% of abnormal plutei, so as to have comparable results with the three PUAs. Embryos were washed twice at different development times: 40 min and 2, 5, 9 and 24 h post-fertilization (hpf). Embryos were grown to the pluteus stage. The number of abnormal embryos was evaluated by fixing embryos in formaldehyde (4% in FSW) and counting under the light microscope.

One-way ANOVA with Tukey's post-hoc test was performed using GraphPad Prism version 4.00 for Windows (GraphPad Software, San Diego, CA, USA).

4.4. Addition of PUAs after Fertilization

Eggs were fertilized without PUAs, according to the procedure reported above. The development of embryos was followed by microscopic examination for different development times after fertilization (10 min pf, 40 min pf, 2, 3, 5 and 8 hpf). PUAs were then added at the same concentrations used for post-recovery experiments. As controls, PUAs were also added 10 min before fertilization. The number of abnormal plutei was evaluated from fixed embryos. One-way ANOVA with Tukey's post-hoc test was performed using GraphPad Prism version 4.00 for Windows (GraphPad Software, San Diego, CA, USA).

4.5. RNA Extraction and cDNA Synthesis

About 30,000 eggs in 200 mL of FSW were treated for 10 min with the three PUAs at the following concentrations: 1.6 μM for decadienal (as in [13]); 3.0 μM for heptadienal; and 4.5 μM for octadienal. Eggs were then fertilized and collected at different developmental times. Samples were collected at 5, 9, 24 and 48 hpf by centrifugation at 1800 relative centrifugal force for 10 min in a swing out rotor at 4 °C. The pellet was washed with phosphate buffered saline and then frozen in liquid nitrogen and kept at −80 °C.

Total RNA was extracted from each developmental stage using TRIzol (Invitrogen, Life Technologies, Carlsbad, CA, USA) according to the manufacturer's instructions. Extraction with chloroform/isoamyl alcohol (24:1) was performed following RNA precipitation by addition of glycogen and isopropyl alcohol. Contaminating DNA was degraded by treating each sample with a DNase RNase-free kit (Roche, Milan, Italy) according to the manufacturer's instructions. The amount of total RNA extracted was estimated by the absorbance at 260 nm and the purity by 260/280 and 260/230 nm ratios, by a NanoDrop spectrophotometer (ND-1000 UV-Vis Spectrophotometer; NanoDrop Technologies, Wilmington, DE, USA). The integrity of RNA was evaluated by agarose gel electrophoresis. Intact rRNA subunits (28S and 18S) were observed on the gel indicating minimal degradation of the RNA. For each sample, 600 ng of total RNA extracted was retrotranscribed with an iScript™ cDNA Synthesis kit (Bio-Rad, Milan, Italy), following the manufacturer's instructions. Synthetized cDNA was used in real-time qPCR experiments without dilution.

To evaluate the efficiency of cDNA synthesis, a PCR was performed with primers of the reference gene, ubiquitin. The reaction was carried out on the C1000 Touch Thermal Cycler GeneAmp PCR System 9700 (Applied Biosystem, Monza, Italy) in a 30 μL final volume with 3 μL 10× PCR reaction buffer (Roche, Milan, Italy), 3 μL 10× 2 mM dNTP, 1 μL 5 U/μL Taq (Roche, Milan, Italy), 100 ng/μL of each oligo, template cDNA and nuclease free water to 30 μL. The PCR program consisted of a denaturation step at 95 °C for 5 min, 35 cycles at 95 °C for 45 s, 60 °C for 1 min and 72 °C for 30 s and a final extension step at 72 °C for 10 min.

4.6. Gene Expression by Real-Time qPCR

For all real-time qPCR experiments, the data from each cDNA sample were normalized using ubiquitin mRNA as the endogenous control level, the level of which remained relatively constant in all developmental stages examined according to Nemer *et al.* ([49]; for more details, see Romano *et al.* [12]). The expression level of seventeen genes, previously analysed by real-time qPCR in Marrone *et al.* [13], were analysed together with fourteen new genes (see Table 1): Pl-p19 (*p19*), Pl-p16 (*p16*; [14]), metallothionein 4 (*MT4*; [21]), metallothionein 5 (*MT5*; [21]), metallothionein 6 (*MT6*; [21]), metallothionein 7 (*MT7*; [21], metallothionein 8 (*MT8*; [21]), blimp (*Blimp*; [16]), ALG-2 interacting protein X/1(*Alix*; [15]), multi-drug resistance protein 1 (*MDR 1*), Wnt5 (*Wnt5*), Wnt6 (*Wnt6*) and Wnt8 (*Wnt8*). We also analysed the catalase gene (*CAT*); since the *CAT* sequence of *P. lividus* is not available, a 156-bp fragment was amplified using specific primers for this gene from *Strongylocentrotus purpuratus*. The amplified fragment using a Taq High Fidelity PCR System (Roche, Milan, Italy) was purified from agarose gel using the QIAquick Gel Extraction kit (Qiagen, Milan, Italy), and the specificity of the PCR product for catalase was checked by DNA sequencing.

Gene sequences were retrieved from NCBI [50]. For each gene, specific primers were designed on the basis of nucleotide sequences of *P. lividus* (see Table 2); primers reported in the indicated references were used for six genes. The same procedure for the amplified fragments used for *CAT* was applied for these genes to check for the specificity of the amplified products.

Table 2. Gene name, acronym, accession numbers, primer sequences and lengths of PCR amplified fragments are reported for the genes analysed. For the genes that have a reference, the lengths of PCR fragments were not reported.

Gene Name	Acronym	Accession Number	Primer	Sequence (5′→3′)	PCR Fragment (bp)
Pl-p19	*p19*	FR693764	Pl_P19_F1	GACAAGCTCGACATCAACAAG	205
			Pl_P19_R1	CTGGAGTCGATGCTGCATCATG	
Pl-p16	*p16*	FR693763	Pl-p16 For	CGGGCAGCGATGACTCA	104
[14]			Pl-p16 Rev	AAATGCCATACCGCTCTTCTGT	
Metallothionein 4	*MT4*		MT4 For	GCTCAAAATCTTCAACATGGCTAATGA	
[21]			MT4 Rev	AGCACTTTCCAGTTTCACAACAAGC	
Metallothionein 5	*MT5*		MT5 For	CGACTTTAGCTCAAATTCATCACCATG	
[21]			MT5 Rev	TCCACAGCATTTACCATCCTTGC	
Metallothionein 6	*MT6*		MT6 For	CACGATTTGTGCTCAATCCTTCAT	
[21]			MT6 Rev	TTTGTGCATGATGTTCCACAGC	
Metallothionein 7	*MT7*		MT7 For	CGTCAAGAGATCAAAATCATCAACCA	
[21]			MT7 Rev	ACAGCACTCGCCAGTAATACAGCAC	

Table 2. *Cont.*

Metallothionein 8 [21]	*MT8*		MT8 For MT8 Rev	GATGGTTGTCGTCGCTCCTAACA TCAAGAAAGGCTGGTATCAAATCTGAC	
Blimp	*Blimp*	HQ322503	Pl_Blimp1_For Pl_Blimp1_Rev	CTGTCTACTCCATGCCGTCC GCCTCCTGCTTCAGATCAGC	161
ALG-2 interacting protein X/1 [15]	*Alix*	HE646599	AL1500for AL1610rev	TACCAGACCATTCTCAACAAT TGCTATTTCCGCTTCGCTTTT	110
Multi drug resistance protein 1	*MDR1*	JQ793791	Pl_MDR1_F2 Pl_MDR1_Rev	GTCAAGGTACTCAATGGGGTC CGGATGTCAATGCCATCAATC	158
Wnt 5	*Wnt5*	HM449806	Pl_Wnt5_F2 Pl_Wnt5_Rev	CACCCAGCCCCTGTGCAGTG CTGCAGTTCCACCTCCTATTC	135
Wnt 6	*Wnt6*	HQ322504	Pl_Wnt6_For Pl_Wnt6_For	CGAATCTGCCGACGATCACG GCATTGTCGTACAGTTCCACC	164
Wnt 8	*Wnt8*	HM449816	Pl_Wnt8_For Pl_Wnt8_For	CTGTAAGTGTCATGGCGTCTC GAGCGAATCGGAGATGACGG	197
Catalase	*CAT*	SPU_000281.1	Sp-CAT_F1 Sp-CAT_R1	GACTTCGTCTTCACCGACGAG GACTCAAAGGGTGCAGCCTTG	156

The specificity of amplification reactions was verified by melting curve analysis. The efficiency of each primer pair was calculated according to standard method curves using the equation $E=10^{-1/slope}$. Five serial dilutions were set up to determine Ct values and reaction efficiencies for all primer pairs. Standard curves were generated for each oligonucleotide pair using Ct values *versus* the logarithm of each dilution factor. PCR efficiencies were calculated for control and target genes and were found to be 2. Diluted cDNA was used as a template in a reaction containing a final concentration of 0.3 mM for each primer and 1× FastStart SYBR Green master mix (total volume of 10 μL) (Applied Biosystems, Monza, Italy). PCR amplifications were performed in a ViiATM7 Real Time PCR System (Applied Biosystems, Monza, Italy) thermal cycler using the following thermal profile: 95 °C for 10 min, one cycle for cDNA denaturation; 95 °C for 15 s and 60 °C for 1 min, 40 cycles for amplification; 72 °C for 5 min, one cycle for final elongation; one cycle for melting curve analysis (from 60 °C to 95 °C) to verify the presence of a single product. Each assay included a no-template control for each primer pair. To capture intra-assay variability, all real-time qPCR reactions were carried out in triplicate. Fluorescence was measured using ViiATM7 software (Applied Biosystems, Monza, Italy). The expression of each gene was analysed and internally normalized against ubiquitin using REST software (Relative Expression Software Tool, Weihenstephan, Germany) based on the Pfaffl method [51,52]. Relative expression ratios above two cycles were considered significant. Experiments were repeated at least twice. Statistical analysis was performed using GraphPad Prism version 4.00 for Windows (GraphPad Software, San Diego, CA, USA).

5. Conclusions

In conclusion, our findings provide molecular evidence for the toxic effects of diatom-derived PUAs and propose novel tools for understanding the cellular mechanisms of the response to aldehyde exposure in benthic organisms. Sea urchins may come in contact with diatom PUAs in the field at the end of a bloom, with the mass sinking of diatoms to the sediment, which represents a major source of organic matter for benthic systems [53]. This is of considerable ecological relevance given the importance of diatom blooms in nutrient-rich aquatic environments. Moreover, our results may be useful in understanding how changes in gene expression levels may be used as an early indicator of stressful conditions in the marine environment. Generally, these studies are fundamental to our understanding of how marine organisms attempt to defend themselves from environmental toxicants, benefitting from the protection provided by an integrated network of genes.

Acknowledgments

We thank the Molecular Biology Service for providing primers for PCR experiments and for PCR product sequencing. We also thank the Fishing Service of our institute for providing sea urchins. Many thanks are also due to Davide Caramiello of the Marine Resources for Research Service for his technical support in sea urchin maintenance and gamete collection. Stefano Varrella has been supported by a Stazione Zoologica Anton Dohrn PhD fellowship.

Author Contributions

M.C., G.R., A.I. conceived and designed the experiments. S.V., N.R., M.C., G.R. performed the experiments. M.C., G.R., A.I., M.B., S.V. analyzed the data. M.C., G.R., A.I. contributed reagents/materials/analysis tools. M.C., G.R., A.I., M.B., S.V. wrote the paper.

References

1. Caldwell, G.S. The Influence of Bioactive Oxylipins from Marine Diatoms on Invertebrate Reproduction and Development. *Mar. Drugs* **2009**, *7*, 367–400.
2. Ianora, A.; Miralto, A. Toxigenic effects of diatoms on grazers, phytoplankton and other microbes: A review. *Ecotoxicology* **2010**, *19*, 493–511.
3. Miralto, A.; Barone, G.; Romano, G.; Poulet, S.A.; Ianora, A.; Russo, G.L.; Buttino, I.; Mazzarella, G.; Laabir, M.; Cabrini, M.; *et al.* The insidious effect of diatoms on copepod reproduction. *Nature* **1999**, *402*, 173–176.
4. D'Ippolito, G.; Iadicicco, I.; Romano, G.; Fontana, A.; Iadicicco, O. Detection of short-chain aldehydes in marine organisms: The diatom *Thalassiosira rotula*. *Tetrahedron Lett.* **2002**, *43*, 6137–6140.
5. Wichard, T.; Poulet, S.A.; Halsband-Lenk, C.; Albaina, A.; Harris, R.; Liu, D.; Pohnert, G. Survey of the chemical defence potential of diatoms: Screening of fifty species for α,β,γ,δ-unsaturated aldehydes. *J. Chem. Ecol.* **2005**, *31*, 949–958.

6. Wendel, T.; Juttner, F. Lipoxygenase-mediated formation of hydrocarbons and unsaturated aldehydes in freshwater diatoms. *Phytochemistry* **1996**, *41*, 1445–1449.
7. Caldwell, G.S.; Olive, P.J.W.; Bentley, M.G. Inhibition of embryonic development and fertilization in broadcast spawning marine invertebrates by water soluble diatom extracts and the diatom toxin 2-*trans*,4-*trans* decadienal. *Aquat. Toxicol.* **2002**, *60*, 123–137.
8. Caldwell, G.S.; Bentley, M.G.; Olive, P.J.W. First evidence of sperm motility inhibition by the diatom aldehyde 2*E*,4*E*-decadienal. *Mar. Ecol. Prog. Ser.* **2004**, *273*, 97–108.
9. Romano, G.; Russo, G.L.; Buttino, I.; Ianora, A.; Miralto, A. A marine diatom-derived aldehyde induces apoptosis in copepod and sea urchin embryos. *J. Exp. Biol.* **2003**, *206*, 3487–3494.
10. Hansen, E.; Even, Y.; Geneviere, A.M. The alpha,beta,gamma,delta-unsaturated aldehyde 2-*trans*-4-*trans*-decadienal disturbs DNA replication and mitotic events in early sea urchin embryos. *Toxicol. Sci.* **2004**, *81*, 190–197.
11. Romano, G.; Miralto, A.; Ianora, A. Teratogenic Effects of diatom metabolites on sea urchin *Paracentrotus lividus* embryos. *Mar. Drugs* **2010**, *8*, 950–967.
12. Romano, G.; Costantini, M.; Buttino, I.; Ianora, A.; Palumbo, A. Nitric oxide mediates the stress response induced by diatom aldehydes in the sea urchin *Paracentrotus lividus*. *PLoS One* **2011**, *6*, e25980.
13. Marrone, V.; Piscopo, M.; Romano, G.; Ianora, A.; Palumbo, A.; Costantini, M. Defensome against toxic diatom aldehydes in the sea urchin *Paracentrotus lividus*. *PLoS One* **2012**, *7*, e31750.
14. Costa, C.; Karakostis, K.; Zito, F.; Matranga, V. Phylogenetic analysis and expression patterns of p16 and p19 in *Paracentrotus lividus* embryos. *Dev. Genes Evol.* **2012**, *222*, 245–251.
15. Romancino, D.P.; Anello, L.; Morici, G.; d'Azzo, A.; Bongiovanni, A.; Di Bernardo, M. Identification and characterization of PlAlix, the Alix homologue from the Mediterranean sea urchin *Paracentrotus lividus*. *Dev. Growth Differ.* **2013**, *55*, 237–246.
16. Lhomond, G.; McClay, D.R.; Gache, C.; Croce, J.C. Frizzled 1/2/7 signaling directs β-catenin nuclearisation and initiates andoderm specification in macromeres during sea urchin embryogenesis. *Development* **2012**, *139*, 816-825.
17. McIntyre, D.C.; Seay, N.W.; Croce, J.C.; McClay, D.R. Short-range Wnt5 signaling initiates specification of sea urchin posterior ectoderm. *Development* **2013**, *140*, 4881–4889.
18. Croce, J.; Range, R.; Wu, S.Y.; Miranda, E.; Lhomond, G.; Peng, J.C.; Lepage, T.; McClay, D.R. Wnt6 activates endoderm in the sea urchin gene regulatory network. *Development* **2011**, *138*, 3297–3306.
19. Hammond, L.M.; Hofmann, G.E. Early developmental gene regulation in *Strongylocentrotus purpuratus* embryos in response to elevated CO_2 seawater conditions. *J. Exp. Biol.* **2012**, *215*, 2445–2454.
20. Bošnjak, I.; Zaja, R.; Klobučar, R.S.; Sver, L.; Franekić, J.; Smital, T. Identification of ABC transporter genes in gonad tissue of two mediterranean sea urchin species: Black, *Arbacia lixula* L., and rocky, *Paracentrotus lividus* L. *Bull. Environ. Contam. Toxicol.* **2013**, *91*, 415–419.

21. Ragusa, M.A.; Costa, S.; Gianguzza, M.; Roccheri, M.C.; Gianguzza, F. Effects of cadmium exposure on sea urchin development assessed by SSH and RT-qPCR: Metallothionein genes and their differential induction. *Mol. Biol. Rep.* **2013**, *40*, 2157–2167.
22. Pohnert, G. Wound-activated chemical defense in unicellular planktonic algae. *Angew. Chem. Int. Ed.* **2000**, *39*, 4352–4354.
23. Wichard, T.; Gerecht, A.; Boersma, M.; Poulet, S.A.; Wiltshire, K.; Pohnert, G. Lipid and fatty acid composition of diatoms revisited: Rapid wound-activated change of food quality parameters influences herbivorous copepod reproductive success. *Chembiochem* **2007**, *8*, 1146–1153.
24. Vidoudez, C.; Nejstgaard, J.C.; Jakobsen, H.H.; Pohnert, G. Dynamics of dissolved and particulate polyunsaturated aldehydes in mesocosms inoculated with different densities of the diatom *Skeletonema marinoi*. *Mar. Drugs* **2011**, *9*, 345–358.
25. Buttino, I.; De Rosa, G.; Carotenuto, Y.; Mazzella, M.; Ianora, A.; Esposito, F.; Vitiello, V.; Quaglia, F.; La Rotonda, M.I.; Miralto, A. Aldehyde-encapsulating liposomes impair marine grazer survivorship. *J. Exp. Biol.* **2008**, *211*, 1426–1433.
26. Ceballos, S.; Ianora, A. Different diatoms induce contrasting effects on the reproductive success of the copepod *Temora stylifera*. *J. Exp. Mar. Bio. Ecol.* **2003**, *294*, 189–202.
27. Taylor, R.L.; Caldwell, G.S.; Dunstan, H.J.; Bentley, M.G. Short-term impacts of polyunsaturated aldehyde-producing diatoms on the harpacticoid copepod, *Tisbe holothuriae*. *J. Exp. Mar. Biol. Ecol.* **2007**, *341*, 60–69.
28. Lauritano, C.; Borra, M.; Carotenuto, Y.; Biffali, E.; Miralto, A.; Procaccini, G.; Ianora, A. First molecular evidence of diatom effects in the copepod Calanus helgolandicus. *J. Exp. Mar. Biol. Ecol.* **2011**, *404*, 79–86.
29. Lauritano, C.; Carotenuto, Y.; Miralto, A.; Procaccini, G.; Ianora, A. Copepod population-specific response to a toxic diatom diet. *PLoS One* **2012**, *7*, e47262.
30. Aluigi, M.G.; Angelini, C.; Corte, G.; Falugi, C. The sea urchin, *Paracentrotus lividus*, embryo as a "bioethical" model for neurodevelopmental toxicity testing: Effects of diazinon on the intracellular distribution of OTX2-like proteins. *Cell Biol. Toxicol.* **2008**, *24*, 587–601.
31. Pinsino, A.; Matranga, V.; Trinchella, F.; Roccheri, M.C. Sea urchin embryos as an *in vivo* model for the assessment of manganese toxicity: Developmental and stress response effects. *Ecotoxicology* **2010**, *19*, 555–562.
32. Russo, R.; Bonaventura, R.; Zito, F.; Schroder, H.C.; Muller, I.; Muller, W.E.G.; Matranga, V. Stress to cadmium monitored by metallothionein gene induction in *Paracentrotus lividus* embryos. *Cell Stress Chaperones* **2003**, *8*, 232–241.
33. Bonaventura, R.; Poma, V.; Costa, C.; Matranga, V. UVB radiation prevents skeleton growth and stimulates the expression of stress markers in sea urchin embryos. *Biochem. Biophys. Res. Commun.* **2005**, *328*, 150–157.
34. Bonaventura, R.; Zito, F.; Costa, C.; Giarrusso, S.; Celi, F.; Matranga, V. Stress response gene activation protects sea urchin embryos exposed to X-rays. *Cell Stress Chaperones* **2011**, *16*, 681–687.
35. Lister, K.N.; Lamare, M.D.; Burritt, D.J. Sea ice protects the embryos of the Antarctic sea urchin *Sterechinus neumayeri* from oxidative damage due to naturally enhanced levels of UV-B radiation. *J. Exp. Biol.* **2010**, *213*, 1967–1975.

36. Lister, K.N.; Lamare, M.D.; Burritt, D.J. Oxidative damage in response to natural levels of UV-B radiation in larvae of the tropical sea urchin *Tripneustes gratilla*. *Photochem. Photobiol.* **2010**, *86*, 1091–1098.
37. Adams, N.L.; Campanale, J.P.; Foltz, K.R. Proteomic responses of sea urchin embryos to stressful ultraviolet radiation. *Integr. Comp. Biol.* **2012**, *52*, 665–680.
38. Byrne, M. Global change ecotoxicology: Identification of early life history bottlenecks in marine invertebrates, variable species responses and variable experimental approaches. *Mar. Environ. Res.* **2012**, *76*, 3–15.
39. Falugi, C.; Aluigi, M.G.; Chiantore, M.C.; Privitera, D.; Ramoino, P.; Gatti, M.A.; Fabrizi, A.; Pinsino, A.; Matranga, V. Toxicity of metal oxide nanoparticles in immune cells of the sea urchin. *Mar. Environ. Res.* **2012**, *76*, 114–121.
40. Roccheri, M.C.; Agnello, M.; Bonaventura, R.; Matranga, V. Cadmium induces the expression of specific stress proteins in sea urchin embryos. *Biochem. Biophys. Res. Commun.* **2004**, *321*, 80–87.
41. Diller, K.R. Stress protein expression kinetics. *Annu. Rev. Biomed. Eng.* **2006**, *8*, 403–424.
42. Rhee, J.-S.; Raisuddin, S.; Lee, K.-W.; Seo, J.S.; Ki, J.-S.; Kim, I.-C.; Park, H.G.; Lee, J.-S. Heat shock protein (Hsp) gene responses of the intertidal copepod *Tigriopus japonicus* to environmental toxicants. *Comp. Biochem. Physiol. C Toxicol. Pharmacol.* **2009**, *149*, 104–112.
43. Campos, A.C.E.; Molognoni, F.; Melo, F.H.M.; Galdieri, L.C.; Carneiro, C.R.W.; D'Almeida, V.; Correa, M.; Jasiulionis, M.G. Oxidative stress modulates DNA methylation during melanocyte anchorage blockade associated with malignant transformation. *Neoplasia* **2007**, *9*, 1111–1121.
44. Pinsino, A.; Roccheri, M.C.; Matranga, V. Manganese overload affects p38 MAPK phosphorylation and metalloproteinase activity during sea urchin embryonic development. *Mar. Environ. Res.* **2014**, *93*, 64–69.
45. Lepage, T.; Sardet, C.; Gache, C. Spatial expression of the hatching enzyme gene in the sea urchin embryo. *Dev. Biol.* **1992**, *150*, 23–32.
46. Ghiglione, C.; Lhomond, G.; Lepage, T.; Gache, C. Structure of the sea urchin hatching enzyme gene. *Eur. J. Biochem*. **1994**, *219*, 845–854.
47. Duboc, V.; Röttinger, E.; Lapraz, F.; Besnardeau, L.; Lepage, T. Left-right asymmetry in the sea urchin embryo is regulated by nodal signaling on the right side. *Dev. Cell* **2005**, *9*, 147–158.
48. Shipp, L.E.; Hamdoun, A. ATP-binding cassette (ABC) transporter expression and localization in sea urchin development. *Dev. Dyn.* **2012**, *241*, 1111–1124.
49. Nemer, M.; Rondinelli, E.; Infante, D.; Infante, A.A. Polyubiquitin RNA characteristics and conditional induction in sea urchin embryos. *Dev. Biol.* **1991**, *145*, 255–265.
50. National Center for Biotechnology Information. Available online: http://www. ncbi.nlm.nih.gov (accessed on 20 January 2013).
51. Pfaffl, M.W. A new mathematical model for relative quantification in real-time RT-PCR. *Nucleic Acids Res.* **2001**, *29*, e45.
52. Pfaffl, M.W.; Horgan, G.W.; Dempfle, L. Relative expression software tool (REST) for group-wise comparison and statistical analysis of relative expression results in real-time PCR. *Nucleic Acids Res.* **2002**, *30*, e36.

53. Vanaverbeke, J.; Franco, M.A.; van Oevelen, D.; Moodley, L.; Provoost, P.; Steyaert, M.; Soetaert, K.; Vincx, M. Benthic responses to sedimentation of phytoplankton on the Belgian Continental Shelf. In *Current Status of Eutrophication in the Belgian Coastal Zone*; Presses Universitaires de Bruxelles: Bruxelles, Belgium, 2006; pp. 73–90.

4

Identification of Plakortide E from the Caribbean Sponge *Plakortis halichondroides* as a Trypanocidal Protease Inhibitor using Bioactivity-Guided Fractionation

Swarna Oli [1], Usama Ramadan Abdelmohsen [2,3], Ute Hentschel [2] and Tanja Schirmeister [1,*]

[1] Institute of Pharmacy and Biochemistry, Johannes Gutenberg-University of Mainz, Staudinger Weg 5, Mainz 55128, Germany; E-Mail: swarnoli@uni-mainz.de

[2] Department of Botany II, Julius-von-Sachs-Institute for Biological Sciences, University of Würzburg, Julius-von-Sachs Platz 3, Würzburg 97082, Germany; E-Mails: usama.ramadan@uni-wuerzburg.de (U.R.A.); ute.hentschel@uni-wuerzburg.de (U.H.)

[3] Department of Pharmacognosy, Faculty of Pharmacy, Minia University, Minia 61519, Egypt

* Author to whom correspondence should be addressed; E-Mail: schirmei@uni-mainz.de

---

Abstract: In this paper, we report new protease inhibitory activity of plakortide E towards cathepsins and cathepsin-like parasitic proteases. We further report on its anti-parasitic activity against *Trypanosoma brucei* with an IC_{50} value of 5 μM and without cytotoxic effects against J774.1 macrophages at 100 μM concentration. Plakortide E was isolated from the sponge *Plakortis halichondroides* using enzyme assay-guided fractionation and identified by NMR spectroscopy and mass spectrometry. Furthermore, enzyme kinetic studies confirmed plakortide E as a non-competitive, slowly-binding, reversible inhibitor of rhodesain.

Keywords: *Plakortis halichondroides*; plakortide E; protease inhibitor; slowly-binding reversible inhibitor; cathepsin; rhodesain

---

1. Introduction

Proteases enable breakdown of proteins via catalytic hydrolysis of peptide bonds [1]. Malfunction in the control of protease activity leads to undesired and unregulated proteolysis which causes many diseases. Therefore, inhibitors of proteases have the potential to provide successful therapeutics for a wide range of diseases [2,3].

Marine sponges of the family Plakinidae are known to be rich sources of structurally unique and biologically active metabolites [4]. Bioactivity-guided fractionation of the crude cyclohexane extract from the sponge *Plakortis halichondroides* yielded a pure endoperoxide metabolite, named plakortide E (Figure 1), which was previously isolated from the same sponge species [5,6]. Plakortide E was previously shown to stimulate sarcoplasmic reticulum (SR) Ca^{2+} ATPase activity [5]. Other endoperoxides from the plakortin family, e.g., six-membered plakortin [7], dihydroplakortin, 3-epiplakortin, plakortide Q [8] and plakortide M [9], are known to be active against Plasmodia, while the five-membered endoperoxide plakortide E was reported to be inactive [10]. In this work, we highlight its new anti-protease and anti-parasitic activities.

2. Results and Discussion

The lyophilized material of the sponge *Plakortis halichondroides* was sequentially extracted with three different solvents and the crude extracts were tested for protease inhibitory activity against the following proteases: Human cysteine proteases cathepsin B [11] and L [12], the related parasite enzyme rhodesain [13] from *Trypanosoma brucei rhodesiense*, and the two cysteine proteases expressed by the SARS coronavirus, namely SARS main protease [14] and SARS papain-like protease [15]. The active crude cyclohexane extract (CY) was further fractionated using column chromatography and finally purified with HPLC to yield the active pure metabolite plakortide E (Figure 1). The purification process was based upon the bioactivity results, *i.e.*, only fractions which showed activity against the enzymes were purified further. The activity of the fractions gradually enhanced with every step of purification process, with the exception of the first cyclohexane extract whichshowed very high inhibition of cathepsins B and L (Figure 2) probably being due to presence of other non polar active compounds.

Figure 1. Structure of plakortide E.

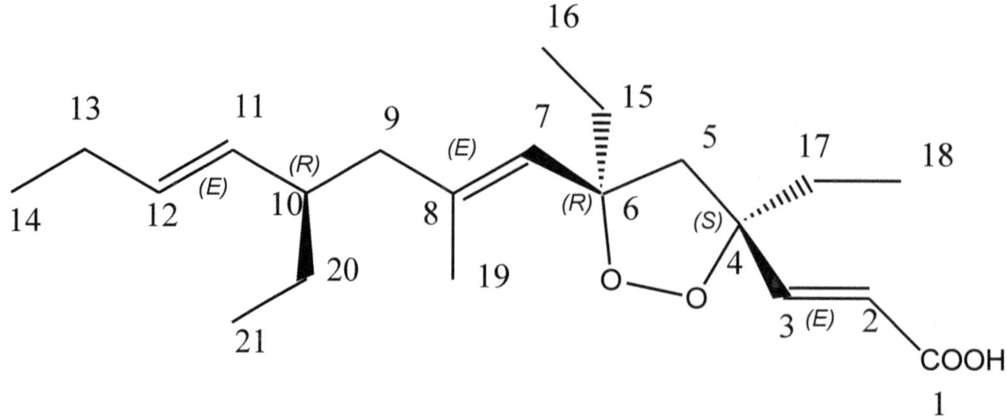

Figure 2. Protease inhibitory activity of the crude cyclohexane extract (CY), the subfractions CYFr III and CY E as well as plakortide E.

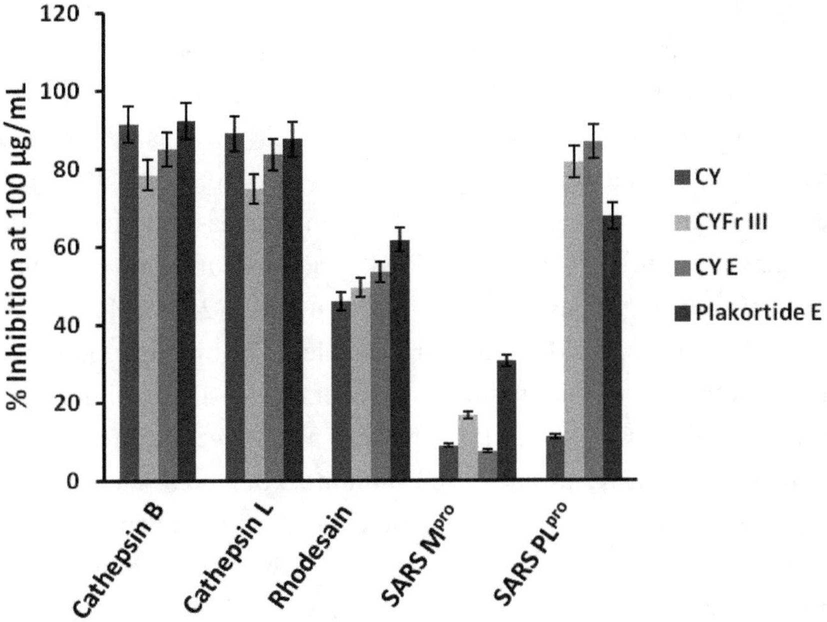

The pure metabolite plakortide E was further tested against the parasite cathepsin-L like protease falcipain-2 [16] from *Plasmodium falciparum*, as well as against the mammalian serine proteases chymotrypsin and the serine protease from Dengue virus (NS2B/NS3 protease) [17]. Inhibition at 100 μg/mL (285.71 μM) was only found with the cathepsins and the cathepsin-like proteases (Figure 3), especially with cathepsin B, L, rhodesain and SARS PL^{pro}.

Figure 3. Percent inhibition of proteases by plakortide E at 100 μg/mL (285 μM), tests were performed in triplicates.

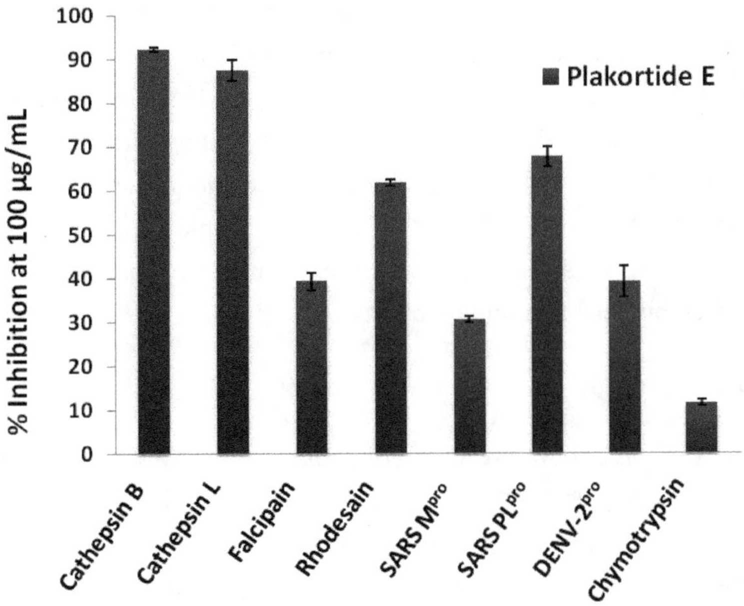

In order to study the inhibition mechanism of plakortide E, enzyme assays [18] with various inhibitor and substrate concentrations were performed with cathepsin B to determine whether the inhibition is competitive with respect to the substrate. The IC_{50} values for cathepsin B were determined at three different substrate concentrations (50, 100, 200 μM). The values (29, 26, 36 μg/mL, 82, 73, 103 μM) were almost similar indicating non-competitive inhibition. For cathepsin L, an IC_{50} of 37 μg/mL (105 μM) at a substrate concentration of [S] = 6.25 μM and for rhodesain an IC_{50} of 44 μg/mL (124 μM) at a substrate concentration of [S] = 30.0 μM was determined. With rhodesain, non-linear progress curves for the substrate hydrolysis were observed in the presence of inhibitor. Therefore, we determined the time-dependency of inhibition by measuring IC_{50} values in correlation to incubation time of enzyme with the inhibitor added, prior to substrate addition. The IC_{50} values decrease with longer incubation times (5 min → 90 μg/mL (257 μM); 30 min → 72 μg/mL (205 μM); 60 min → 27 μg/mL (77 μM)) indicating covalent inhibition or other mechanisms leading to slow binding. To address whether the inhibition is reversible or irreversible, further kinetic studies were performed as described by Copeland [19]. Dilution assays were performed and compared with K11777 [20], an irreversible vinylsulfone-based inhibitor of rhodesain. The enzyme rhodesain (100 fold concentration as used in the normal enzyme assays) was preincubated with the 10 fold IC_{50} concentration of the compound for one hour, which allows the formation of enzyme-inhibitor complexes and leads to total block of enzyme activity. The above complex was then diluted 100 fold by adding assay buffer and substrate. Thus, the enzyme concentration was reduced to the one used in the normal assays and the inhibitor concentration was reduced to 1/10 of the IC_{50}. If reversible, the inhibitor will dissociate from the complex and enzyme activity is recovered. In case of irreversible inhibition, dissociation of the complex cannot occur and no enzyme activity will be detected. These assays showed reactivation of the enzyme in the case of plakortide E to about 50% activity, while in contrast no enzyme activity was detected with K11777. This indicates a reversible inhibition by plakortide E. In summary, plakortide E was determined to be a non-competitive, reversible, inhibitor of cysteine proteases, in case of rhodesain slow-binding was observed.

Furthermore, plakortide E was tested against the parasites *Leishmania major* promastigotes and the trypomastigote forms of *Trypanosoma brucei*, *Candida albicans*, and was also tested for its cytotoxicity against J774.1 macrophages. The compound exhibited trypanocidal activity with an IC_{50} value of 5 μM (after 48 h and also after 72 h). This may at least in parts be due to the protease inhibiting properties of plakortide E. It did not show activity against *Leishmania* which also express a variety of cathepsin-like proteases [21]. No activity against *Candida*, and no cytotoxic effects against macrophages at 100 μM were observed. Since *Leishmania* promastigote forms express less cysteine proteases than the amastigote forms, the cysteine-protease inhibiting properties of the compound may not be sufficient for detectable leishmanicidal activity.

3. Experimental Section

The sponge *Plakortis halichondroides* was collected by SCUBA diving at depths of 30 m in Bahamas in July 2008 (GPS: 26°27′3.25″N, 77°54′14.59″W). Sponge tissues were cut into small pieces and preserved at −80 °C until extraction. The frozen material was then dried by lyophilization. The lyophilized material (640 g) was subsequently macerated and sequentially extracted with cyclohexane (CY), methylene dichloride (DCM), and finally methanol (MeOH). After filtration, the crude extracts were concentrated under reduced pressure. The crude cyclohexane extract (15.27 g) was chromatographed on a silica gel (200 g) column and eluted with an isocratic solvent (cyclohexane/methylene dichloride/methanol/formic acid (2:1:1:0.05)). The eluted fractions were combined based upon TLC results to yield five fractions (CYFr I–V). Further fractionation of the fraction CYFr III by silica gel column chromatography using the solvent system (cyclohexane/methylene dichloride (90:10) with increasing polarity (chloroform/methanol (10:90)) afforded seven subfractions (CY A–G). The subfraction CY E was subjected to preparative HPLC using a RP 18 column (eluent methanol/water with 0.1% formic acid 70:30, flow 8 mL/min) affording 3 fractions (CY M, N and P). The fraction CY N was further purified using preparative HPLC using RP 18 column (methanol/water amended with 0.1% formic acid 70:30, flow 8 mL/min, and the retention time of the peak was observed at 40 min) to yield the pure bioactive compound **1**. The compound **1** was identified as plakortide E, by means of MS and NMR spectral data (Table 1) and comparison to previously published NMR data [5,6,22]. Enzyme assays [18,21,23–25] and parasite growth assays [21,23–26] were performed as described previously.

Table 1. NMR-spectroscopic data of plakortide E (**1**) in $CDCl_3$ (^1H: 400 MHz; ^{13}C: 100 MHz, δ in ppm).

Position	δ_C	Multiplicity	δ_H	Multiplicity	*J* (Hz)
1	171.76	C			
2	119.75	CH	6.09	d	15.8
3	152.31	CH	6.96	d	15.8
4	87.31	C			
5	56.11	CH_2	2.55 2.45	d d	12.0 12.0
6	89.39	C			
7	126.71	CH	5.12	s	
8	136.82	C			
9	46.67	CH_2	2.00 1.85	m m	
10	42.68	CH	1.99	m	
11	132.91	CH	5.05	ddt	15.2, 8.3
12	132.10	CH	5.34	dt	15.2, 6.3, 6.3
13	25.73	CH_2	1.96	m	
14	14.19	CH_3	0.94	t	7.4
15	32.36	CH_2	1.88 1.64	m m	
16	9.01	CH_3	0.89	t	7.4
17	30.88	CH_2	1.76	m	
18	8.95	CH_3	0.92	t	7.5
19	17.92	CH_3	1.62	d	1.2
20	27.79	CH_2	1.34 1.10	m m	
21	11.72	CH_3	0.81	t	7.4

NMR spectra (Table 1) were obtained with a BRUKER (Bruker Biospin GmbH, Rheinstetten Germany), Typ Advance 400 spectrometer. Mass spectra were measured using a Bruker micrOTOF 88 mass spectrometer. Column chromatography was performed using silica gel (0.063–0.200 mm mesh, Merck, Darmstadt, Germany). TLC was conducted on pre-coated silica gel 60 F254 plates (0.20 mm, Merck, Darmstadt, Germany); spots were detected using vanillin spray reagent, UV 254 nm and iodinevapours. Reagents were purchased from Sigma-Aldrich (Munich, Germany) or Fluka (Munich, Germany). Solvents were purchased from Roth (Karlsruhe, Germany) or Merck (Darmstadt, Germany). High performance liquid chromatography was performed on a Varian ProStar (Rheinfelden, Schweiz), consisting of an analytical/preparative HPLC Upscale Linear system (0.05–50 mL/min at 275 bar pressure with scale-mast), a preparative autosampler and a 2-channel UV detector. The detection wavelengths were 254 nm and 230 nm.

Plakortide E: The product was obtained as colourless viscous oil (0.018 g). $CDCl_3$ was used as solvent for measuring NMR spectra. ESI-MS found: 373.23404 $[M + Na]^+$, calcd. for $C_{21}H_{34}O_4$, 350.49. The specific rotation of plakortide E was $[\alpha]^D{}_{22} = +60.7°$, $c = 0.00313$ in $CHCl_3$.

Enzyme assays and *in vitro* tests for antiparasitic activity were performed as published previously: for cathepsin-like cysteine proteases see [18,23–26], for SARS M^{pro} see [27], for SARS PL^{pro} see [15], for Dengue virus protease see [28], for assays against *T. brucei* see [24,29–32], for assays on macrophages see [33], for assays on *L. major* promastigotes see [21], for assays on *C. albicans* see [34,35].

4. Conclusions

Plakortide E, obtained from the marine sponge *Plakortis halichondroides*, was identified as a new protease inhibitor. Plakortide E showed selectivity towards the cathepsin-like cysteine proteases, with a non-competitive, reversible, and, in the case of rhodesain, a slow-binding inhibitory mode of action. The anti-protease activity of the compound may contribute to its anti-parasitic activity against *Trypanosoma brucei*, as rhodesain and also the cathepsin B like protease TbCatB [13] are known to be essential for the parasite's growth and pathogenicity.

Acknowledgments

We would like to thank Cornelia Heindl and Anna Kucharski from University of Wuerzburg, Germany and Ulrike Nowe, Sabine Maehrlein, Nicole Heindl from University of Mainz for performing the enzyme assays. We gratefully acknowledge Antje Fuss and Svetlana Sologub (SFB 630 TP Z1, University of Wuerzburg) for performing the parasite and toxicity tests. We thank Joe Pawlik (UNC Wilmington, USA) for excellent organisation of the Bahamas 2008 expedition, during which the material was collected. Financial support is provided by the Deutsche Forschungsgemeinschaft, SFB 630 (TPs A4, A5).

Author Contributions

Swarna Oli, isolation and characterization of plakortide E, manuscript preparation, structure elucidation; Usama Ramadan Abdelmohsen, compound purification; Ute Hentschel, manuscript preparation; Tanja Schirmeister, manuscript preparation, analysis of assay results.

References

1. Turk, B. Targeting proteases: Successes, failures and future prospects. *Nat. Rev. Drug Discov.* **2006**, *5*, 785–799.
2. Powers, J.C.; Asgian, J.L.; Ekici, Ö.D.; James, K.E. Irreversible inhibitors of serine, cysteine and threonine proteases. *Chem. Rev.* **2002**, *102*, 4639–4750.
3. Rawlings, N.D.; Morton, F.R. The MEROPS batch BLAST: A tool to detect peptidases and their non-peptidase homologues in a genome. *Biochimie* **2008**, *90*, 243–259.
4. Sata, N.; Abinsay, H.; Yoshida, W.Y.; Horgen, F.D.; Sitachitta, N.; Kelly, M.; Scheuer, P.J. Lehualides A–D, metabolites from a Hawaiian sponge of the genus *Plakortis*. *J. Nat. Prod.* **2005**, *68*, 1400–1403.
5. Patil, A.D.; Freyer, A.J.; Bean, M.F.; Carte, B.K.; Westley, J.W.; Johnson, R.K.; Lahouratate, P. The plakortones, novel bicyclic lactones from the sponge *Plakortis halichondroides*: Activators of cardiac SR-Ca^{2+} pumping ATPase. *Tetrahedron* **1996**, *52*, 377–394.
6. Chen, Y.; McCarthy, P.J.; Harmody, D.K.; Schimoler-O'Rourke, R.; Chilson, K.; Selitrennikoff, C.; Pomponi, S.A.; Wright, A.E. New bioactive peroxides from marine sponges of the family plakiniidae. *J. Nat. Prod.* **2002**, *65*, 1509–1512.
7. Fattorusso, C.; Campiani, G.; Catalanotti, B.; Persico, M.; Basilico, N.; Parapini, S.; Taramelli, D.; Campagnuolo, C.; Fattorusso, E.; Romano, A.; *et al.* Endoperoxide derivatives from marine organisms: 1,2-Dioxanes of the plakortin family as novel antimalarial agents. *J. Med. Chem.* **2006**, *49*, 7088–7094.
8. Campagnuolo, C.; Fattorusso, E.; Romano, A.; Taglialatela-Scafati, O.; Basilico, N.; Parapini, S.; Taramelli, D. Antimalarial polyketide cycloperoxides from the marine sponge *Plakortis simplex*. *Eur. J. Org. Chem.* **2005**, *23*, 5077–5083.
9. Hu, J.-F.; Gao, H.-F.; Kelly, M.; Hamann, M.T. Plakortides I–L, four new cyclic peroxides from an undescribed Jamaican sponge *Plakortis* sp. (Homosclerophorida, Plakinidae). *Tetrahedron* **2001**, *57*, 9379–9383.
10. Fattorusso, E.; Parapini, S.; Campagnuolo, C.; Basilico, N.; Taglialatela-Scafati, O.; Taramelli, D. Activity against *Plasmodium falciparum* of cycloperoxide compounds obtained from the sponge *Plakortis simplex*. *J. Antimicrob. Chemother.* **2002**, *50*, 883–888.
11. Palermo, C.; Joyce, J.A. Cysteine cathepsin proteases as pharmacological targets in cancer. *Trends Pharm. Sci.* **2008**, *29*, 22–28.
12. Lankelma, J.M.; Voorend, D.M.; Barwari, T.; Koetsveld, J.; Van der Spek, A.H.; de Porto, A.P.N.A.; Van Rooijen, G.; Van Noorden, C.J.F. Cathepsin L target in cancer treatment? *Life Sci.* **2010**, *86*, 225–233.
13. Kerr, I.D.; Wu, P.; Marion-Tsukamaki, R.; Mackey, Z.B.; Brinen, L.S. Crystal structures of TbCatB and rhodesain, potential chemotherapeutic targets and major cysteine proteases of *Trypanosoma brucei*. *PLoS Negl. Trop. Dis.* **2010**, *4*, doi:10.1371/journal.pntd.0000701.
14. Anand, K.; Ziebuhr, J.; Wadhwani, P.; Mesters, J.R.; Hilgenfeld, R. Coronavirus main proteinase (3CLpro) structure: Basis for design of anti-SARS drugs. *Science* **2003**, *300*, 1763–1767.

15. Han, Y.-S.; Chang, G.-G.; Juo, C.-G.; Lee, H.-J.; Yeh, S.-H.; Hsu, J.T.-A.; Chen, X. Papain-like protease 2 (PLP2) from Severe Acute Respiratory Syndrome Coronavirus (SARS-CoV): Expression, purification, characterization, and inhibition. *Biochemistry* **2005**, *44*, 10349–10359.
16. Rosenthal, P. Falcipains and other cysteine proteases of malaria parasites. In *Cysteine Proteases of Pathogenic Organisms*; Robinson, M., Dalton, J., Eds.; Springer: San Francisco, California, USA, 2011; Volume 712, pp. 30–48.
17. Bessaud, M.; Pastorino, B.A.M.; Peyrefitte, C.N.; Rolland, D.; Grandadam, M.; Tolou, H.J. Functional characterization of the NS2B/NS3 protease complex from seven viruses belonging to different groups inside the genus *Flavivirus*. *Virus Res.* **2006**, *120*, 79–90.
18. Breuning, A.; Degel, B.; Schulz, F.; Buchold, C.; Stempka, M.; Machon, U.; Heppner, S.; Gelhaus, C.; Leippe, M.; Leyh, M.; *et al.* Michael acceptor based antiplasmodial and antitrypanosomal cysteine protease inhibitors with unusual amino acids. *J. Med. Chem.* **2010**, *53*, 1951–1963.
19. Copeland, R.A. A Guide for Medicinal Chemists and Pharmacologists. In *Evaluation of Enzyme Inhibitors in Drug Discovery*; John Wiley & Sons, Inc.: Hoboken, NJ, USA, 2005.
20. Yang, P.-Y.; Wang, M.; He, C.Y.; Yao, S.Q. Proteomic profiling and potential cellular target identification of K11777, a clinical cysteine protease inhibitor, in *Trypanosoma brucei*. *Chem. Commun.* **2012**, *48*, 835–837.
21. Ponte-Sucre, A.; Vicik, R.; Schultheis, M.; Schirmeister, T.; Moll, H. Aziridine-2,3-dicarboxylates, peptidomimetic cysteine protease inhibitors with antileishmanial activity. *Antimicrob. Agents Chemother.* **2006**, *50*, 2439–2447.
22. Sun, X.Y.; Tian, X.Y.; Li, Z.W.; Peng, X.S.; Wong, H.N. Total synthesis of plakortide E and biomimetic synthesis of plakortone B. *Chemistry* **2011**, *17*, 5874–5880.
23. Vicik, R.; Busemann, M.; Gelhaus, C.; Stiefl, N.; Scheiber, J.; Schmitz, W.; Schulz, F.; Mladenovic, M.; Engels, B.; Leippe, M.; *et al.* Aziridide-based inhibitors of cathepsin L: synthesis, inhibition activity, and docking studies. *ChemMedChem* **2006**, *1*, 1126–1141.
24. Vicik, R.; Hoerr, V.; Glaser, M.; Schultheis, M.; Hansell, E.; McKerrow, J.H.; Holzgrabe, U.; Caffrey, C.R.; Ponte-Sucre, A.; Moll, H.; Stich, A.; *et al.* Aziridine-2,3-dicarboxylate inhibitors targeting the major cysteine protease of *Trypanosoma brucei* as lead trypanocidal agents. *Bioorg. Med. Chem. Lett.* **2006**, *16*, 2753–2757.
25. Ehmke, V.; Heindl, C.; Rottmann, M.; Freymond, C.; Schweizer, W.B.; Brun, R.; Stich, A.; Schirmeister, T.; Diederich, F. Potent and selective inhibition of cysteine proteases from *Plasmodium falciparum* and *Trypanosoma brucei*. *ChemMedChem* **2011**, *6*, 273–278.
26. Ettari, R.; Zappala, M.; Micale, N.; Giofre, S.; Schirmeister, T.; Grasso, S. Peptidomimetics containing a vinyl ketone warhead as falcipain-2 inhibitor. *Eur. J. Med. Chem.* **2011**, *46*, 2058–2065.
27. Käppler, U.; Stiefl, N.; Schiller, M.; Vicik, R.; Breuning, A.; Schmitz, W.; Rupprecht, D.; Schmuck, C.; Baumann, K.; Ziebuhr, J.; *et al.* A new lead for non-peptidic active-site-directed inhibitors of the SARS coronavirus main protease discovered by a combination of screening and docking methods. *J. Med. Chem.* **2005**, *48*, 6832–6842.

28. Steuer, C.; Heinonen, K.H.; Kattner, L.; Klein, C.D. Optimization of assay conditions for dengue virus protease: Effect of various polyols and nonionic detergents. *J. Biomol. Screen.* **2009**, *14*, 1102–1108.
29. Baltz, T.; Baltz, D.; Giroud, C.; Crocket, J. Cultivation in a semi-defined medium of animal infective forms of *Trypanosoma brucei*, *T. equiperdum*, *T. evansi*, *T. rhodesiense* and *T. gambiense*. *EMBO J.* **1985**, *4*, 1273–1277.
30. Räz, B.; Iten, M.; Grether-Bühler, Y.; Kaminsky, R.; Brun, R. The Alamar Blue assay to determine drug sensitivity of African trypanosomes (*T.b. rhodesiense* and *T.b. gambiense*) *in vitro*. *Acta Trop.* **1997**, *68*, 139–147.
31. Merschjohann, K.; Sporer, F.; Steverding, D.; Wink, M. *In vitro* effect of alkaloids on bloodstream forms of *Trypanosoma brucei* and *T. congolense*. *Planta Med.* **2001**, *67*, 623–627.
32. Huber, W.; Koella, J.C. A comparison of three methods of estimating EC_{50} in studies of drug resistance of malaria parasites. *Acta Trop.* **1993**, *55*, 257–261.
33. Ahmed, S.A.; Gogal, R.M.; Walsh, J.E. A new rapid and simple non-radioactive assay to monitor and determine the proliferation of lymphocytes: An alternative to [^{3}H] thymidine incorporation assay. *J. Immunol. Methods* **1994**, *170*, 211–224.
34. Degel, B.; Staib, P.; Rohrer, S.; Scheiber, J.; Martina, E.; Büchold, C.; Baumann, K.; Morschhäuser, J.; Schirmeister, T. *Cis*-configured aziridines are new pseudo-irreversible dual-mode inhibitors of *Candida albicans* secreted aspartic protease 2. *ChemMedChem* **2008**, *3*, 302–315.
35. Büchold, C.; Hemberger, Y.; Heindl, C.; Welker, A.; Degel, B.; Pfeuffer, T.; Staib, P.; Schneider, S.; Rosenthal, P.J.; Gut, J.; *et al.* New *cis*-configured aziridine-2-carboxylates as aspartic acid protease inhibitors. *ChemMedChem* **2011**, *6*, 141–152.

5

Mode of Action of Diterpene and Characterization of Related Metabolites from the Soft Coral, *Xenia elongata*

Eric H. Andrianasolo [1], Liti Haramaty [1], Eileen White [2], Richard Lutz [1] and Paul Falkowski [1,*]

[1] Center for Marine Biotechnology, Institute of Marine and Coastal Sciences, Rutgers, The State University of New Jersey, New Brunswick, NJ 08901, USA; E-Mails: andriane@marine.rutgers.edu (E.H.A.); haramaty@marine.rutgers.edu (L.H.); rlutz@imcs.marine.rutgers.edu (R.L.)

[2] The Cancer Institute of New Jersey, 195 Little Albany Street, New Brunswick, NJ 08903, USA; E-Mail: epwhite@cinj.rutgers.edu

* Author to whom correspondence should be addressed; E-Mail: falko@imcs.marine.rutgers.edu

---

Abstract: Chemical and biological investigation of the cultured marine soft coral *Xenia elongata* led to the isolation of two new diterpenes (**2**, **3**). Their structures were elucidated using a combination of NMR and mass spectrometry. Biological evaluations and assessments were determined using the specific apoptosis induction assay based on genetically engineered mammalian cell line D3 deficient in Bak and Bax and derived from a mouse epithelial cell. The diterpenes induce apoptosis in low micromolar concentrations. The results indicate that the previously isolated compound (**1**) affects cell in a manner similar to that of HSP90 and HDAC inhibitors and in a manner opposite of PI3 kinase/mTOR inhibitors. Compound (**3**) inhibits selectively HDAC6 in high micromolar concentrations.

Keywords: structural and functional characterization of marine drug; marine biotechnology; bioactive compounds and bioproducts; drug discovery and development

---

1. Introduction

Natural products are the most reliable and rich source of new anticancer entities. Over the past 30 years, nearly 63% of anticancer drugs introduced to the pharmaceutical development market are natural products or can be traced from a natural products origin [1,2]. The soft coral of the genus *Xenia elongata* has produced interesting biological molecules, specifically apoptosis inducer compounds [3,4]. These molecules embed a framework of nine-membered rings and a particular arrangement of functional groups with multiple stereocenters. These structural features limit the range of chemical reactions that are applicable to their synthesis. There is little in the existing synthetic literature to define an effective strategy for the synthesis of these molecules [5]. The first total synthesis of an optically active xenicane diterpene has not yet been achieved [5,6]. In our ongoing effort to discover and develop new marine natural product biomedicinals, and in order to explore the diversity of natural products from the soft coral *Xenia elongata*, chemical and biological investigation of this organism were undertaken. Two novel diterpenes (**2, 3**) were isolated and, remarkably, these new diterpenes induce apoptosis and, more importantly, **3** is 10-fold more active than the most active and previously isolated compound **1** (Figure 1). The detail mode of action of **1** by analysis of MCF-7 breast cancer treated with **1** was also reported here. HDAC inhibitory assay of **3** was also displayed and discussed.

Figure 1. Structure of compounds **1**, **2** and **3**.

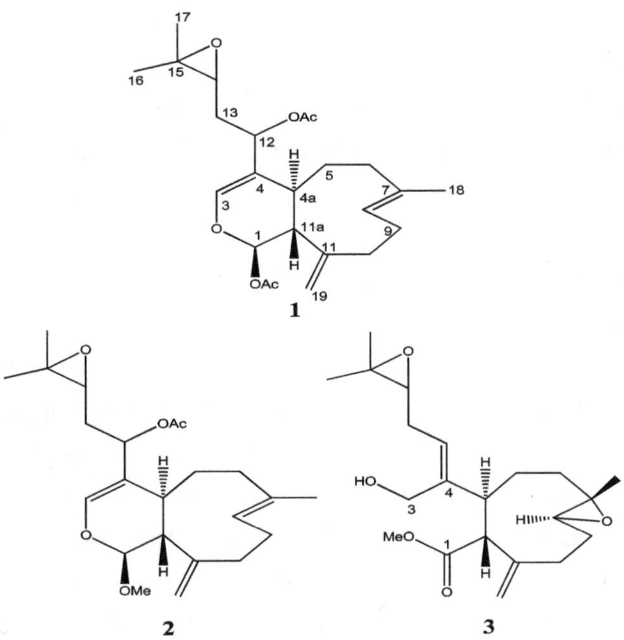

2. Results and Discussion

Live colonies of *Xenia elongata* were grown in the coral laboratory at the Marine Biotechnology Center, Institute of Marine and Coastal Sciences, Rutgers University. For extraction, 200 g of the colonies were isolated and frozen. A methanol (or dichloromethane) soluble fraction was extracted from frozen coral tissue, lyophilized and dissolved in DMSO. The extract that induced apoptosis was subsequently fractionated and purified by analytical RPHPLC. Using this strategy, compounds with proapoptotic activity were isolated from the whole tissue extracts. Chemical structures of the two compounds (**2** and **3**) were ascertained by standard spectroscopic techniques as described below.

The molecular formula of **2** was established as $C_{23}H_{34}O_5$ on the basis of HRESIMS (*m/z* 413.2299 $[M + Na]^+$ calcd. for $C_{23}H_{34}O_5Na$, 413.2298). This indicated a difference of 28 mass units and probably a carbonyl group, compared to compound **1** [3]. The ^{1}H NMR (Supplementary Figure S1) of **2** indicated clearly the existence of the following functional groups: an 1-methoxydihydropyran moiety [δ 5.35 (d, J = 1.4 Hz), H-1 and 6.51 (s), H-3], a terminal methylene [δ 5.05 (s) and 5.15 (s), H-19, H-19′], two methyls α to oxygen [δ 1.33 (s), H-16, H-17], an epoxy signal at [δ 2.75 (dd, J = 5.9, 6.2 Hz), H-14] and a vinyl methyl group [1.60 (s), H-18]. The NMR spectra of compounds **1** and **2** were compared in $CDCl_3$ and were found to be similar, with the exception of the carbon at C-1 [δ 99.3] (Supplementary Figure S2) for **2** instead of C-1 [δ 91.7] for **1** in addition to that **2** has one less carbonyl signal, suggesting that a methoxy group is attached to C-1 in **2** instead of acetate group in **1**. The deshielding effect of the methoxy group to the carbon at C-1 position is more pronounced compared to the deshielding effect of an acetate group to the carbon at the same position.

The relative stereochemistry of **2** was established with the same method as described in our previous work [3] which is based on ROESY data (Supplementary Figure S3), coupling constant analyses and chemical shifts comparison to **1** and xeninculin [7], tsitsixenicin A [8,9] and related compounds [7,9]. The coupling constant (J = 12 Hz) between H-4a and H-11a suggests a trans ring junction. The small coupling constant (J = 1.4 Hz) between H-1 and H-11a would favor the α-position for H-1 if the six-member ring was in a quasi-boat conformation. A closer look of the crystal structure of xenicin and essentially the six-member ring, which is also present in **2**, suggests that H-1, if it is in the α-position, should display ROESY cross peak correlation to H-4a. In order to verify this suggestion, a high scan ROESY experiment was performed. Careful analysis of the ROESY data, particularly in the cross region of H-1 and H-4a, revealed unambiguously that these two protons display a ROESY cross peak correlation (Figure 2). Based on this result, it appears that the configurations at C-1, C-4a, and C-11a in compound **2** are essentially the same as those in **1** and xeninculin, in which the protons H-1 and H-4a are on the α face of the ring and the proton H-11a is oriented on the β-side. This observation confirmed that the stereochemistry of this proton at H-1 position is well conserved and common to any molecule with dihydropyran moiety in the xenicane diterpene class. Previous attempts to establish the stereochemistry of an acetate group at C-12 in xenicane diterpenes by chemical transformations and spectroscopy have failed [9,10], and therefore the stereochemistry at C-12 remains unassigned. This information suggested that the acetate group attached to C-12 is very reactive and probably leads to further transformation of the molecule as described by the proposed rearrangement in Scheme 1. From the above results, the structure of **2** was formulated as shown.

Figure 2. Selected ROESY cross peaks for compounds (**2**) and (**3**).

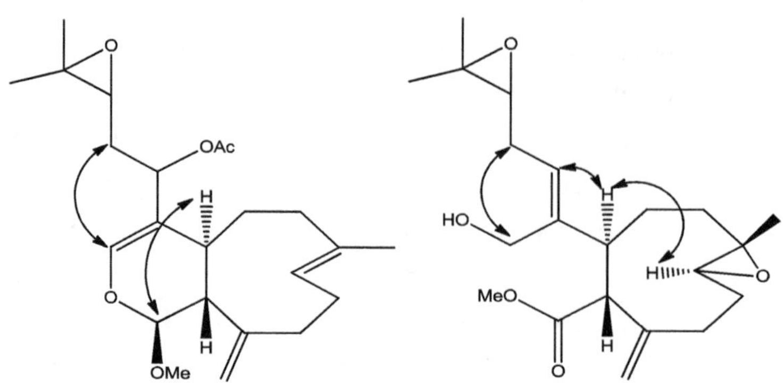

A molecular formula of $C_{21}H_{32}O_5$ for **3** was determined from HRESIMS data (Supplementary Figure S4 Low resolution). ^{1}H NMR (Supplementary Figure S5) of **3** indicated the presence of two epoxy signals at [δ 2.80 (dd, J = 11.5, 3.2 Hz), H-8] and [δ 2.76 (dd, J = 7.2, 6.1 Hz), H-14], one methyl α to oxygen [δ 1.16 (s), H-18] and a terminal methylene [δ 5.00 (s) and 5.20 (s), H-19, H-19′]. The presence of two equivalent protons at δ [3.72, d(6)], H-3 for an isolated methylene attached to oxygen indicated that **3** has a primary alcohol functional group at C-3 which also revealed that the dihydropyran moiety presents in **1** and **2** is absent in **3**. The NMR data of **3** were analogous to those of diterpene xenitacin isolated from the Formosan soft coral *Xenia umbellata* [4] except that xenitacin had an acetate group attached to C-3 instead of a hydroxy group for **3** at C-3 (Supplementary Figure S6).

The relative stereochemistry of **3** was investigated with the aid of a ROESY spectrum. The *Z*-configuration was assigned to $\Delta^{4(12)}$ double bond based on the observed ROESY cross peak between H-12 [δ 5.45] and H-4a [δ 3.29]. The configuration of the two chiral centers at C-4a and C-11a was deduced by ROESY data and comparison with those of xenitacin [4]. The large coupling constant (J = 12 Hz) between H-4a [δ 3.29] and H-11a [δ 3.45] suggests that they have a configuration opposite to each other [4]. The observed ROESY cross peak between H-4a [δ 3.29] and H-8 [δ 2.80] (Figure 2, Supplementary Figure S7) suggests that they have similar configuration to xenitacin [4]. Therefore the structure of **3** was established as shown. Key HMBC and selected COSY correlations were established for **2** and **3** in order to assign all carbons framework in their structures (Figure 3 and Supplementary Figures S8–S13).

Figure 3. Key HMBC and selected COSY correlations for compounds (**2**) and (**3**).

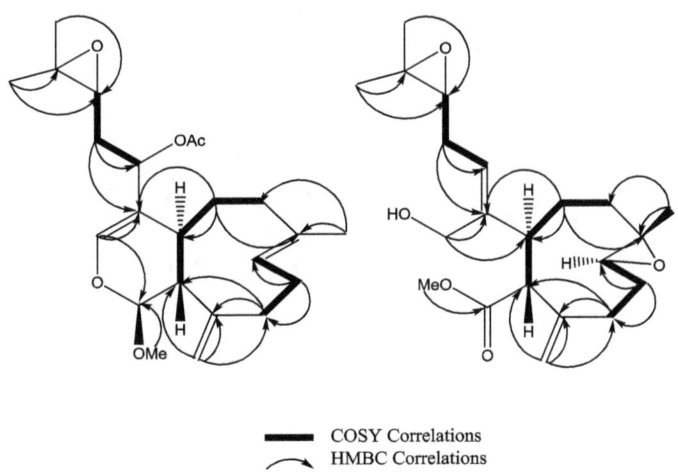

In order to assess the pro-apoptotic activities of these compounds as a marker for their potential anticancer efficacy, methanol soluble fractions of *Xenia* extracts were tested for apoptosis induction in an MTT assay using the apoptosis competent W2 and apoptosis resistant D3 cells, as described in our previous work [3]. Compounds **2** and **3** induce measurable cell death in W2 but not in D3 cells (Supplementary Figures S14 and S15). To quantify apoptosis induction, viability was determined using an MTT assay [11]. Measurements were taken at time 0 and 48 h after addition of compounds in different concentrations. Viability was calculated as the difference between time 0 (addition of compound) and 48 h. Apoptosis induction was defined as at least 20% death of W2 cells and a 10% or higher growth of D3 cells by MTT assay. The optimal concentration of the compounds for apoptosis induction was determined based on the dose response of the two cells lines by MTT assay. The results reveal a significant induction of apoptosis by the whole tissue extract and by individual compounds (11.5 μM for **2** and 1.2 μM for **3**), with the most remarkable induction shown by compound **3** (1.2 μM) which is ten times more active than compound **1** previously characterized [3]. Treatment of W2 and D3 cells with 0.1 μM staurosporine, a protein kinase inhibitor and potent inducer of apoptosis, was included as a positive control for apoptosis induction and for comparison. Thus, the diterpenes we have isolated and characterized effectively and specifically activate apoptosis in immortalized mammalian epithelial cells.

There are efforts underway to synthesize optically active xenicane diterpene [12] and work is in progress for the synthesis of diterpene from *Xenia elongata*. One synthetic compound (Supplementary Figure S16) from this work which has the same structure of **3** but does not have the side chain was also tested for comparison of activity. The result revealed that this synthetic compound did not induce apoptosis, which suggested that the side chain is critical for the activity. It is also reported before that a similar structure, xenitacin, isolated from the Formosan soft coral *Xenia umbellata* [4], is the most active compound in its group in a cytotoxicity assay against P-388, HT-29 and A549 cells [4]. These bioactivity results may suggest that a conversion or rearrangement toward the most active structure may undergo within the organism itself or even in the cell in order to increase the efficacy of affecting or killing cancer cells. A proposed conversion or rearrangement of **2** to **3** is displayed (Scheme 1). It is also possible that **1** will convert to **2** by enzymatic reaction of deacetylation followed by methylation and then **2** will be converted to **3**.

Scheme 1. Proposed rearrangement of **2** under acidic conditions, to give **3**.

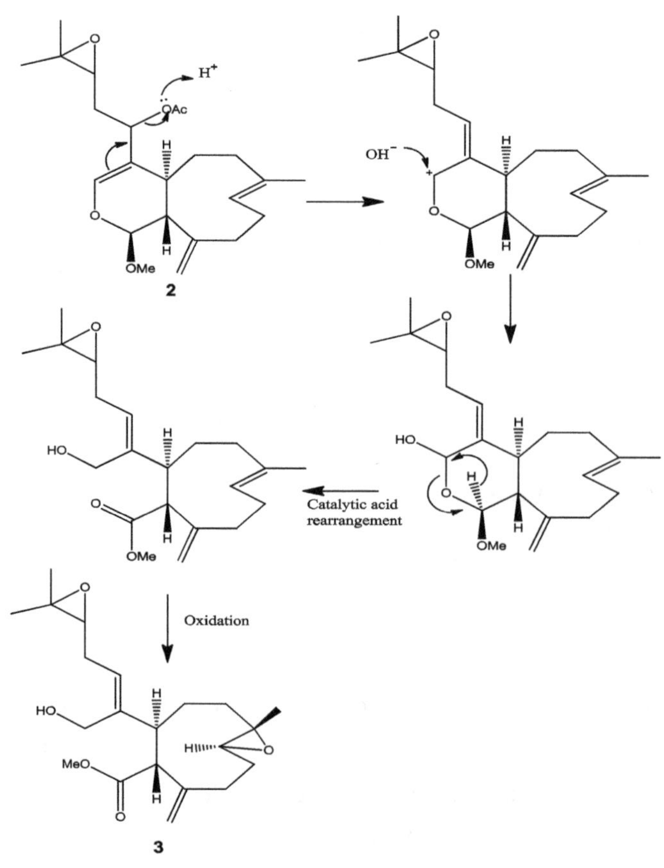

In order to determine the mode of action of diterpenes from *Xenia elongata*, a connectivity analysis for MCF-7 breast cancer cells treated with compound **1** was performed. The objective is to analyze changes in gene expression in a breast cancer cell line exposed to a soft coral diterpene, which is compound **1** in this experiment, and to compare the resulting signature to perturbagen signatures in the Connectivity Map [13].

To prepare a 0.12 M stock solution in DMSO, 1.6 mg of **1** was used. Proliferating MCF-7 breast cancer cells were treated in triplicate with either a 120 μM final concentration of **1** (final DMSO concentration 0.1%), or were left untreated for 8 h. Cells were harvested before the RNA was extracted then purified, reverse transcribed, and hybridized to an Affymetrix array by the CINJ Transcriptional Profiling Core. The resulting gene expression data were analyzed.

Expression levels of 22,277 genes in three untreated and three treated samples were obtained using an Affymetrix array. For this initial analysis, the 11,400 genes were used for which all six samples provided clearly measurable signals. (Genes for which any sample was determined to be "A" or "M" were excluded, and only those for which all six were called "P" by the Affymetrix software were kept).

The package "siggenes" [14] was used in the bioconductor set of gene expression packages for the R statistical system. Two different methods were considered for obtaining the most differentially expressed genes. The first method is known as "significance analysis of arrays" or "SAM" and was first proposed by Tusher *et al.* [15]. This method computes a modified t-statistic comparing treated to control samples for each gene. It then identifies differentially expressed genes so that the false discovery rate (FDR) is a pre-specified value. The second method, known as "Empirical Bayes Analysis of Microarrays" or "EBAM" [16] also computes the modified t-statistic. Unlike SAM, however, it computes the posterior probability that each gene is differentially expressed. Each of the two methods can be used to create a list of genes that are differentially expressed. The two lists will generally be different, although there is usually some overlap.

For the SAM method, the parameter delta was set to 1.15. This produced an FDR of 3.58%, and resulted in a list of 215 genes, of which an estimated 20 genes are falsely declared to be differentiated. A SAM plot is given in Figure 4, with some of the genes labeled. A complete list is given in Supplementary Table S1.

For the EBAM method, the required posterior probability was set to 0.7, which resulted in an FDR of 15.4%, to generate a list of about the same number of genes (211). An EBAM plot of posterior probabilities is given in Figure 5, and a complete list of differentially expressed genes is given in Supplementary Table S2. A list of the 61 genes that are present on both lists is given in Supplementary Table S3. One gene, with code "36711_at," is clearly at the top of both lists. This protein is a novel MAFF (v-maf musculoaponeurotic fibrosarcoma (avian) oncogene family, protein F)-LIKE protein.

The diterpene connectivity scores for the 453 "instances" (representing treatment of various cell lines with 164 different compounds) are shown in Supplementary Tables S4 and S5. Common drug targets among the mostly highly connected instances include the proteasome, calmodulin, and histone deacetylases (HDACs). The presence of structurally distinct compounds that yield high connectivity scores suggests that the diterpene **1** acts as a proteasome and calmodulin inhibitor. Conversely, instances associated with the most highly ranked negative scores (corresponding to a gene expression profile similar to a reversal of the diterpene signature) include compounds that target the PI3 kinase/mTOR pathway. Additional inferences can be made from "permuted results" provided by the connectivity analysis (Table 1). In particular, analysis of permuted results supports the idea that the diterpene **1** is affecting cells in a manner similar to that of HSP90 and HDAC inhibitors and in a manner opposite to that of PI3 kinase/mTOR inhibitors. Note that the proteasome inhibitors are not included in the permuted results because they are represented by single instances.

Figure 4. SAM Plot, with some genes labeled.

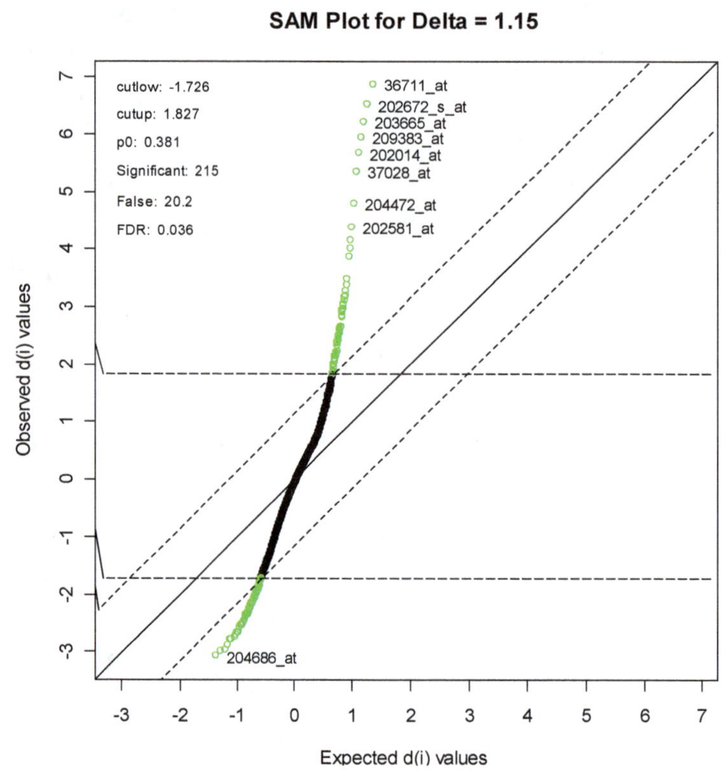

Figure 5. EBAM plot. The points above the dotted line correspond to genes with a posterior probability of 0.9 of being differentially expressed.

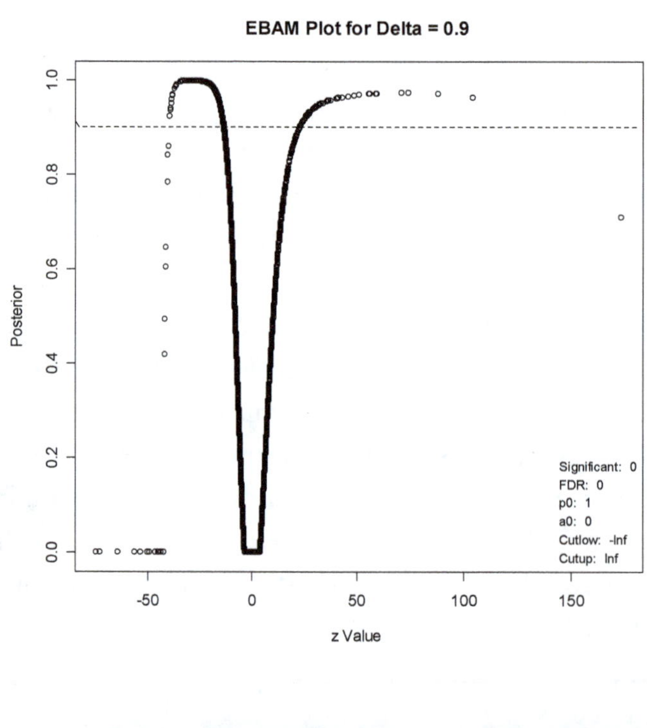

Table 1. Permuted results.

Rank	CMAP name	Score	N	Enrichment	*P* value	Target
1	sirolimus	−0.254	10	−0.770	0.0000	mTOR
2	17-allylamino-geldanamycin	0.496	18	0.611	0.0000	HSP90
3	trichostatin A	0.565	12	0.714	0.0000	HDACs
4	LY-294002	−0.038	17	−0.506	0.0001	PI3 kinase
5	geldanamycin	0.544	6	0.759	0.0004	HSP90
6	5253409	0.748	2	0.971	0.0009	?
7	5224221	0.758	2	0.967	0.0015	?
8	vorinostat	0.684	2	0.956	0.0032	HDAC
9	tetraethylene-pentamine	−0.084	6	−0.662	0.0055	proteases
10	wortmannin	−0.248	8	−0.565	0.0071	PI3 kinase

Each row in the table summarizes the performance of all instances produced with the same perturbagen. The table displays the cmap name for those perturbagens, the arithmetic mean of the connectivity scores for each of those instances (labeled "score"), the number of those instances (labeled "*n*"), a measure of the enrichment of those instances in the list of all instances ordered by descending order of connectivity score and up score (labeled "enrichment") and a permutation *p*-value for that enrichment score (labeled "*p*"). Enrichment scores and *p*-values are not provided for perturbagens represented by only one instance or where the mean of the connectivity scores for their instances is zero.

In order to verify the predicted mode of action of diterpene from connectivity analysis, HDAC inhibitory assays were performed. Given the availability of compound material and the structural similarity of all isolated diterpenes, only compound **3** was subjected to enzymatic inhibition test against the class I (HDAC1, HDAC2, HDAC3 and HDAC8), the class II A (HDAC4, HDAC5, HDAC7 and HDAC9) and the class II B (HDAC6) enzymes. HDAC inhibitors Trichostatin A (TSA) and TMP269 (TMP) were used as reference compounds. The results showed that compound **3** up to 50 or 100 μM did not significantly inhibit the class I (HDAC1, Supplementary Table S6; HDAC2, Supplementary Table S7; HDAC3, Supplementary Table S8; HDAC8, Supplementary Table S9) as well as the class II A (HDAC4, Supplementary Table S10; HDAC5, Supplementary Table S11; HDAC7, Supplementary Table S12; HDAC9, Supplementary Table S13), though compound **3** inhibited class II B HDAC6 with IC_{50} around 80 μM (Figure 6, Supplementary Table S14). The predicted mode of action of diterpene was verified. HDAC6 appears to be the key deacetylase regulating Hsp90; treatment of breast cancer cells with Hsp90 inhibitors or HDAC inhibitor (vorinostat) induces breast cancer cell differentiation [17]. Cell death signaling is complex and it is clear that there are multiple alternate pathways, especially in cancer cell types. The HDAC6 inhibition pathway is among several pathways that the diterpene may trigger simultaneously, thus explaining the difference of potency between the apoptosis induction assay and the HDAC6 inhibition assay.

Figure 6. Inhibition of HDAC6 by compound **3**, IC_{50} around 80 μM.

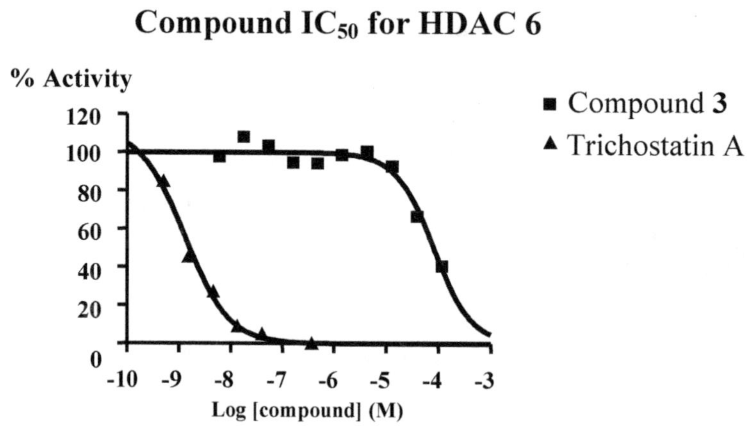

	Compound **3**	Trichostatin A
HILLSLOPE	-1.190	-1.036
EC50	8.093e-005	1.288e-009

3. Experimental Section

3.1. General Experimental Procedures

Optical rotations were measured on JASCO P 1010 polarimeter. FT-IR spectra were obtained employing Hewlett Packard 8452A and Nicolet 510 instruments. All NMR spectra were recorded on a Bruker Avance DRX300 and DPX400 spectrometers. Spectra were referenced to residual solvent signal with resonances at $\delta_{H/C}$ 7.26/77.1 ($CDCl_3$). ESI MS data were acquired on a Waters Micromass LCT Classic mass spectrometer and Varian 500-MS LC Ion Trap. HPLC separations were performed using Waters 510 HPLC pumps, a Waters 717 plus autosampler, and Waters 996 photodiode array detector. All solvents were purchased as HPLC grade.

3.2. Extraction and Isolation Procedures

The organism was purchased from Ocean Reef Aquariums (ORA). Live colonies of *Xenia elongata* were kept in optimal growing conditions (Salinity 34 g/L, temperature 25 °C, light [min. 250 μE m^{-2} s^{-1} and max. 500 μE m^{-2} s^{-1}], pH 8.2–8.4) in the coral laboratory at the Marine Biotechnology Center, Institute of Marine and Coastal Sciences, Rutgers University. A voucher specimen is available as collection number SC/XE/Apr-04 System 1A. The soft coral was stored at −80 °C after addition of liquid nitrogen before workup. The material (80 g) was extracted three times, first with CH_2Cl_2 and then with MeOH, to give a nonpolar crude organic extract (259.2 mg) and a polar crude organic extract (537.9 mg). A portion of these two extracts (30 mg each) was tested for apoptosis induction. The polar crude organic extract was found active and subjected to fractionation by solid phase extraction cartridge (Reverse-phase C18) to give three fractions using a stepwise gradient of H_2O-MeOH as solvent system. The fraction eluting with 15% H_2O and 85% MeOH had apoptosis induction activity. This fraction was further chromatographed on analytical RP HPLC (Phenomenex luna C8, 250 × 4.60 mm) using a gradient elution (starting with 80% water and 20% CH_3CN, flow rate 1 mL/mn) to yield successively 3.1 mg of **2** (t_R = 27 mn), 2.6 mg of **3** (t_R = 33 min). The extract/compound weight percents are 0.57% and 0.48% for **2** and **3**, respectively.

Compound **2**: $[\alpha]_D^{22.5}$ −58 (*c* 0.87, $CHCl_3$); IR ν_{max} (neat) 3075, 2970, 2930, 2860, 1736, 1670, 1608, 1445, 1370, 1236, 1155, 1015, 950, 870, 835, 790 cm^{-1}; ^1H NMR and ^{13}C NMR, see Table 2; HRESIMS *m*/*z* 413.2299 (calcd. for $C_{23}H_{34}O_5Na$, 413.2298).

Table 2. NMR Spectroscopic Data of Compound **2** ^1H (300 MHz, $CDCl_3$) and ^{13}C (75 MHz, $CDCl_3$).

	Compound 2		
Position	δ_C	δ_H	**HMBC**
1	99.3 CH	5.35, d (1.4)	3, 4a
3	142.7 CH	6.51, s	1, 4, 4a, 12
4	112.6 qC		
4a	35.7 CH	2.19, dd (8, 12)	3, 4, 5, 11
5	31.0 CH_2	1.50, m 1.95, m	4a, 6
6	40.1 CH_2	2.08, m 2.22, m	5, 7, 8
7	135.5 qC		
8	124.6 CH	5.25, t (5.8)	9
9	24.7 CH_2	2.45, m 2.08, m	8, 10
10	35.7 CH_2	2.30, m	11, 19
11	150.0 qC		
11a	55.1 CH	1.70, dd (1.2,12)	1, 4a, 11
12	70.8 CH	4.85,dd (5.8, 6.1)	3, 4, 4a, 13, 14, CO
13	31.9 CH_2	1.98, m 2.00, m	12, 14
14	62.8 CH	2.75, dd (5.9, 6.2)	13
15	60.1 qC		
16	18.9 CH_3	1.33, s	14, 15, 17
17	24.7 CH_3	1.33, s	14, 15, 16
18	17.7 CH_3	1.60, s	6, 7, 8
19	119.8 CH_2	5.05, s 5.15, s	10, 11, 11a
CH_3 of CH_3COO	21.0 CH_3	2.09, s	CO
CO of CH_3COO	170.9 qC		
CH_3 of CH_3O	51.5 CH_3	3.52, s	1

Compound **3**: $[\alpha]^{22.5}_D$ +12 (*c* 0.13, $CHCl_3$); IR ν_{max} (neat) 3075, 2970, 1740, 1730, 1600, 1370, 1240 cm^{-1}; ^1H NMR and ^{13}C NMR, see Table 3; HRESIMS *m*/*z* 365.2325 (calcd. for $C_{21}H_{33}O_5$, 365.2328).

Table 3. NMR Spectroscopic Data of Compound **3** ^{1}H (300 MHz, $CDCl_3$) and ^{13}C (75 MHz, $CDCl_3$).

Compound 3			
Position	δ_C	δ_H	**COSY**
1	172.4 qC		
3	65.5 CH_2	3.72, d(6)	
4	138.1 qC		
4a	38.3 CH	3.29, m	5, 11a
5	28.7 CH_2	1.55, m 1.73, m	4a, 6
6	38.3 CH_2	1.17, m 1.99, m	5
7	59.7 qC		
8	63.8 CH	2.80, dd (11.5, 3.2)	9
9	27.2 CH_2	2.13, m 1.43, m	8, 10
10	26.2 CH_2	2.13, m 2.34, m	9
11..	143.0 qC		
11a	60.6 CH	3.45, d (12.0)	4a
12	126.6 CH	5.45, dd (10.1, 7.2)	13
13	28.1 CH_2	2.30, m 2.38, m	12, 14
14	63.3 CH	2.76, dd (7.2, 6.1)	13
15	58.8 qC		
16	19.0 CH_3	1.29, s	
17	24.6 CH_3	1.32, s	
18	18.8 CH_3	1.21, s	
19	121.1 CH_2	5.00, s 5.20, s	
CH_3O	51.6 CH_3	3.50, s	

3.3. Biological Evaluation–Apoptosis Induction

Apoptosis induction in the presence of compounds (**2**, **3**) was carried out in the following manner. W2 (apoptosis competent) and D3 (apoptosis defective) cells were plated in 96- and 6-well plates and incubated for 24 h, after which they were evenly spread at about 50% confluency. At this time, compounds dissolved in DMSO and diluted in growth medium (DMEM) were added to the cells at various concentrations. DMSO concentration was kept at 0.5% in all wells. Plates were incubated for 24, 48 and 72 h. Cell viability was determined using a modification of the MTT assay [11], where the reduction of yellow tertazolium salt (MTT—3-(4,5-dimethylthiazol-2-yl)-2,5) to purple formazan indicates mitochondrial activity, and thus cell viability. Cells were incubated with 0.5 mg/mL MTT for 3 h. Supernatant was aspirated and DMSO was added to dissolve the formazan crystals. After 30 min incubation at 37 °C, with shaking, absorbance was read at 570 nm on a Spectra MAX 250 (Molecular Devices) plate reader. Differential growth from time 0 to 48 h was calculated. Starurosporine, an apoptosis inducer, and DMSO were used as positive and negative controls, respectively.

3.4. Connectivity Analyses

Cells were harvested and RNA extracted using TRIzol Reagent, (Invitrogen, Carlsbad, CA, USA) and RNeasy kit (Qiagen Sciences, MD, USA). RNA was purified, reverse transcribed, and hybridized to an Affymetrix U133 2.0 array by the CINJ Transcriptional Profiling Core. The resulting gene expression data were analyzed. Two methods were used to identify the most differentially expressed genes in the treated *versus* control groups. The first method (significance analysis of arrays) was set at a false discovery rate of 3.58%, and yielded 215 genes. A subset of these 215 genes, including 145 with at least a 3-fold difference in expression (58 upregulated by treatment, 87 downregulated), was defined as the diterpene signature, and was used for the connectivity analysis supplementary (Supplementary Table S5). The diterpene signature was compared to 453 gene expression profiles present in the Connectivity Map [13]. For each of the 453 profiles, a connectivity score was calculated (based on the Kolmogorov–Smirnov statistic), which represents the relative similarity of the profile to the diterpene signature.

3.5. HDAC Inhibition Assay

3.5.1. Class I

In singlet 10-dose IC_{50} mode with 3-fold serial dilution starting at 50 μM against 4 HDACs (HDAC1, HDAC2, HDAC3 and HDAC8), 0.5 mg of Compound **3** was tested.

HDAC reference compound Trichostatin A (TSA) and TMP 269 were tested in a 10-dose IC_{50} with 3-fold serial dilution starting at 10 μM.

Substrate for HDAC1, 2 and 3: Fluorogenic peptide from p53 residues 379–382 (RHKK(Ac)AMC.

Substrate for HDAC8: Fluorogenic peptide from p53 residues 379–382 (RHK(Ac)K(Ac)AMC).

Data pages include raw data, % Enzyme activity (relative to DMSO controls), and curve fits. Curve fits were performed where the enzyme activities at the highest concentration of compounds were less than 65%. IC_{50} values were calculated using the GraphPad Prism 4 program based on a sigmoidal dose-response equation. The blank (DMSO) value was entered as 1.0×10^{-12} of concentration for curve fitting.

3.5.2. Class II A and Class II B

In singlet 10-dose IC_{50} mode with 3-fold serial dilution starting at 100 μM against 5 HDACs [II A (HDAC4, HDAC5, HDAC7, and HDAC9) IIB (HDAC6)], 0.5 mg of Compound **3** was tested.

HDAC reference compounds Trichostatin A (TSA) and TMP269 (TMP) were tested in a 10-dose IC_{50} with 3-fold serial dilution starting at 10 μM.

Substrate for HDAC 6: Fluorogenic peptide from p53 residues 379–382 (RHKK(Ac)AMC.

Substrate for HDAC 4,5,7 and 9: Fluorogenic HDAC Class2a Substrate (Trifluoroacetyl Lysine, Ac-LGK(TFA)-AMC).

IC_{50} values were calculated using the GraphPad Prism 4 program based on a sigmoidal dose-response equation. The blank (DMSO) value was entered as 1.0×10^{-12} of concentration for curve fitting.

3.5.3. Class II B for IC_{50} Calculation

0.5 mg of Compound **3** was tested in duplicate 10-dose IC_{50} mode with 3-fold serial dilution starting at 120 μM against 1 HDAC (HDAC6).

HDAC reference compounds Trichostatin A (TSA) was tested in a 10-dose IC_{50} with 3-fold serial dilution starting at 10 μM.

Substrate for HDAC6: Fluorogenic peptide from p53 residues 379–382 (RHKK(Ac)AMC.

IC_{50} values were calculated using the GraphPad Prism 4 program based on a sigmoidal dose-response equation. The blank (DMSO) value was entered as 1.0×10^{-12} of concentration for curve fitting.

4. Conclusions

The compounds described here represent novel natural products with specific proapoptotic, and therefore potential anticancer, activities. This work describes the mode of action of diterpene from the soft coral *Xenia elongata* using connectivity analysis for MCF-7 cells treated with a coral diterpene. The coral diterpene is affecting cells in a manner similar to that of HSP90 and HDAC inhibitors and in a manner opposite to that of PI3 kinase/mTOR inhibitors. The vast majority of human solid tumors are of epithelial origin, and defects in apoptosis, mostly upstream of Bax and Bak, play important roles in both tumor suppression and mediation of chemotherapeutic response. Consequently, efforts are increasingly focused on developing drugs that can re-activate the apoptotic pathway. The compounds identified here induce apoptosis upstream of Bax and Bak and may have a potential for use as anticancer agents that exploit the apoptosis pathway in tumor cells. Furthermore, the diterpene selectively inhibits HDAC6 and represents a new model structure of selective HDAC inhibitors which will contribute to the development of HDAC practical isoform selective [18]. Future work will likely focus on the identification of the protein targeted by the coral diterpene and the optimization of essential pharmocophores.

Acknowledgments

We thank Murugesan. Gounder and Eric. Rubin from Robert Wood Johnson Medical School and Cancer Institute of New Jersey for the connectivity analysis. We also thank Dirk. Moore from Biometrics Cancer Institute of New Jersey for gene expression analysis. We acknowledge Michael Drahl and Tony Wang for making the synthetic compound available. This research was funded by Rutgers University through an Academic Excellence award to Paul Falkowski and Richard Lutz.

References

1. Newman, D.J.; Cragg, G.M. Natural products as sources of new drugs over the last 25 years. *J. Nat. Prod.* **2007**, *70*, 461–477.
2. Simmons, T.L.; Andrianasolo, E.H.; McPhail, K.; Flatt, P.; Gerwick, W.H. Marine natural products as anticancer drugs. *Mol. Cancer Ther.* **2005**, *4*, 333–342.
3. Andrianasolo, E.H.; Haramaty, L.; Degenhardt, K.; Mathew, R.; White, E.; Lutz, R.; Falkowski, P. Induction of apoptosis by diterpenes from the soft coral *Xenia elongata*. *J. Nat. Prod.* **2007**, *70*, 1551–1557.
4. Duh, C.-Y.; El-Gamal, A.A.H.; Chiang, C.-Y.; Chu, C.-J.; Wang, S.-K.; Dai, C.-F. New cytotoxic xenia diterpenoids from the formosan soft coral *Xenia umbellata*. *J. Nat. Prod.* **2002**, *65*, 1882–1885.

5. Pollex, A.; Hiersemann, M. Catalytic asymmetric claisen rearrangement in natural product synthesis: Synthetic studies toward (−)-xeniolide F. *Org. Lett.* **2005**, *7*, 5705–5708.
6. Mushti, C.S.; Kim, J.-H.; Corey, E.J. Total synthesis of antheliolide A. *J. Am. Chem. Soc.* **2006**, *128*, 14050–14052.
7. Kashman, Y.; Groweiss, A. New diterpenoids from the soft corals Xenia macrospiculata and Xenia obscuronata. *J. Org. Chem.* **1980**, *45*, 3814–3824.
8. Davies-Coleman, M.T.; Beukes, D.R. Ten years of marine natural products research at Rhodes University. *S. Afr. J. Sci.* **2004**, *100*, 539–543.
9. Hooper, G.J.; Davies-Coleman, M.T.; Schleyer, M. New diterpenes from the South African soft coral *Eleutherobia aurea*. *J. Nat. Prod.* **1997**, *60*, 889–893.
10. Coval, S.J.; Scheuer, P.J.; Matsumoto, G.K.; Clardy, J. Two new xenicin diterpenoids from the octocoral Anthelia edmondsoni. *Tetrahedron* **1984**, *40*, 3823–3828.
11. Mosmann, T. Rapid colorimetric assay for cellular growth and survival: Application to proliferation and cytotoxicity assays. *J. Immunol. Meth.* **1983**, *65*, 55–63.
12. Drahl, M.A.; Akhmedov, N.G.; Williams, L.J. Selective conversion of an enantioenriched cyclononadienone to the xeniolide, xenibellol, and florlide cores: An integrated routing strategy. *Tetrahedron Lett.* **2011**, *52*, 325–328.
13. Connectivity MAP 02. Available online: http://www.broad.mit.edu/cmap/ (accessed on 12 December 2007).
14. Schwender, H.; Krause, A.; Ickstadt, K. Identifying Interesting Genes with siggenes. *RNews* **2006**, *6*, 45–50.
15. Tusher, V.G.; Tibshirani, R.; Chu, G. Significance analysis of microarrays applied to the ionizing radiation response. *Proc. Natl. Acad. Sci. USA* **2001**, *98*, 5116–5121.
16. Efron, B.; Tibshirani, R.; Storey, J.D.; Tusher, V. Empirical bayes analysis of a microarray experiment. *J. Am. Statist. Assoc.* **2001**, *96*, 1151–1160.
17. Thomas, S.; Munster, P.N. Histone deacetylase inhibitor induced modulation of anti-estrogen therapy. *Cancer Lett.* **2009**, *280*, 184–191.
18. Su, H.; Altucci, L.; You, Q. Competitive or noncompetitive, that's the question: Research toward histone deacetylase inhibitors. *Mol. Cancer Ther.* **2008**, *7*, 1007–1012.

6

CS5931, a Novel Polypeptide in *Ciona savignyi*, Represses Angiogenesis via Inhibiting Vascular Endothelial Growth Factor (VEGF) and Matrix Metalloproteinases (MMPs)

Ge Liu [1,2], Ming Liu [3,*], Jianteng Wei [4], Haijuan Huang [5], Yuyan Zhang [5], Jin Zhao [6], Lin Xiao [7], Ning Wu [1], Lanhong Zheng [8] and Xiukun Lin [9,*]

[1] Institute of Oceanology, Chinese Academy of Sciences, 7 Nanhai Road, Qingdao 266071, China; E-Mails: liug878@163.com (G.L.); Wuning2283720@126.com (N.W.)
[2] The University of Chinese Academy of Sciences, Beijing 100049, China
[3] Key Laboratory of Marine Drugs, Ministry of Education, School of Medicine and Pharmacy, Ocean University of China, Qingdao 266003, China
[4] Lanzhou Institute of Chemical Physics, Chinese Academy of Sciences, Lanzhou 730000, China; E-Mail: Weijt@163.com
[5] Qingdao University of Science and Technology, Qingdao 266042, China; E-Mails: qkhuanghaijuan@126.com (H.H.); zhangyuyanlcs@163.com (Y.Z.)
[6] Department of Biotechnology, Zhengzhou University, Zhengzhou 450001, China; E-Mail: zhaojinwang2136@126.com
[7] Qingdao Agricultural University, Qingdao 266109, China; E-Mail: xiaolin_qd@163.com
[8] Yellow Sea Fisheries Research Institute, Chinese Academy of Fishery Sciences, Qingdao 266071, China; E-Mail: zhenglanhong@126.com
[9] Department of Pharmacology, Capital Medical University, Beijing 100069, China

* Authors to whom correspondence should be addressed; E-Mails: lmouc@hotmail.com (M.L.); linxiukun@yahoo.com (X.L.)

Abstract: CS5931 is a novel polypeptide from *Ciona savignyi* with anticancer activities. Previous study in our laboratory has shown that CS5931 can induce cell death via mitochondrial apoptotic pathway. In the present study, we found that the polypeptide could inhibit angiogenesis both *in vitro* and *in vivo*. CS5931 inhibited the proliferation, migration and formation of capillary-like structures of HUVECs (Human Umbilical Vein Endothelial Cell) in a dose-dependent manner. Additionally, CS5931 repressed spontaneous angiogenesis of the zebrafish vessels. Further studies showed that CS5931 also blocked vascular endothelial growth factor (VEGF) production but without any effect on its mRNA expression. Moreover, CS5931 reduced the expression of matrix metalloproteinases (MMP-2 and MMP-9) both on protein and mRNA levels in HUVEC cells. We demonstrated that CS5931 possessed strong anti-angiogenic activity both *in vitro* and *in vivo*, possible via VEGF and MMPs. This study indicates that CS5931 has the potential to be developed as a novel therapeutic agent as an inhibitor of angiogenesis for the treatment of cancer.

Keywords: CS5931; anti-angiogenesis; zebrafish; HUVECs; VEGF and MMPs

---

1. Introduction

Angiogenesis, the sprouting or intussusception of pre-existing blood vessels to form new vessels, involves many coordinated endothelial cell activities, including proliferation, migration, alignment, and cord formation [1,2]. It is well documented that angiogenesis is a prerequisite of solid tumor growth, since the tumor progression and metastasis depend on the existence of a functional blood supply system [3–6]. The angiogenesis process was regulated by many angiogenic factors; VEGF (vascular endothelial growth factor) is one of the major regulators of angiogenesis and can stimulate the growth of blood vessels directly; MMPs (matrix metalloproteinases), particularly MMP-2 and MMP-9, involved in extracellular matrix degradation [7], are crucial for the endothelial cell migration, organization and hence, angiogenesis [8,9]. Therefore, VEGF and MMPs are closely related to the angiogenesis, and targeting these factors has become a promising anticancer strategy [6,10]. In recent years, a number of anti-angiogenic agents have progressed to clinical studies [7,11]. However, most of the anti-angiogenic candidates do not target tumor cells specifically and can cause endothelial dysfunction and vessel pruning in healthy tissues, and what is more, can induce drug resistance of cancer and stromal cells [12,13]. To overcome these defects, the continuous effort in searching for novel anti-angiogenic agents remains pressing.

The ocean is an important source of biologically active substances and numerous novel peptides with anticancer activity have been isolated from marine organisms [14–16]. Many of them have been reported to display anti-angiogenic activity. For example, Somocystinamide A (ScA), a lipopeptide derived from *L.majuscula*, could inhibit angiogenesis by inducing apoptosis in angiogenic endothelial cells [17]. Aplidine, a cyclic depsipeptide isolated from the Mediterranean tunicate *Aplidium albicans*, also displayed anti-angiogenic activity both *in vitro* and *in vivo* [5].

In our previous study, a novel polypeptide, CS5931, with potent antitumor activity, was purified from *Ciona savignyi*. We found that CS5931 polypeptide displayed significant cytotoxicity against several cancer cells and induced apoptosis in HCT-8 cells via mitochondrial-mediated pathway [18]. Sequence analysis reveals that there is high homology between CS5931 and human granulin A in their primary and 3D structure [19]. However, little is known about its mode of action. Our primary study shows that CS5931 also inhibits the proliferation of HUVEC cells, suggesting the possibility of its anti-angiogenic activity. In the present study, we evaluated the effects of recombinant CS5931 polypeptide on proliferation, migration and cord formation of HUVECs, and the effect of the polypeptide on the formation of new vessels in zebrafish embryos was also examined.

2. Results and Discussion

2.1. CS5931 Inhibits the Proliferation of HUVECs

MTT assay was performed to determine if CS5931 could inhibit the proliferation of HUVECs. As shown in Figure 1A, the polypeptide caused a significant growth-inhibiting effect on HUVECs compared with the untreated HUVECs; the proliferation rate decreased by 74.3, 83.1 and 90.2% after incubation for 24 h, while it decreased by 81.0, 87.4 and 95.1% after incubation for 48 h, when treating the cells with CS5931 at a concentration of 40, 80 and 160 μg/mL, respectively.

Figure 1. Effect of CS5931 on HUVEC proliferation. The cells were incubated in the absence or presence of certain concentrations of CS5931 at 37 °C for 24 h, 48 h (**A**) and 10 h (**B**) and cell viability was determined by MTT assay as described in the Materials and Methods section. (**C**) Cells were stained by PI/Hoechst 33258 after they were treated with 0 (1), 10 (2), 20 (3), 30 (4), 40 μg/mL (5) CS5931 for 24 h and pictures were taken using a fluorescence microscopy. The apoptotic cell with bright blue fluorescence was indicated by an arrow. Results are normalized to untreated cells. All experiments were repeated more than three times. Values represent the means ± SD of triplicate measurements. * $P < 0.05$, ** $P < 0.01$ *versus* medium control.

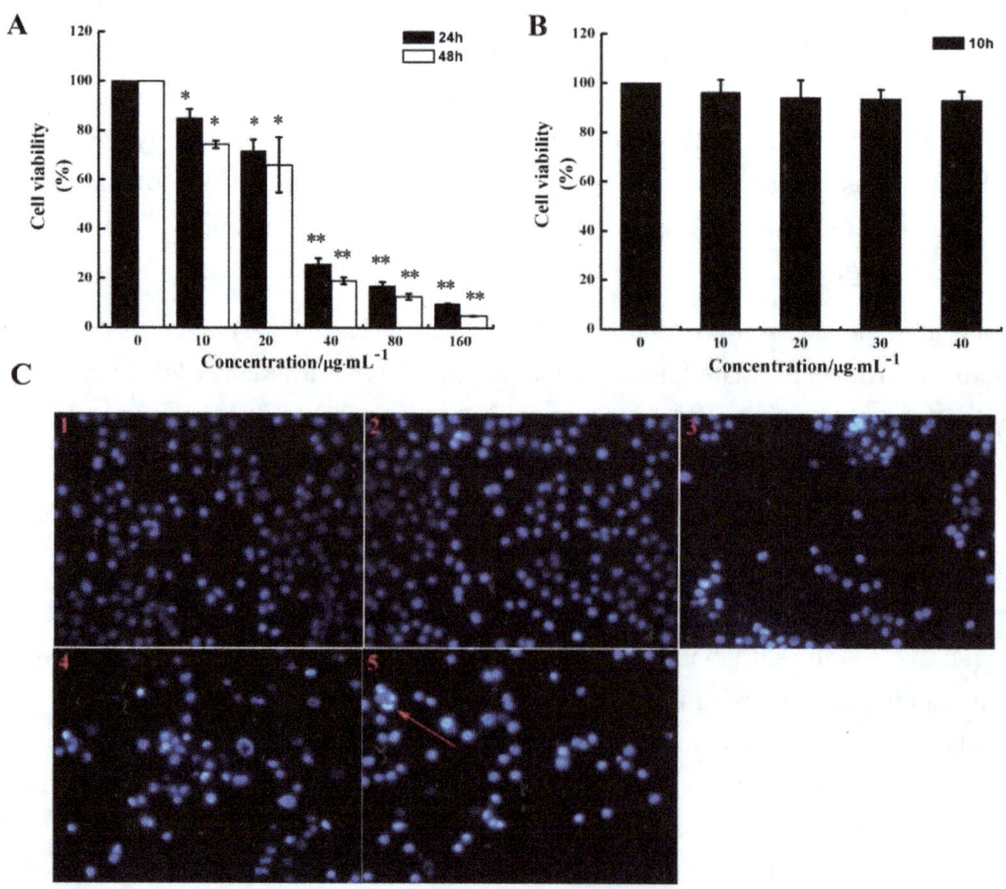

A previous study has shown that CS5931 induced cell death via mitochondrial-mediated pathway in human colon cancer HCT116 cells. In order to determine if the polypeptide can induce apoptosis in HUVEC, PI/Hoechst 33258 staining assay was performed. Our results showed that treatment with higher concentration of CS5931 for 24 h could induce endothelial cell apoptosis to some extent (Figure 1C). These results revealed that the repression of HUVEC growth induced by CS5931 is in a dose and time dependent manner (Figure 1A,B) and the polypeptide could inhibit the proliferation of HUVEC cells via apoptosis pathway when treating the cells for more than 24 h.

It is reported that some cytotoxic antitumor drugs can affect endothelial cell functions and angiogenesis [20,21]. However, not all of them are true anti-angiogenic agents because they need a higher drug concentration to achieve inhibitory effect in endothelial cells than that in tumor cells [22,23]. CS5931 inhibits endothelial cell proliferation, with an IC_{50} similar to that observed for tumor cells [19], indicating that the anti-angiogenic effect might indeed occur in tumors and contribute to the final anticancer activity.

Figure 2. CS5931 inhibits the migration of HUVECs. Cells were treated without (1) or with 10 (2), 20 (3), 30 (4) and 40 μg/mL (5) of CS5931. After incubation for 8 h, cell migration was analyzed using scratch-wound assay (**A**) as well as Transwell assay (**B**). (**C**) Quantitative evaluations of HUVECs migration induced by CS5931 in the scratch-wound assay and Transwell assay (**D**). Results are normalized to untreated cells. All experiments were repeated more than three times. Values represent the means ± SD of triplicate measurements. * $P < 0.05$, ** $P < 0.01$ *versus* medium control.

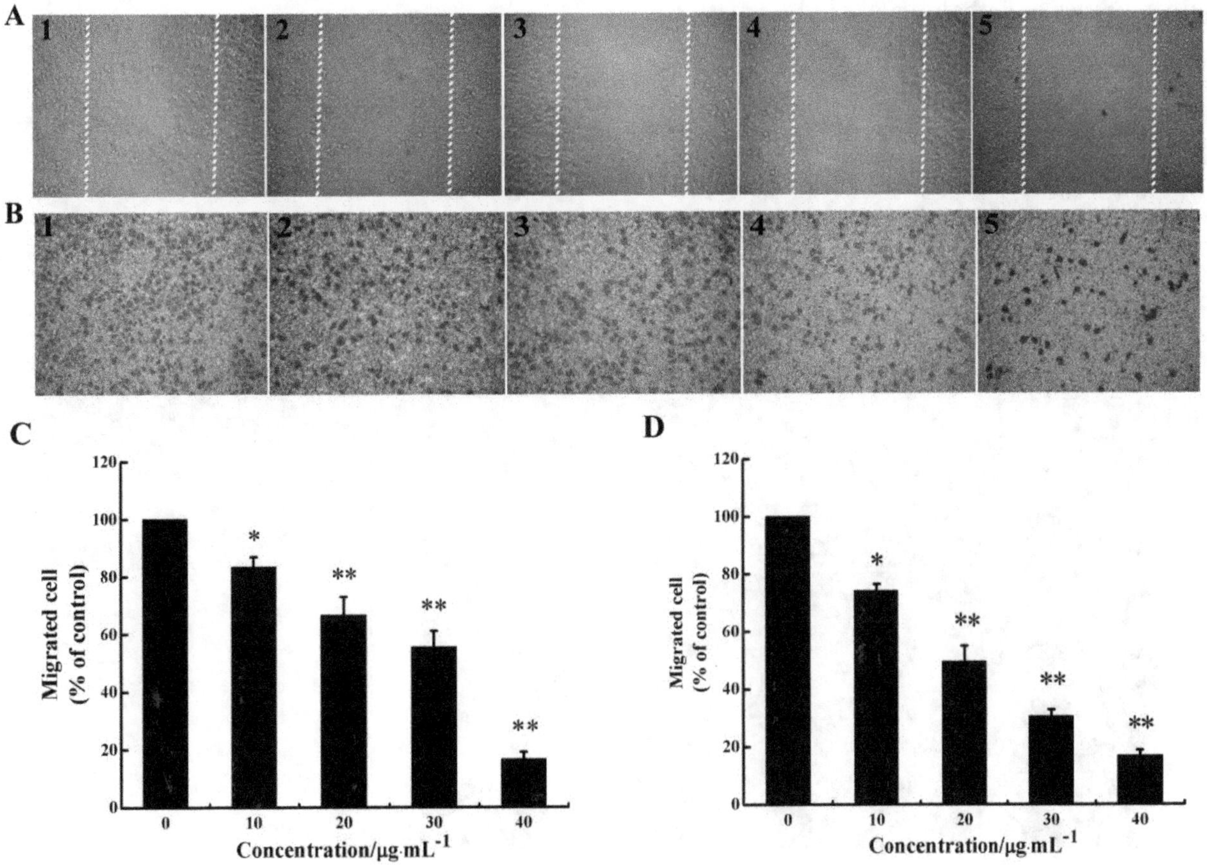

2.2. CS5931 Represses the Migration of HUVECs

As migration of endothelial cell is necessary for angiogenesis, we investigated the effect of CS5931 on directional cell motility using a scratch-wound assay and Transwell assay. The results of scratch-wound assay showed that the wound healing was gradually reduced as the concentration of the polypeptide increased in a dose-dependent relationship (Figure 2A,C). The results of Transwell assay revealed that treatment with CS5931 resulted in a concentration-dependent suppression of cell migration; the inhibition rate of the cells was 25.8, 50.3, 69.4 and 83.1%, when treating the cells with CS5931 at a concentration of 10, 20, 30 and 40 μg/mL, respectively (Figure 2B,D). Both the findings indicated that CS5931 could prevent HUVECs migration significantly. Since the inhibition of cell migration by CS5931 occurred at exposure times at which cell proliferation was not obviously affected (Figure 1B) and cell apoptosis did not happen, the results suggested that CS5931 might indeed exert its anti-angiogenic effect by affecting HUVEC migration.

2.3. CS5931 Disrupts the Cord Formation of HUVECs

Since tube formation is an important process in angiogenesis, we then explored the effects of CS5931 on capillary-like tube structure formation ability. Three-dimensional layer of Matrigel experiment was performed and the results showed that the capillary-like tube formation was inhibited significantly when treating the cells with CS5931 (Figure 3B–E); higher concentration of CS5931 abrogated the cord formation completely (Figure 3E). In contrast, the capillary-like tube structure network could be clearly found in cells untreated with the polypeptide (Figure 3A). These findings demonstrated that CS5931 could suppress endothelial cell cord formation.

Figure 3. Effect of CS5931 on the formation of capillary-like structures of HUVECs. HUVECs were seeded on the surface of the Matrigel in a 96-well plate and treated without (**A**) or with 10 (**B**), 20 (**C**), 30 (**D**) and 40 μg/mL (**E**) of CS5931. After incubation for 6 h, capillary tube formation was examined using an inverted microscope.

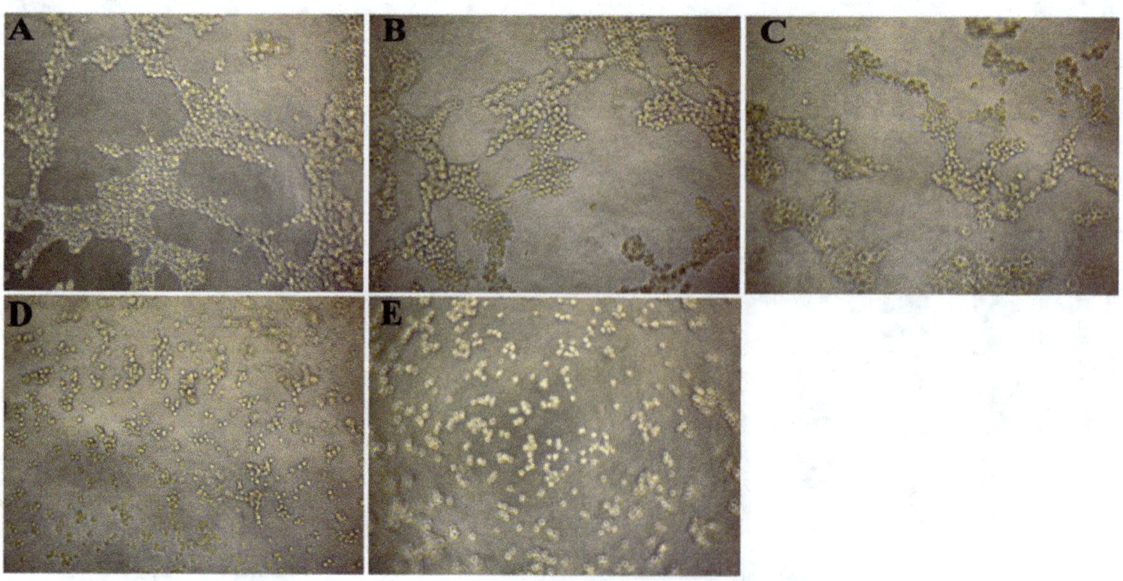

The antimotility activity of CS5931 was apparently sufficient to confer true anti-angiogenic activity. This was further supported by the finding that CS5931 prevented cord formation *in vitro*, an assay that mimics the final events during angiogenesis, when endothelial cells become aligned and organized in a capillary-like structure. Once again, the inhibitory effect occurred before endothelial cell proliferation was dramatically affected and apoptosis has not happened, suggesting that CS5931 might indeed prevent the process of angiogenesis.

2.4. CS5931 Blocks Vessel Formation in Zebrafish Embryos in Vivo

Zebrafish has been used as an ideal model to evaluate the angiogenic activity of anticancer agents. In the present study, SIV (subintestinal vessel) formation of zebrafish embryos was measured to assess the anti-angiogenic activity of CS5931 *in vivo*.

As seen in Figure 4A, CS5931 prevented SIV formation in zebrafish embryos significantly; treatment with 10, 20, 30 μg/mL CS5931 reduced SIVs growth by 27.6, 47.2 and 82.8%, respectively (Figure 4B). Our results confirmed that CS5931 significantly inhibited angiogenesis both *in vitro* and *in vivo*. Cisplatin, a cytoxic antineoplastic drug, can inhibit endothelial cell proliferation *in vitro* but not affect angiogenesis *in vivo* [24–27], implying that drugs affecting endothelial cell proliferation are not necessarily anti-angiogenic. Our study revealed that the polypeptide also affected the growth of SIVs in zebrafish embryos. The results suggest that CS5931, unlike cisplatin, not only affects angiogenesis of HUVECs *in vitro*, but also impairs the development of angiogenesis *in vivo*.

Figure 4. CS5931 affects subintestinal vessel (SIV) formation in zebrafish embryos. (**A**) Lateral view of stained zebrafish embryos at 72 hpf. Embryos were treated without (1), or with 10 (2), 20 (3), and 30 μg/mL (4) of CS5931. Vessels of zebrafish embryos were observed using a dissecting stereomicroscope. SIVs were indicated by an arrow. (**B**) Quantification of growth repression of SIVs induced by CS5931. Results are normalized to untreated embryos. All experiments were repeated more than three times. Values represent the means ± SD of triplicate measurements. * $P < 0.05$, ** $P < 0.01$ *versus* medium control.

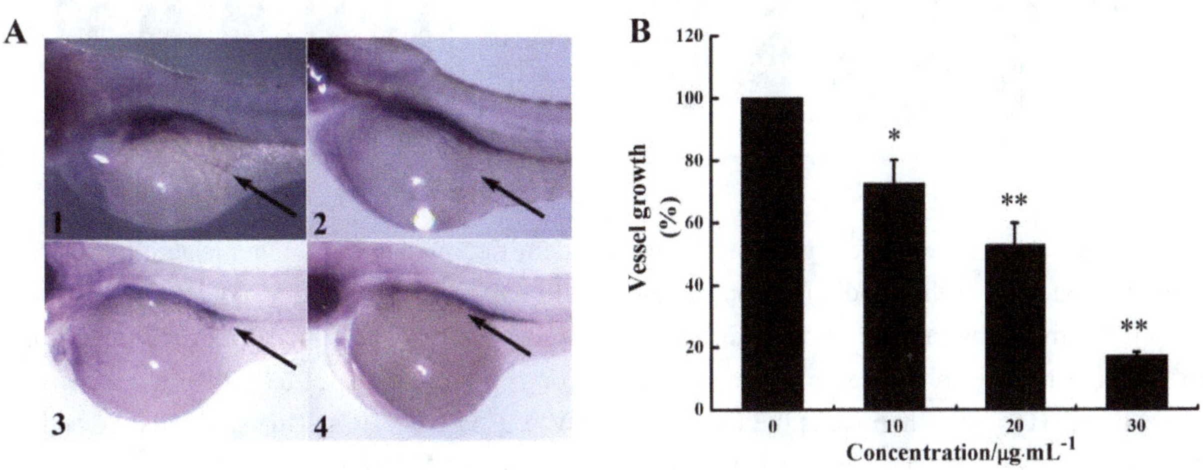

2.5. CS5931 Reduces VEGF Expression in HUVECs

Since VEGF is one of the foremost regulators of angiogenesis, we investigated whether CS5931 could decrease VEGF expression of HUVECs both at mRNA and protein levels using RT-PCR and Western Blotting. As shown in Figure 5A,B, the VEGF protein level decreased significantly in cells treated with CS5931. To further investigate if CS5931 affected the expression of VEGF mRNA, we also assessed the VEGF mRNA expression in HUVECs. The results showed that after treatment of the cells with CS5931 for 24 h, the mRNA expression of VEGF remained basically unchanged (Figure 5C,D). These results suggested that the effect of CS5931 on VEGF expression occurs at the translational level. However, the exact mechanism of action still remains unknown; whether the polypeptide affects the translational efficiency of VEGF mRNA or increases its degradation still needs to be addressed.

Figure 5. Effect of CS5931 on the expression of VEGF protein and mRNA. Cells were treated without (1) or with 10 (2), 20 (3), 30 (4) and 40 μg/mL (5) of CS5931, respectively. After incubation for 24 h, the VEGF expression was analyzed by Western Blotting (**A**) and RT-PCR analysis (**C**). (**B**) and (**D**) represent the quantitative evaluation of the expression of VEGF. Results are normalized to untreated cells. All experiments were repeated more than three times. Values represent the means ± SD of triplicate measurements. * $P < 0.05$, ** $P < 0.01$ *versus* medium control.

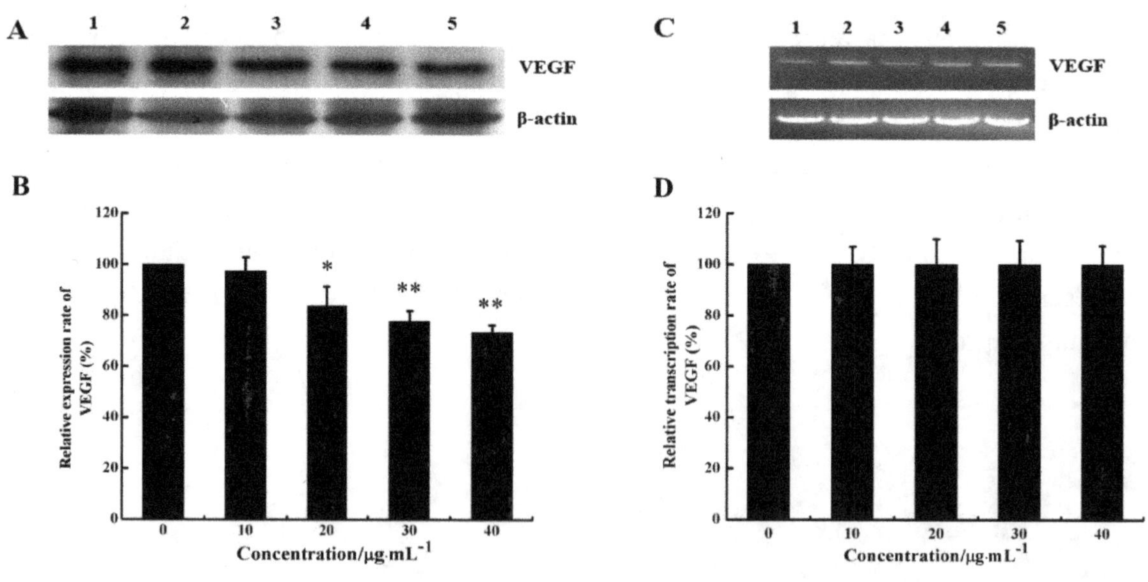

Several peptides from marine organisms have been identified with anti-angiogenic activity. Aplidine, a peptide isolated from the Mediterranean tunicate *Aplidium albicans*, displays potent anti-angiogenic activity [5] with an obvious inhibitory effect on VEGF production [28]. A polypeptide, PG155, isolated from Shark Cartilage in our laboratory possesses strong anti-angiogenic activity both *in vitro* and *in vivo* [29]; PG155 also suppresses the expression of VEGF in HUVECs. This study also reveals that CS5931 displays its anti-angiogenic activity via the repression of VEGF expression.

2.6. CS5931 Down-Regulates the Secretion and mRNA Expression of MMPs

Given that the process of cell invasion requires enzymatic digestion of the matrix, we investigated whether CS5931 affected the production of matrix-degrading metalloproteinases in endothelial cells. Gelatin zymography was carried out to investigate whether CS5931 affects the gelatinolytic activity of MMP-2 and MMP-9 secreted from endothelial cells. As shown in Figure 6A, CS5931 suppressed MMP-2 and MMP-9 secretion in a dose-response pattern. The quantitative analysis of the gelatinolytic activity indicated that, when cells were treated with 10, 20, 30, and 40 μg/mL CS5931, MMP-2 activity was reduced by 21.6, 44.6, 45.6 and 46.8%, while MMP-9 activity was reduced by 11.9, 13.5, 17.8 and 43.6%, respectively (Figure 6B). These results indicated that CS5931 inhibited the secretion of MMP-2/9 in endothelial cells.

Figure 6. Effect of CS5931 on the secretion of matrix metalloproteinases (MMPs) protein and mRNA expression. HUVECs were treated without (1) or with 10 (2), 20 (3), 30 (4) and 40 μg/mL (5) of CS5931, respectively. After incubation for 24 h, the secretion of MMPs protein was analyzed by gelatin zymography (**A**) and MMPs mRNA expression was analyzed by RT-PCR (**C**). (**B**) and (**D**) represent the quantitative analysis of MMPs secretion and mRNA expression. Results are normalized to untreated cells. All experiments were repeated more than three times. Values represent the means ± SD of triplicate measurements. * $P < 0.05$, ** $P < 0.01$ *versus* medium control.

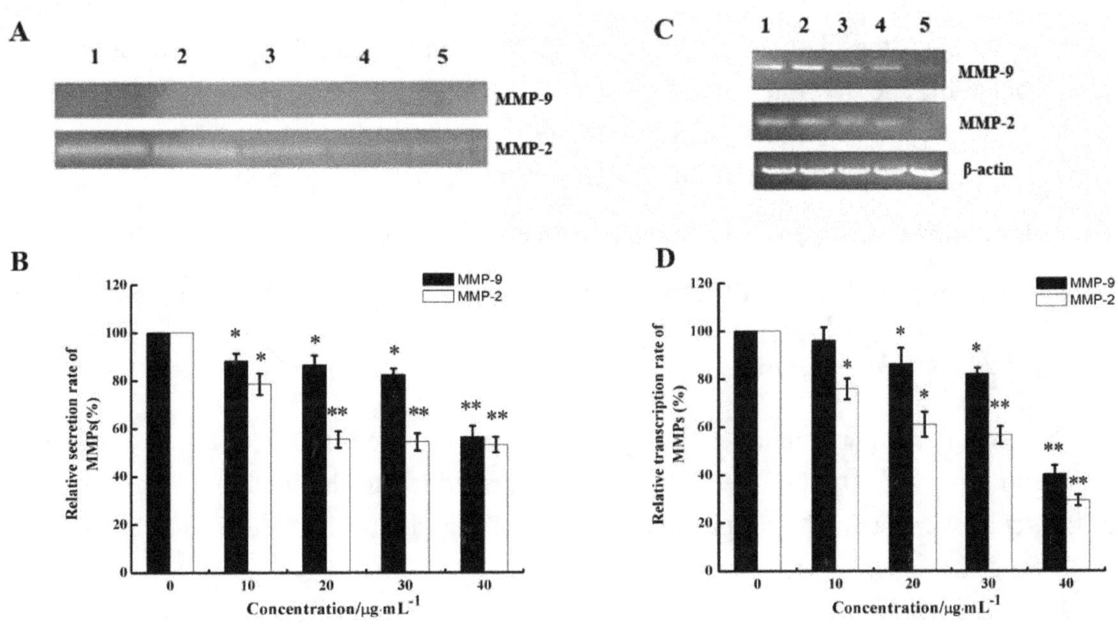

The mRNA levels of MMP-2/9 were further examined by semi-quantitative RT-PCR after HUVECs were treated with certain concentrations of CS5931 for 24 h. As shown in Figure 6C, CS5931 could inhibit the gene expression of MMP-2/9 in a concentration-dependent manner. The inhibition rate of MMP-2 and MMP-9 was about 70.7% and 59.7%, respectively, when treating the cells with CS5931 at a concentration of 40 μg/mL for 24 h (Figure 6D). These results revealed that CS5931 not only inhibited the expression of MMP-2/9 mRNA, but also eliminated the secretion of MMP-2/9 and MMPs, which are also involved in the anti-angiogenic activity induced by CS5931.

3. Experimental Section

3.1. Materials and Reagents

The recombinant CS5931 polypeptide was prepared in our laboratory according to the method described by Zhao *et al.* [19]. Transwell inserts with 8 μm pores polycarbonate filters was purchased from Corning company (Corning Costar, Cambridge, MA, USA). Matrigel was provided by BD company (Becton Dickinson, Bedford, MA, USA). All other reagents used in the experiment were analytical grade.

3.2. Cells and Cell Culture

HUVECs were purchased from American Type Culture Collection and cultured at 37 °C in a humidified atmosphere of 5% CO_2 and 95% air in RPMI-1640 (GIBCO, Invitrogen, Grand Island, NY, USA) supplemented with 10% fetal bovine serum (FBS; GIBCO, Invitrogen, Grand Island, NY, USA). All experiments were carried out with the same batch of HUVECs.

3.3. HUVEC Proliferation Assay

The anti-proliferative effects of CS5931 on the growth of HUVECs were examined *in vitro* using MTT assay [30]. In brief, cells (3.5×10^3/well) were plated in the 96-well plate in RPMI-1640 medium containing 10% FBS and cultured at 37 °C for 24 h. Then, CS5931 with certain concentrations were added to the medium. After incubation for 48 h, MTT solution (5 mg/mL, 20 μL) was added into each well and incubated for another 4 h. Then, 150 μL DMSO was added, and the plate was gently agitated until the color reaction was uniform and the OD490 was determined by a microplate reader with subtracted background absorbance. The inhibition ratio was calculated as follows:

$$\text{Inhibition ratio (\%)} = (OD_{\text{control group}} - OD_{\text{experimental value}})/OD_{\text{control group}} \times 100\%$$

The IC_{50} value was expressed as the concentration of drugs causing 50% inhibition.

3.4. PI/Hoechst 33258 Staining Assay

HUVECs were counted and plated in 24-well plates and cultured at 37 °C for 24 h. CS5931 with certain concentrations (0, 10, 20, 30 and 40 μg/mL) were added to the medium. After incubation for another 24 h, cells were stained with propidium iodide (PI) and Hoechst 33258 and then examined by fluorescence microscopy.

3.5. Scratch-Wound Cell Migration Assay

HUVECs were counted and plated in 96-well plates. After growth to 90% confluence, "scratch" wounds were created by scraping cell monolayers with sterile disposable yellow tips [31–33]. After washing the scratched cell monolayers with PBS twice, RPMI-1640 medium supplemented with 2% FBS containing certain concentrations of CS5931 was added. After incubation for 10 h, 3 fields of each wound were selected and photographed with an inverted phase-contrast microscope. The number of HUVECs migrated from the edge of the injured monolayer was quantified by measuring the distance between the wound edges before and after injury. The migration rate at the start of the experiment was measured and arbitrarily defined as 100%. After incubation for 10 h, the width of the remaining wound was measured and the migration inhibition rate was calculated using the following formula:

Inhibition rate (%) = [1 − (average size in drug treated group/average size in control group)] × 100%

3.6. Transwell Migration Assay

Migration of HUVECs was determined in a Transwell Boyden chamber (Corning Costar, Cambridge, MA, USA) containing a polycarbonate filter with a pore size of 8 μm as described previously [29]. In brief, cell suspension (5×10^5 cell/mL, 100 μL) containing certain concentrations of CS5931 (0, 10, 20, 30 and 40 μg/mL) was added to the upper compartment of each well. The lower compartment contained 0.6 mL of RPMI-1640 medium supplemented with 20% FBS. After incubation for 8 h at 37 °C, the filter was removed and fixed with 95% ethanol and stained with 0.1% crystal violet. HUVECs remained on the upper side of the filter were removed with the cotton swabs. Cells on the lower surface of the filter (migrated) were counted manually with a digital camera in five random fields and the average number of total fields was calculated. The inhibition rate of migration was calculated as using the following formula:

Inhibition rate = [1 − (migrated cells treated/migrated cells control)] × 100%

3.7. HUVEC Tube Formation Assay

The ability of HUVEC to form capillary-like structures on a 3D layer of basement membrane (Matrigel) was tested as described previously [26]. In brief, ice-cold Matrigel (BD, Bedford, MA, USA) diluted with serum-free RPMI-1640 medium was layered in a 96-well plate and incubated at 37 °C for 30 min to allow polymerisation. HUVECs (3×10^4 cell/well) were plated onto the Matrigel layer in the culture medium containing certain concentrations of CS5931 (10, 20, 30 and 40 μg/mL) for 6 h. Photographs from randomly chosen fields were taken using a digital camera when cords were formed.

3.8. Alkaline Phosphatase Staining for Visual Inspection in Zebrafish Embryos

Zebrafish embryos were generated by natural pairwise mating as described in the zebrafish handbook [34]. Embryos at 24 h post-fertilization (hpf) were manually dechorionated with 1 mg/mL trypsin for 10 min at room temperature immediately prior to drug treatment. Embryos treated with CS5931 were maintained in 24-well plates (20 embryos/well) for an additional 48 h.

Vessel staining was performed as described previously [29,35,36]. Briefly, at 72 hpf, embryos were fixed with 4% paraformaldehyde for 2 h at room temperature, then washed three times in PBS and dehydrated by immersing in 25, 50, 75 and 100% methanol in PBT buffer and rehydrated stepwise to 100% PBT. The embryos were then equilibrated in NTMT buffer (0.1 M Tris-HCl at pH 9.5, 50 mM $MgCl_2$, 0.1 M NaCl, and 0.1% Tween-20) at room temperature for 45 min. The staining reaction was started by incubating embryos with NBT/BCIP solution for about 45 min and the reaction was stopped by adding PBST. Embryos were then immersed in a solution of 5% formamide and 10% hydrogen peroxide in PBS for 30 min to remove endogenous melanin in the pigment cells and allow full visualization of the stained vessels. Embryos were then examined on a dissecting stereomicroscope

(Zeiss, Jena, Germany). Images were collected and stored using digital camera and total SIV vessel length was determined by point-to-point measurement using Image J software.

3.9. Western Blotting Analysis

To determine the effect of CS5931 on VEGF protein, Western Blotting was performed as previously described [37]. Briefly, HUVECs were placed in a 6-well plate for 24 h, and CS5931 with certain concentrations (0, 10, 20, 30 and 40 μg/mL) was added and incubated for another 24 h. Cells were collected by trypsin digestion method and then lysed in the ice-cold RIPA lysis buffer supplemented with a protease inhibitor cocktail. The protein concentration of the lysates was quantified using the BCA protein assay (Pierce, Thermo Scientific, Waltham, MA, USA). Cell lysates (50 μg per lane) were resolved by 15% SDS-PAGE and transferred to nitrocellulose membranes. After incubation in blocking solution (7.5% non-fat milk) for 2 h, nitrocellulose membranes were incubated with 1:1000 dilution primary antibodies to human VEGF and β-actin proteins overnight at 4 °C. Membranes were incubated with 1:2000 dilution of HRP-conjugated second antibody for 1 h at room temperature and then the signal was detected with the enhanced chemiluminescence system and the relative intensity was analyzed using Image J software.

3.10. Production of Matrix Metalloproteinases (MMPs)

HUVEC cells were seeded in a 6-well plate for 24 h. CS5931 with certain concentrations were added. After incubation for another 24 h, the MMPs were analyzed using gelatin zymography assay as described previously [5,9,38]. Briefly, the presence of MMP-2/9 in the serum-free supernatants was analyzed by electrophoresis on an 8% non-denaturing SDS-PAGE containing 0.1% gelatin. After electrophoresis, the gel was washed with 2.5% Triton-X100 four times for 15 min to remove SDS and incubated in 50 mM Tris-HCl buffer, pH 7.6, containing 0.15 M NaCl, 10 mM $CaCl_2$, and 0.02% (w/v) Brij-35 for 42 h at 37 °C. Gels were then stained with 0.5% Coomassie brilliant blue R250 in 25% methanol and 10% acetic acid, and destained in the same solution without Coomassie blue. Gelatinolytic activity was visualized by negative staining and the results were analyzed using gel documentation system and Image J software.

3.11. Semi-Quantitative Reverse Transcription and Polymerase Chain Reaction (RT-PCR)

Total cellular RNA was isolated from HUVEC cells and RT-PCR was performed using PrimeScript RT reagent kit with gDNA Eraser (TaKaRa Bio Group, Otsu, Japan), according to manufacturer's protocol. Briefly, cDNA was synthesized using total RNA (2 μg), Oligo dT Primer, and dNTP Mixture. The sequences of the primers used are shown in Table 1.

3.12. Statistical Analysis

Data was expressed as mean ± standard deviation (SD). Statistics was analyzed using Student's *t*-test. *P* values less than 0.05 were considered statistically significant.

Table 1. Primers.

	Primers	Sequences	Base pairs
P1	β-actin primer forward	5′-ACACTGTGCCCATCTAGGAGG -3′	21
P2	β-actin primer reverse	5′-AGGGGCCGGACTCGTCATACT-3′	21
P3	VEGF primer forward	5′-TTGCTGCTCTACCTCCAC-3′	18
P4	VEGF primer reverse	5′-AATGCTTTCTCCGCTCTG-3′	18
P5	MMP-2 primer forward	5′-GGCCCTGTCACTCCTGAGAT-3′	20
P6	MMP-2 primer reverse	5′-GGCATCCAGGTTATCGGGGA-3′	20
P7	MMP-9 primer forward	5′-CGGAGCACGGAGACGGGTAT-3′	20
P8	MMP-9 primer reverse	5′-TGAAGGGGAAGACGCACAGC-3′	20

PCR reactions were carried out in 50 μL of reaction mixture containing 25 μL Premix Taq DNA polymerase buffer, 0.5 μL of each primer (20 μM), and 2 μL cDNA. PCR was performed under the following conditions: an initial denaturation step at 94 °C for 5 min, followed by 35 cycles at 94 °C for 30 s, 60 °C for 30 s, 72 °C for 1 min, and a final extension step at 72 °C for 10 min. PCR fragments were separated on a 2% agarose gel. After ethidium bromide staining, the gels were illuminated at UV transilluminator and photographed and the relative intensity was analyzed using Image J software.

4. Conclusions

Taken together, our study confirmed that the marine polypeptide CS5931 inhibits angiogenesis both *in vitro* and *in vivo*. VEGF and MMPs play an important role in the anti-angiogenic activity of the polypeptide. Considering its low molecular weight as well as its high production in genetic preparation, CS5931 possesses the potential to be developed as a novel anticancer agent. Further study is in progress in our laboratory to address if the polypeptide can inhibit tumor growth in xenograft nude mice model.

Acknowledgments

This work was supported by National Innovative Drugs Development program of China (No. 2014ZX-091022043-0001) and 863 High Technology Project (No. 2014AA093503). This work was also supported by the National Natural Science Foundation of China (No. 81001396, No. 81273550 and No. 41306157) and International Cooperation project of China-Israel (2013DFG33000). The authors are grateful to all members of the laboratory for their continuous technical advice and helpful discussion.

Author Contributions

X.L. and M.L. contributed to the study concept and design, and the manuscript preparation. G.L. performed the experimental studies *in vitro* and *in vivo*, and analyzed the data. W.J., H.H., Y.Z. and J.Z. helped acquire data and statistical analysis. L.X., N.W. and L.Z. revised the article critically for intellectual content.

References

1. Wang, Z.; Zheng, L.; Yang, S.; Niu, R.; Chu, E.; Lin, X. *N*-acetylchitooligosaccharide is a potent angiogenic inhibitor both *in vivo* and *in vitro*. *Biochem. Biophys. Res. Commun.* **2007**, *357*, 26–31.
2. Ellis, L.M.; Liu, W.; Ahmad, S.A.; Fan, F.; Jung, Y.D.; Shaheen, R.M.; Reinmuth, N. Overview of angiogenesis: biologic implications for antiangiogenic therapy. *Semin. Oncol.* **2001**, *28*, 94–104.
3. Carmeliet, P.; Jain, R.K. Angiogenesis in cancer and other diseases. *Nature* **2000**, *407*, 249–257.
4. Folkman, J. Angiogenesis in cancer, vascular, rheumatoid and other disease. *Nature Med.* **1995**, *1*, 27–30.
5. Taraboletti, G.; Poli, M.; Dossi, R.; Manenti, L.; Borsotti, P.; Faircloth, G.T.; Broggini, M.; D'Incalci, M.; Ribatti, D.; Giavazzi, R. Antiangiogenic activity of aplidine, a new agent of marine origin. *Br. J. Cancer* **2004**, *90*, 2418–2424.
6. Jubb, A.M.; Oates, A.J.; Holden, S.; Koeppen, H. Predicting benefit from anti-angiogenic agents in malignancy. *Nature Rev. Cancer* **2006**, *6*, 626–635.
7. Taraboletti, G.; Margosio, B. Antiangiogenic and antivascular therapy for cancer. *Curr. Opin. Pharmacol.* **2001**, *1*, 378–384.
8. Fidler, I.J.; Singh, R.K.; Yoneda, J.; Kumar, R.; Xu, L.; Dong, Z.; Bielenberg, D.R.; McCarty, M.; Ellis, L.M. Critical determinants of neoplastic angiogenesis. *Cancer J.* **2000**, *6*, S225–S236.
9. Li, A.; Varney, M.L.; Valasek, J.; Godfrey, M.; Dave, B.J.; Singh, R.K. Autocrine role of interleukin-8 in induction of endothelial cell proliferation, survival, migration and MMP-2 production and angiogenesis. *Angiogenesis* **2005**, *8*, 63–71.
10. Folkman, J. What is the evidence that tumors are angiogenesis dependent? *J. Natl. Cancer Inst.* **1990**, *82*, 4–7.
11. Brower, V. Tumor angiogenesis-new drugs on the block. *Nat. Biotechnol.* **1999**, *17*, 963–968.
12. Loges, S.; Schmidt, T.; Carmeliet, P. Mechanisms of resistance to anti-angiogenic therapy and development of third-generation anti-angiogenic drug candidates. *Genes Cancer* **2010**, *1*, 12–25.
13. Deplanque, G.; Harris, A. Anti-angiogenic agents: Clinical trial design and therapies in development. *Eur. Cancer* **2000**, *36*, 1713–1724.
14. Suarez-Jimenez, G.-M.; Burgos-Hernandez, A.; Ezquerra-Brauer, J.-M. Bioactive peptides and depsipeptides with anticancer potential: Sources from marine animals. *Mar. Drugs* **2012**, *10*, 963–986.
15. Jha, R.K.; Zi-rong, X. Biomedical compounds from marine organisms. *Mar. Drugs* **2004**, *2*, 123–146.
16. Zheng, L.; Lin, X.; Wu, N.; Liu, M.; Zheng, Y.; Sheng, J.; Ji, X.; Sun, M. Targeting cellular apoptotic pathway with peptides from marine organisms. *Biochim. Biophys. Acta* **2013**, *1836*, 42–48.
17. Wrasidlo, W.; Mielgo, A.; Torres, V.A.; Barbero, S.; Stoletov, K.; Suyama, T.L.; Klemke, R.L.; Gerwick, W.H.; Carson, D.A.; Stupack, D.G. The marine lipopeptide somocystinamide A triggers apoptosis via caspase 8. *Proc. Natl. Acad. Sci. USA* **2008**, *105*, 2313–2318.
18. Cheng, L.; Wang, C.; Liu, H.; Wang, F.; Zheng, L.; Zhao, J.; Chu, E.; Lin, X. A Novel Polypeptide Extracted From *Ciona savignyi* Induces Apoptosis Through a Mitochondrial-Mediated Pathway in Human Colorectal Carcinoma Cells. *Clin. Colorectal Cancer* **2012**, *11*, 207–214.

19. Zhao, J.; Wei, J.; Liu, M.; Xiao, L.; Wu, N.; Liu, G.; Huang, H.; Zhang, Y.; Zheng, L.; Lin, X. Cloning, characterization and expression of a cDNA encoding a granulin-like polypeptide in *Ciona savignyi*. *Biochimie* **2013**, *95*, 1611–1619.
20. Hanahan, D.; Bergers, G.; Bergsland, E. Less is more, regularly: metronomic dosing of cytotoxic drugs can target tumor angiogenesis in mice. *J. Clin. Investig.* **2000**, *105*, 1045–1047.
21. Kerbel, R.; Folkman, J. Clinical translation of angiogenesis inhibitors. *Nature Rev. Cancer* **2002**, *2*, 727–739.
22. Schirner, M. Antiangiogenic chemotherapeutic agents. *Cancer Metastasis Rev.* **2000**, *19*, 67–73.
23. Miller, K.D.; Sweeney, C.J.; Sledge, G.W. Redefining the target: chemotherapeutics as antiangiogenics. *J. Clin. Oncol.* **2001**, *19*, 1195–1206.
24. Nicoletti, M.; Lucchini, V.; Massazza, G.; Abbott, B.; D'Incalci, M.; Giavazzi, R. Antitumor activity of taxol (NSC-125973) in human ovarian carcinomas growing in the peritoneal cavity of nude mice. *Ann. Oncol.* **1993**, *4*, 151–155.
25. Yamamoto, T.; Sudo, K.; Fujita, T. Significant inhibition of endothelial cell growth in tumor vasculature by an angiogenesis inhibitor, TNP-470 (AGM-1470). *Anticancer Res.* **1994**, *14*, 1–3.
26. Belotti, D.; Vergani, V.; Drudis, T.; Borsotti, P.; Pitelli, M.R.; Viale, G.; Giavazzi, R.; Taraboletti, G. The microtubule-affecting drug paclitaxel has antiangiogenic activity. *Clin. Cancer Res.* **1996**, *2*, 1843–1849.
27. Prestayko, A.; D'Aoust, J.; Issell, B.; Crooke, S. Cisplatin (*cis*-diamminedichloroplatinum II). *Cancer Treat. Rev.* **1979**, *6*, 17–39.
28. Broggini, M.; Marchini, S.; Galliera, E.; Borsotti, P.; Taraboletti, G.; Erba, E.; Sironi, M.; Jimeno, J.; Faircloth, G.; Giavazzi, R. Aplidine, a new anticancer agent of marine origin, inhibits vascular endothelial growth factor (VEGF) secretion and blocks VEGF-VEGFR-1 (flt-1) autocrine loop in human leukemia cells MOLT-4. *Leukemia* **2003**, *17*, 52–59.
29. Zheng, L.; Ling, P.; Wang, Z.; Niu, R.; Hu, C.; Zhang, T.; Lin, X. A novel polypeptide from shark cartilage with potent anti-angiogenic activity. *Cancer Biol. Ther.* **2007**, *6*, 775–780.
30. Taraboletti, G.; Micheletti, G.; Rieppi, M.; Poli, M.; Turatto, M.; Rossi, C.; Borsotti, P.; Roccabianca, P.; Scanziani, E.; Nicoletti, M.I. Antiangiogenic and antitumor activity of IDN 5390, a new taxane derivative. *Clin. Cancer Res.* **2002**, *8*, 1182–1188.
31. Mottet, D.; Bellahcène, A.; Pirotte, S.; Waltregny, D.; Deroanne, C.; Lamour, V.; Lidereau, R.; Castronovo, V. Histone deacetylase 7 silencing alters endothelial cell migration, a key step in angiogenesis. *Circ. Res.* **2007**, *101*, 1237–1246.
32. van Mil, A.; Grundmann, S.; Goumans, M.-J.; Lei, Z.; Oerlemans, M.I.; Jaksani, S.; Doevendans, P.A.; Sluijter, J.P. MicroRNA-214 inhibits angiogenesis by targeting Quaking and reducing angiogenic growth factor release. *Cardiovasc. Res.* **2012**, *93*, 655–665.
33. Dimmeler, S.; Dernbach, E.; Zeiher, A.M. Phosphorylation of the endothelial nitric oxide synthase at Ser-1177 is required for VEGF-induced endothelial cell migration. *FEBS Lett.* **2000**, *477*, 258–262.
34. Westerfield, M. *The Zebrafish Book: A Guide for the Laboratory Use of Zebrafish (*Brachydanio rerio*)*; Institution of Neuro Science: Eugene, OR, USA, 1993; pp. 231–236.
35. He, Z.-H.; He, M.-F.; Ma, S.-C.; But, P.P.-H. Anti-angiogenic effects of rhubarb and its anthraquinone derivatives. *J. Ethnopharmacol.* **2009**, *121*, 313–317.

36. Serbedzija, G.N.; Flynn, E.; Willett, C.E. Zebrafish angiogenesis: A new model for drug screening. *Angiogenesis* **1999**, *3*, 353–359.
37. Wei, J.; Liu, M.; Liu, H.; Wang, H.; Wang, F.; Zhang, Y.; Han, L.; Lin, X. Oleanolic acid arrests cell cycle and induces apoptosis via ROS-mediated mitochondrial depolarization and lysosomal membrane permeabilization in human pancreatic cancer cells. *J. Appl. Toxicol.* **2013**, *33*, 756–765.
38. Wang, H.; Wei, J.; Wu, N.; Liu, M.; Wang, C.; Zhang, Y.; Wang, F.; Liu, H.; Lin, X. Mere15, a novel polypeptide from *Meretrix meretrix*, inhibits adhesion, migration and invasion of human lung cancer A549 cells via down-regulating MMPs. *Pharm. Biol.* **2013**, *51*, 145–151.

7

Extraction of Fucoxanthin from Raw Macroalgae excluding Drying and Cell Wall Disruption by Liquefied Dimethyl Ether

Hideki Kanda [1,2,*], Yuichi Kamo [1], Siti Machmudah [1,3], Wahyudiono [1] and Motonobu Goto [2]

1 Department of Chemical Engineering, Nagoya University, Furo-cho, Chikusa-ku, Nagoya 464-8603, Japan; E-Mails: kamo.yuuichi@d.mbox.nagoya-u.ac.jp (Y.K.); machmudah.siti@d.mbox.nagoya-u.ac.jp (S.M.); wahyudiono@b.mbox.nagoya-u.ac.jp (W.)

2 Japan Science and Technology Agency, Chiyoda, Tokyo 102-0076, Japan; E-Mail: mgoto@nuce.nagoya-u.ac.jp

3 Department of Chemical Engineering, Sepuluh Nopember Institute of Technology, Kampus ITS Sukolilo, Surabaya 60111, Indonesia

* Author to whom correspondence should be addressed; E-Mail: kanda@nuce.nagoya-u.ac.jp

---

Abstract: Macroalgae are one of potential sources for carotenoids, such as fucoxanthin, which are consumed by humans and animals. This carotenoid has been applied in both the pharmaceutical and food industries. In this study, extraction of fucoxanthin from wet brown seaweed *Undaria pinnatifida* (water content was 93.2%) was carried out with a simple method using liquefied dimethyl ether (DME) as an extractant in semi-continuous flow-type system. The extraction temperature and absolute pressure were 25 °C and 0.59 MPa, respectively. The liquefied DME was passed through the extractor that filled by *U. pinnatifida* at different time intervals. The time of experiment was only 43 min. The amount of fucoxanthin could approach to 390 μg/g dry of wet *U. pinnatifida* when the amount of DME used was 286 g. Compared with ethanol Soxhlet and supercritical CO_2 extraction, which includes drying and cell disruption, the result was quite high. Thus, DME extraction process appears to be a good method for fucoxanthin recovery from *U. pinnatifida* with improved yields.

Keywords: dimethyl ether; supercritical CO_2; fucoxanthin; wet extraction

---

1. Introduction

Undaria pinnatifida is consumed as one of the most popular, traditional seaweeds, particularly in East Asian countries such as Japan and Korea. In the current food industry, *U. pinnatifida* is mainly manufactured in the form of dry particles by using coarse grinder. Dry *U. pinnatifida* particles swell in hot water, and are used as an ingredient in popular foods such as bean paste soup in Japan. The main functional constituent of *U. pinnatifida* is fucoxanthin [1,2], which is a xanthophylls pigment contained in the chloroplasts of brown macroalgae. Fucoxanthin also had ability to exert effects in humans, such as antidiabetic effect [3,4], cholesterol metabolism [5,6], anti-obesity effect [1], anti-oxidant properties [2,7,8], anti-proliferative effect on cancer cells [9,10], and anti-inflammatory [11]. Extraction of bioactive compounds from macroalgae has great potential and its applications will continue to grow in the following years [12]. Pressurized solvent extraction has been used, for example, to isolate carotenoids from brown macroalgae, such as *Eisenia bicyclis* [13], *Cystoseira abies-marina*, and *Himanthalia elongata* [14]. The results showed that ethanol at high temperatures provide high recoveries of fucoxanthin and other oxygenated carotenoids. Moreover, many studies of biomaterials production have reported that cell disruption is a significant important factor to extract organic components contained in cells. For example, microwave assisted extraction of fucoxanthin from *U. pinnatifida* cells [15–18]. However, fucoxanthin exhibited sensitivity towards some factors such as light and pH, the least stable in acidic pH condition and higher concentration of ascorbic acid supplementation exerted stabilization role on fucoxanthin [19], and carotenoids are thermally decomposed in hot-drying.

Liquefied DME was used as an extractant to enhance extraction of fucoxanthin from *U. pinnatifida*. DME is the simplest form of ether [20], with the following characteristics. (i) DME has a low normal boiling point (−24.8 °C) [21], therefore, DME is not present in the final products at normal temperatures; (ii) Relative permittivity of DME is 1.08 and 5.34 at 30.5 °C, in gaseous and liquid states, respectively. Liquefied DME has high affinity to oily substances [22] and partial miscibility with water [23]; (iii) DME has been approved as a safe extraction solvent for the production of foodstuff and food ingredients by the European Food Safety Authority (EFSA) [24], by the Food Standards Australia New Zealand, and by the United States [25]. The panels of EFSA consider the intended use of dimethyl ether as an extraction solvent to remove fat from animal protein raw materials. Considering (a) that the defatted animal protein is submitted to vacuum which assures that most of the volatile dimethyl ether is eliminated from final animal protein products (b) that the maximum residual limit of dimethyl ether is of 9 μg/kg of extracted animal proteins and (c) that these proteins are used at a level of up to 2% in the final food, the Panel considered that there is no safety concern [24]; (iv) DME exhibits resistance to autoxidation, unlike other alkyl ethers [26]. Owing to these characteristics, the authors successfully extracted lipids and water from wet vegetal biomasses by liquefied DME. In the DME-based extraction technique, drying, cell disruption, and heating of solvent were not needed to improve extraction yields [27].

In this work, the effect of pressure, temperature, and ethanol as co-solvent on the recovery of fucoxanthin from *U. pinnatifida* using supercritical CO_2 would be also presented. However, the lipids content in the extract is not presented, although supercritical CO_2 is well known as an effective technique to extract lipids. Several experimental studies of extraction by supercritical CO_2 have reported that supercritical CO_2 extraction is an environmentally-friendly for lipids and carotenoids extraction from micro- and macroalgae. Mendes *et al.* carried out supercritical CO_2 extraction of carotenoids and other lipids on whole and crushed *Chlorella vulgaris* [28]. Machmudah *et al.* and Kitada *et al.* also extracted carotenoids and astaxanthin from microalgae of *Haematococcus pluvialis* [29] and *C. vulgaris* [30] with supercritical CO_2. The results showed that supercritical CO_2 is more selective for carotenes than the usual organic liquid extraction and it is preferred for handling temperature-sensitive molecules. Extraction using supercritical CO_2 with ethanol as the co-solvent has also been performed with brown seaweed at 0.8 to 30 MPa and 303 to 333 K [31]. The significant amount of oil from brown seaweed was extracted within 50 min. Roh *et al.* reported that the oil extraction yield of brown seaweed was high at higher pressure, and the amounts of fucoxanthin extracted at 20 MPa of CO_2 pressure and 323 K of extraction temperature showed much higher than other extraction conditions [31]. As a comparison, the extraction yield from *U. pinnatifida* was also tested by using the conventional solvent ethanol with a Soxhlet extractor. As explained before that fucoxanthin has been implicated as important dietary nutrients having antioxidant potential. It is one of the reasons why some researchers had evaluated the *in vitro* antioxidant activity of fucoxanthin extracted from algae. For example, Roh *et al.* suggested that fucoxanthin content should be considered as an important feature of brown seaweed, as some of its nutritive and pharmacological effects could be attributed to their presence in plant material [31]. Billakanti *et al.* also reported that the yield of fucoxanthin from *U. pinnatifida* could be improved by using an enzyme-assisted extraction process followed by DME extraction. The results showed that extraction of fucoxanthin from *U. pinnatifida* using supercritical CO_2 with ethanol as a co-solvent and enzymatic pre-treatment followed by DME extraction have no effect on the fucoxanthin properties [20]. Therefore, in this study the antioxidant activity of fucoxanthin was not conducted. Isolation and purification of fucoxanthin from extraction products were also not carried out.

2. Results and Discussion

2.1. Ethanol Soxhlet Extraction

It was well known that the extraction of organic compounds using a range of organic solvents from matrices (soils, sewage sludges, vegetables, and plants) has historically been carried out by using Soxhlet extraction. The apparatus for Soxhlet extraction consisted of a solvent reservoir, extractor body, an electric heat source and a water-cooled reflux condenser. Since the temperature plays a significant role for extraction, increased temperatures can disrupt the strong solute-matrix interactions caused by van der Waals forces, hydrogen bonding, and dipole attractions of the solute molecules and active sites on the matrix [32]. Therefore, the use of ethanol as an extract solvent at elevated temperature is enhanced solubility and mass transfer effects, and disruption of surface equilibria. Kim *et al.* reported that ethanol provided the best fucoxanthin extraction yield from *Phaeodactylum tricornutum* (15.71 mg/g freeze-dried sample weight) [33]. They explained that fucoxanthin content in the extracts produced by the different methods was quite constant (15.42–16.51 mg/g freeze-dried sample weight) but increased steeply based on the percentage of ethanol in water, emphasizing the importance of ethanol in the extraction. Kanazawa *et al.* also reported that 191 μg/g wet sample weight was extracted from *Laminalia japonica* with ethanol at 40 °C [34].

In this work, by using ethanol solvent, 50 μg of fucoxanthin has been extracted from U. pinnatifida 6 g. This result might be due to some contribution from the physical properties of *U. pinnatifida*. The size of *U. pinnatifida* was 0.5–3 cm directly fed without pre-treatment. The particle size is other factors that high influence the mass transfer during Soxhlet process. Bigger particles present lower ratios of surface area to volume. Conversely, smaller particles present higher ratios of surface area to volume, which enhance the contact between solvent and solid matrix and diminish the diffusion path of the particle to reach the surface, resulting in a high extraction yield. In addition, there was possibility that the degradation of fucoxanthin in ethanol solution might have occurred during Soxhlet extraction process (12 h).

2.2. Liquefied DME Extraction

As shown in Figure 1, fucoxanthin has been extracted from *U. pinnatifida* with small amount of DME (24 g). It should be noted that the lipids content in the extract is not determined. The increasing amount of DME used was followed by increasing the amount of fucoxanthin extracted. When the amount of DME used was 286 g (at the last time point of extraction process), the residue of the *U. pinnatifida* was almost perfectly dewatered as like as such kind of dry paper. The color of *U. pinnatifida* was also changed to light green color. The amount of fucoxanthin could approach up to 390 μg/g dry of *U. pinnatifida* at this point.

Figure 1. Yield of fucoxanthin in the extract obtained from wet *U. pinnatifida* by liquefied dimethyl ether (DME).

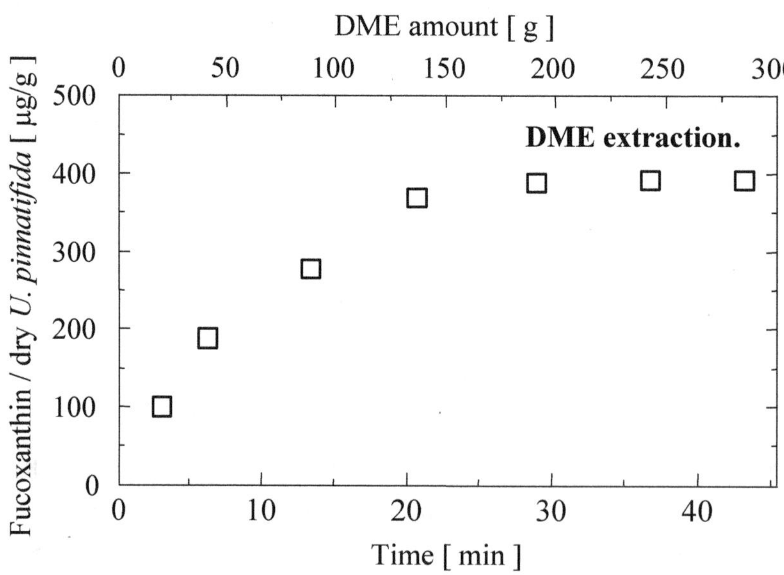

Compared with other, ethanol soxhlet, the fucoxanthin amount by liquefied DME was quite high. It is well known that liquefied DME could dissolve a wide range of polar and non-polar substances. They are also good solvents for many hydrogen-bonded substances. To dissolve hydrogen bonded substances, high solvation energy was needed to break the hydrogen bonds. In this case, DME has ability to act hydrogen bond acceptors, forming hydrogen bonds with hydrogen-bonding solutes. Therefore, liquefied DME can enter into *U. pinnatifida* cells and goes out together with *U. pinnatifida*

components include fucoxanthin. In addition, the cumulative fucoxanthin extracted looks like complete after 191 g of DME consumption, however, the cumulative fucoxanthin extracted seemed clearly to be enhanced with increasing the DME consumption. As mentioned before, despite liquefied DME had high dissolving ability, liquefied DME could also generate a lower viscosity of the analytes in the matrix and, accordingly, a better diffusion rate of the solute from the solid phase to the solvent. Consequently, the fucoxanthin in *U. pinnatifida* might be extracted easily.

2.3. Supercritical CO_2 Extraction

For supercritical CO_2 extraction system, the extraction of fucoxanthin was performed in a semi-continuous flow-type. Physically, the appearance of the extracts was oily. In general, they increased with the increase of pressure at constant temperature. This tendency happened due to direct increase of density and hence solvating power of supercritical CO_2 [35,36]. Figure 2a,b showed the extraction curves of fucoxanthin from *U. pinnatifida* for different extraction times and under different conditions in terms of pressure (10–40 MPa; 60 °C) and temperatures (40–70 °C; 40 MPa). Initially, the yield of fucoxanthin increased significantly up to 150 min extraction time. For example, the yield of fucoxanthin is about 47 μg/g dry weight of *U. pinnatifida* dried after 30 min at 40 MPa and 60 °C. Then, it increased gradually to 58 μg/g dry weight of *U. pinnatifida* at 150 min extraction time. Over 150 min extraction time, the yield of fucoxanthin was not increase significantly. At the same extraction temperature, the same trend was found at 10, 20, and 30 MPa. As previously mentioned, the effect of pressure can be attributed to the increase in solvent power and by the strengthening of intermolecular physical interactions. It is also evident that an increase in pressure had effect on the solubility of fucoxanthin in supercritical CO_2. This result is in agreement with similar trends reported by other authors [29,37]. Machmudah *et al.* reported that the total extract and the astaxanthin extracted were not significantly affected by the increasing pressure at 30–50 MPa, and after which the dramatic increase was observed at the pressure of 55 MPa. They explained that the dependency on the pressure was expected as the CO_2 density increases at higher pressure, and therefore the solvent power to dissolve the substances increases [29]. At constant pressure (40 MPa), the yield of fucoxanthin also increased with increasing extraction temperature. It increased rapidly till 150 min extraction time then remained constant after 200 min extraction time in each extraction temperature. The highest extraction rate of fucoxanthin was reached when the extraction temperature is 60 °C. The maximum fucoxanthin content obtained was 60 μg/g dry weight of *U. pinnatifida* after 270 min extraction time. As shown in the Figure 2b, the amount of fucoxanthin increased with an increase in temperature. These results indicated that the fucoxanthin extractions are dependent on solute vapor pressure which increased with an increase in temperature. Instead of that, the increasing temperature contributed to the decomposition of cell walls, and as a result fucoxanthin and extractable compounds availability for extraction was increased [29]. In addition, due to the fast extraction rate of fucoxanthin from *U. pinnatifida*, this figure also showed that an increase in temperature likely had a stronger effect on the solubility than an increase in pressure. At 70 °C, the yield of fucoxanthin is lower than at 60 °C. In this case, degradation of fucoxanthin might also take place, considering that carotenoids in general are considered to be heat-sensitive [33].

Figure 2. (**a**) Effect of pressures on the amount of fucoxanthin as a function of time at 60 °C of extraction temperature; (**b**) Effect of temperatures on the amount of fucoxanthin as a function of time at 40 MPa of extraction pressure; (**c**) Effect of entrainer flow rate on the amount of fucoxanthin as a function of time at 60 °C and 40 MPa.

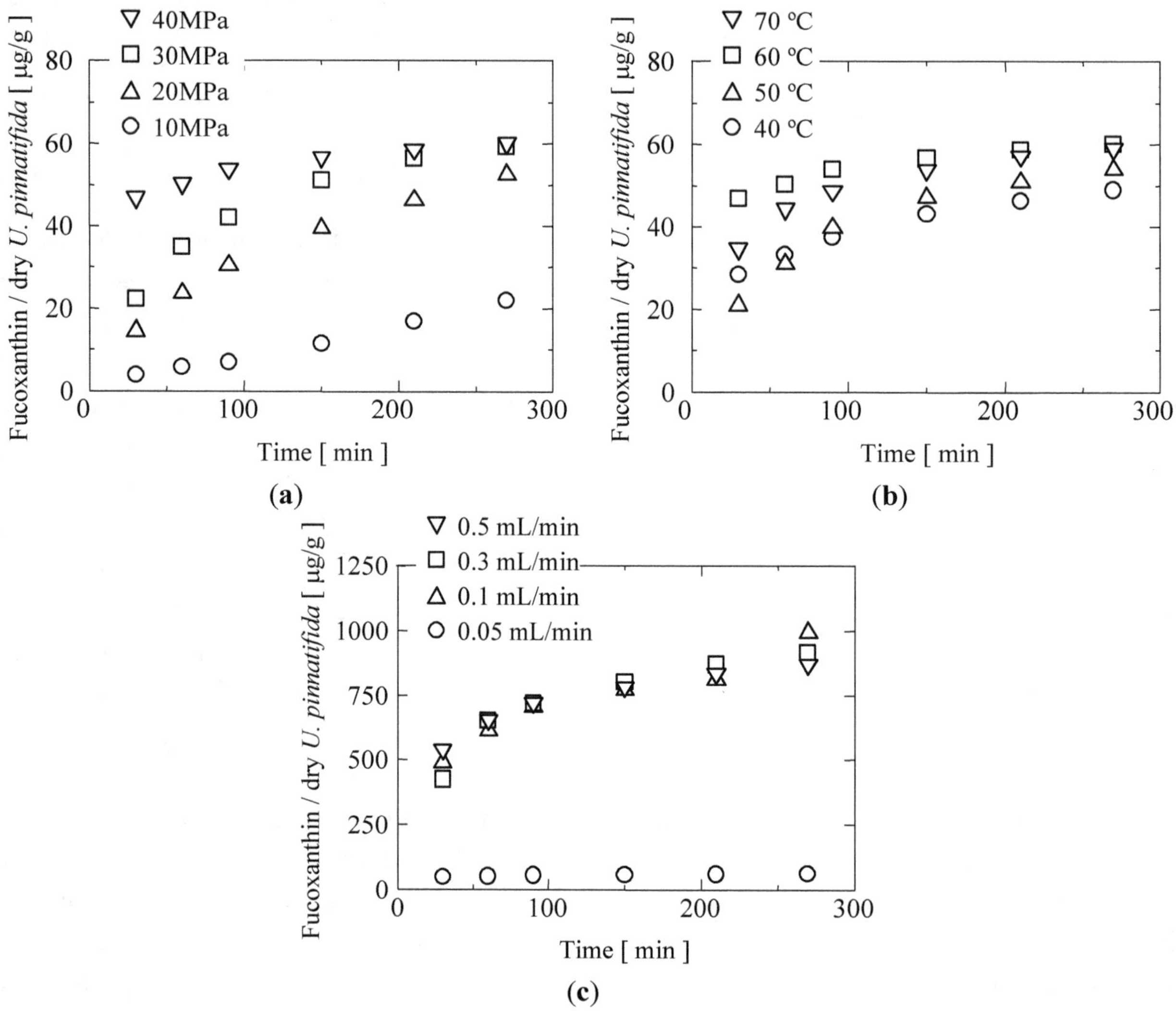

In this work, the amounts of fucoxanthin extracted by supercritical CO_2 extraction with and without ethanol as entrainer were compared. The experiments were conducted under 60 °C and 40 MPa with various entrainer concentrations (1.64%–14.29% in volume). This condition was selected because at that condition the extraction rate of fucoxanthin was high. When the concentration of 1.67% ethanol was employed, there was no significant effect on the amount of fucoxanthin extracted. As shown in Figure 2c, the enhancement of ethanol concentration had high effect on the amount of fucoxanthin extracted. The recovery of fucoxanthin could reach 10-fold. This considerable increase in extraction efficiency was because the ethanol added could enhance the solvent power of supercritical CO_2 and caused swelling of the matrix, thus increasing the internal volume and the surface area for the contact with supercritical CO_2 [29]. Machmudah *et al.* explained that the addition of ethanol in supercritical CO_2 also could cause decomposition of the *H. pluvialis* cellular wall, with the result that the astaxanthin availability for extraction increased [29].

2.4. Comparison of the Results of Three Extraction Techniques

As shown in Table 1, different extraction techniques, such as soxhlet, DME, and supercritical CO_2 extraction have been used to isolate fucoxanthin from the *U. pinnatifida* plant, however none of them can be considered as an optimal method for this purpose. Except DME extraction technique, pre-treatment of *U. pinnatifida,* such as drying, was needed. The main disadvantages of soxhlet extraction are the long time required and the large amount of solvent wasted, which is costly and cause environmental problems. Due to the long extraction time at the boiling point of the ethanol solvent, the possible degradations of fucoxanthin due to local overheating effects might occur. This process was performed to disrupt the cell membrane in *U. pinnatifida,* thus extraction efficiency of fucoxanthin increases. Supercritical CO_2 as known is attractive extraction method for food industries because the solvent is safe, nontoxic and easily removable, the method is fast and extraction parameters can be changed in a wide range of pressure and temperature. However, CO_2 is non-polar fluid and it is the main disadvantage in its use for the isolation of antioxidants. As explained before, to improve the fucoxanthin yield, ethanol was introduced to increase the polarity of CO_2. However, respecting to separation process, it is not only expensive but also difficult. On the contrary, the cell disruption of *U. pinnatifida* is not required when the DME was employed as extractant due to the properties of DME (see "Introduction"). That is why this technique of fucoxanthin extraction is very simple and versatile.

Table 1. Best recoveries of fucoxanthin obtained using different extraction techniques.

Extraction Techniques	Time (h)	Temperature (°C)	Pressure (MPa)	Yield of Fucoxanthin (μg/g)
Ethanol soxhlet	12	78	*	50
Liquefied DME	0.72	25	*	390
Supercritical CO_2	3	60	40	60.12
	3	70	40	59.51
Supercritical CO_2 with entrainer (3.23%)	3	60	40	994.53

* not determined.

3. Experimental Section

3.1. Materials

U. pinnatifida was obtained from Fukui Prefecture, Japan and used as a starting material. Most of *U. pinnatifida* that did not meet strict quality standards are usually discarded as wastes, a situation, which is becoming an environmental concern. The appearance of *U. pinnatifida* is shown in Figure 3. The elemental composition of *U. pinnatifida* is given in Table 2. The apparatus of elemental analysis was conducted by a CHN analyzer (MT-6 Elemental analyzer, Yanaco New Science Inc., Kyoto, Japan) based on flash combustion, which converts all organic substances into combustion gases (H_2O, CO_2, N_2). DME and CO_2 were purchased from Tamiya, Inc. Japan and Sogo Kariya Sanso, Inc. Japan, respectively. Fucoxanthin (94.0%) purchased from Wako Pure Chemicals Industries Ltd. (Osaka, Japan) was used as standard. It was stored at −60 °C in freezer (NF-75SF3, Nihon Freezer Co. Ltd., Tokyo, Japan). The analytical reagents used were acetonitrile, chloroform, and ethanol (HPLC-grade) from Wako Pure Chemicals Industries Ltd. (Osaka, Japan). All chemicals were used as received.

Figure 3. Wet *U. pinnatifida* sample.

Table 2. Elemental analysis of *U. pinnatifida* by CHN analyzer, on average.

Materials	Ultimate Analysis (wet weight%, DAF)			
	C (±0.2)	H (±0.1)	N (±0.1)	O * (±0.3)
Original algae	33.0	5.5	3.5	58.0
Residue	36.1	5.9	4.7	53.3
Extract	55.6	8.6	1.4	34.4

* Oxygen content was determined by difference.

3.2. Analytical Methods

There are many carotenoids could be extracted from plant material. However, in this work, fucoxanthin was subjected as the target of carotenoid compounds recovery from *U. pinnatifida*, and determined quantitatively by using HPLC (high-performance liquid chromatography). The organic components in extracts were recovered with 2 mL of chloroform, and all the solutions were filtered using a disposable filter of 0.45 μm pore size prior to HPLC analysis. The separation of fucoxanthin was carried out according to a previous published method [30] with modification. Initially, the pure compound of fucoxanthin dissolved in chloroform as a standard was injected in the HPLC system to construct calibration curve in 5 point. After separation process in the HPLC column, the amount of fucoxanthin leaving the column will determine the intensity of the signal produced in the detector. By comparing the time it takes for the peak to show up (the retention time) with the retention times for fucoxanhin, the amount of fucoxanthin in the extract can be identified (Figure 4). This analysis can be performed with good precision; therefore, other techniques analysis was not conducted. The HPLC instrument used was an ultraviolet–visible spectroscopy detector (UV-970; JASCO, Tokyo, Japan) equipped with an ODS-3 column (GL Sciences, Tokyo, Japan; 5 μm; 250 mm × 4.6 mm) and operated at 40 °C. The mobile phase contained acetonitrile and water (80/20, v/v) with flow rate of 1.0 mL/min. Preparation of standard curve and detection were accomplished at wavelength of 445 nm.

Figure 4. HPLC chromatogram of fucoxanthin extract from *U. pinnatifida.*

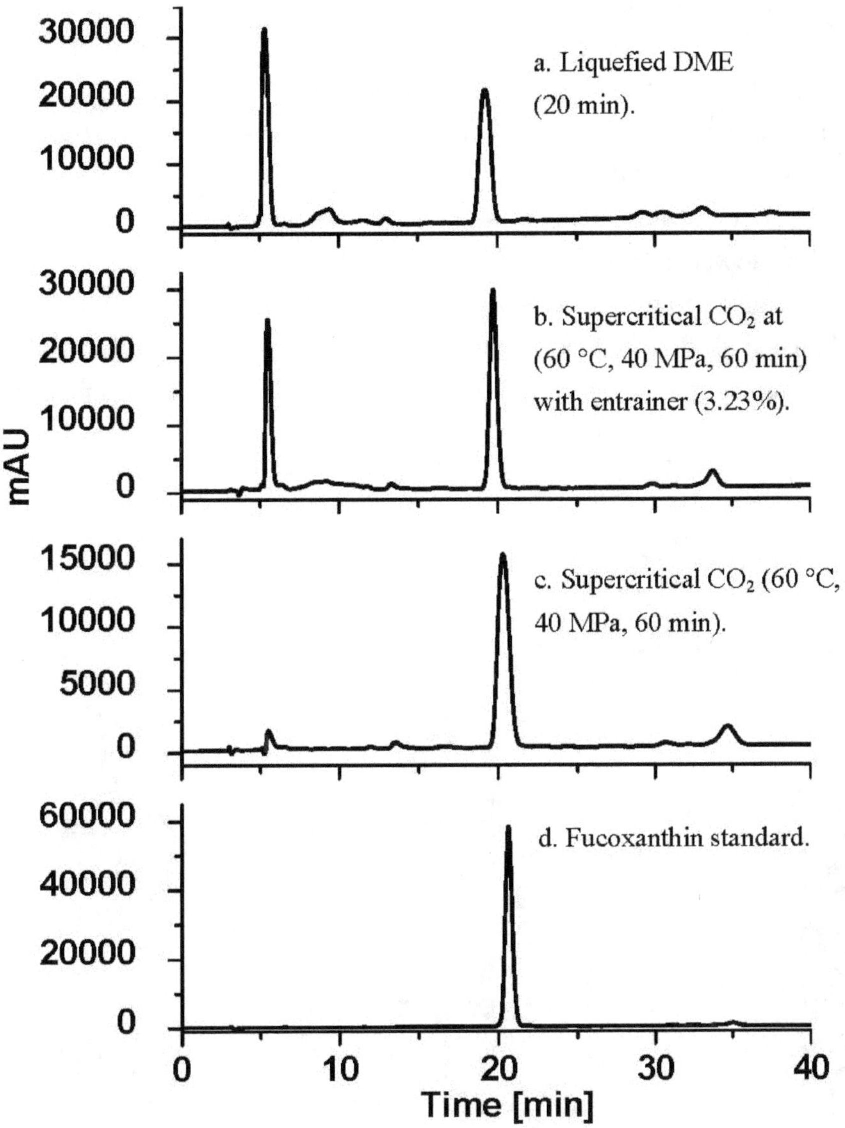

3.3. Ethanol Soxhlet Extraction

In this extraction, the heating mantle temperature was set at 78 °C for 12 h. The amount of dried *U. pinnatifida* and ethanol used in flask were 6 g and 200 mL, respectively. The flask was then removed from the mantle, and the liquid extracts were transferred to bottle that covered with aluminium foil and refrigerated until analysis. In this work (DME extraction, supercritical CO_2 extraction and ethanol soxhlet extraction), the *U. pinnatifida* was fed without any pre-treatments such as microparticulation.

3.4. DME Extraction

Figure 5 depicted the apparatus scheme that used to evaluate the extraction efficiency of the DME extraction. The main apparatus consisted of extractor (HPG-10-5; Taiatsu Techno Corp., Saitama, Japan; volume: 10 cm^3), needle valve to control the DME flow rate, and extract storage tank (HPG-96-3;

Taiatsu Techno Corp., Saitama, Japan; volume: 96 cm^3). Briefly, they were connected in series included feed materials. The extractor and DME storage tank were made of pressure-resistant glass coated with polycarbonate. 4.40 g of the raw *U. pinnatifida* (water content: 93.2%) was loaded into the lower half of the extractor, and the upper half was loaded with colorless glass beads. The DME flow rate was 10 ± 2 cm^3 min^{-1}. The extraction temperature and absolute pressure were 25 ± 1 °C and 0.59 ± 0.02 MPa, respectively. After passing liquefied DME through the extractor at different time intervals, the DME was evaporated by opening the reducing valve of the storage vessel. Next, liquefied DME pass through the lower half of extractor, color of glass beads in the upper half was changed to olive-green color. Because the glass beads and liquefied DME were colorless, the olive-green color indicated the probable presence of fucoxanthin. The extraction test was finished when the color of glass beads was changed again to colorless. The total time needed was less than 43 min.

Figure 5. Schematic diagram of DME extraction.

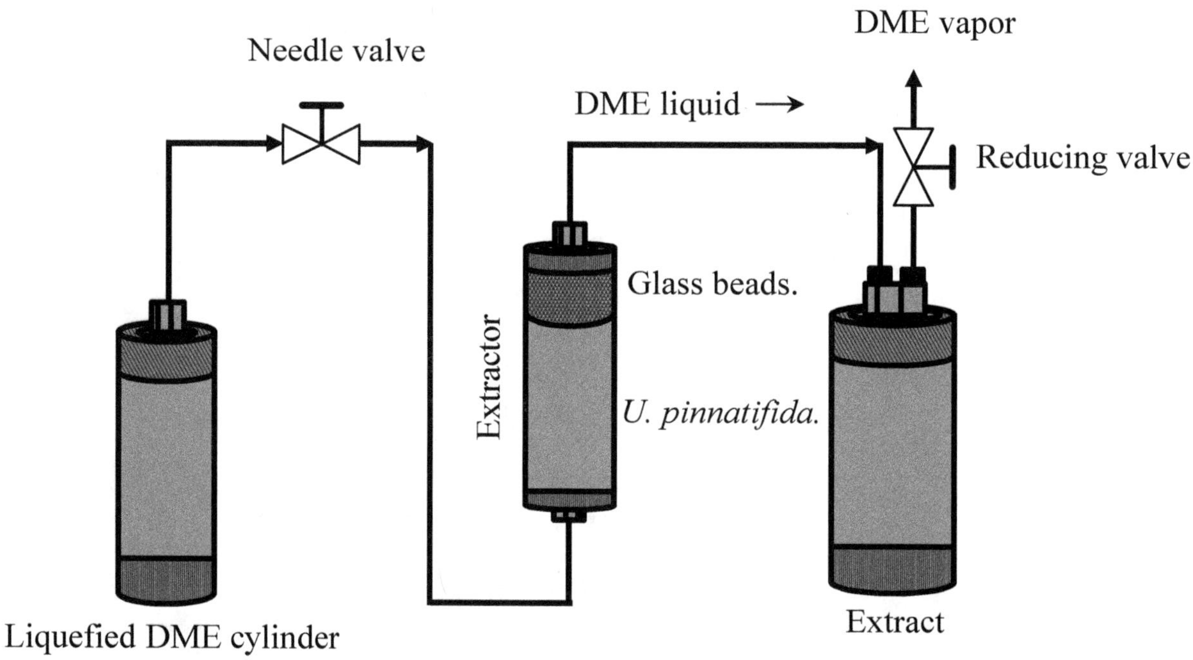

3.5. Supercritical CO_2 Extraction

The supercritical CO_2 extraction was carried out in an apparatus with schematic diagram as shown in Figure 6. The apparatus includes a high-pressure pump for CO_2 (PU-2086; Jasco, Hachioji, Japan), a heating chamber (WFO-400; EYELA, Tokyo, Japan), a 25 mL extraction cell (Thar Technologies, Inc., Pittsburgh, PA, USA) and back pressure regulator (AKICO, Tokyo, Japan). In this work, the extraction of fucoxanthin from *U. pinnatifida* by supercritical CO_2 was conducted at temperatures of 40–70 °C and pressures of 20–40 MPa using a semi-continuous flow-type system with CO_2 flow rate of 3 mL min^{-1}. In each experiment, 3.0 g of dry *U. pinnatifida* sample was loaded into the extraction vessel, filled with glass beads at the bottom and top of the extraction vessel. The extraction vessel was placed in the heating chamber to maintain the operating temperature. The extracts were collected every 1 h for 5 h, weighed and analyzed immediately after collection. In all experiments, including ethanol soxhlet extraction and DME extraction, the extraction products were directly stored in the refrigerator at −60 °C. The bottles used for the collection of extracts were wrapped in aluminum foil. These processes were maintained until analysis. Co-solvent effect of ethanol on the supercritical CO_2 extraction was also examined. Ethanol flow rates that supplied by using high-pressure pump (PU-2080; Jasco, Hachioji, Japan) were between 0.05 and 0.5 mL min^{-1}. In the case of co-solvent used, the pressure and temperature selected were 40 MPa and 60 °C, respectively.

Figure 6. Schematic diagram of supercritical CO_2 extraction.

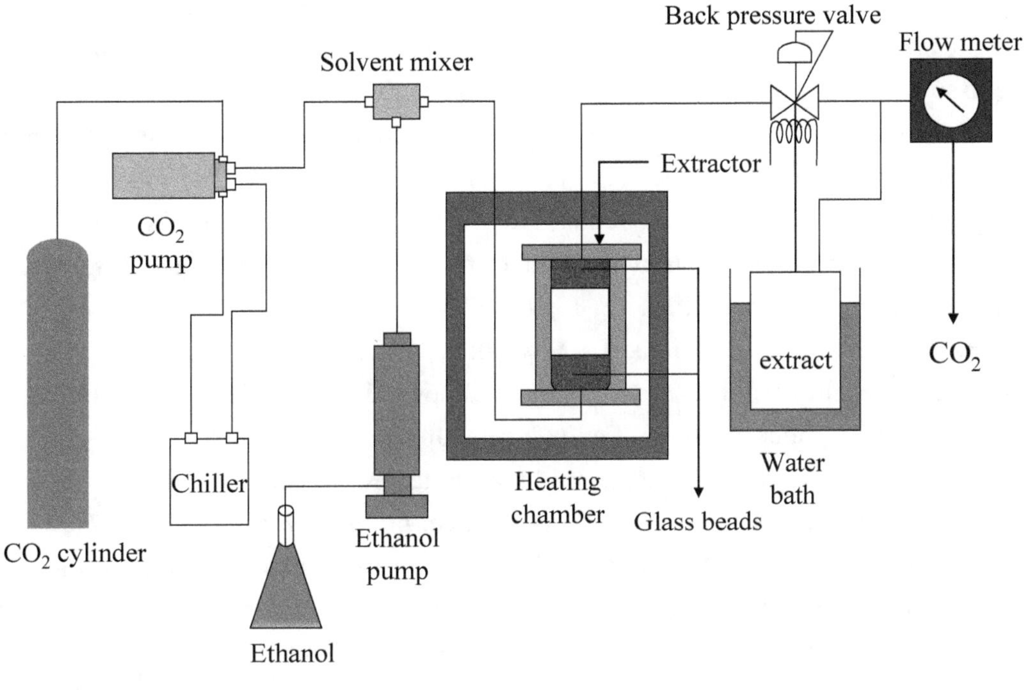

4. Conclusions

Liquefied DME extraction of fucoxanthin from *U. pinnatifida* was studied at temperature of 25 °C and pressure of 0.59 MPa using a semi-continuous flow-type system, a simple and environmentally friendly extraction method. Under these conditions, DME has ability to act hydrogen bond acceptors, forming hydrogen bonds with hydrogen-bonding solutes. Therefore, liquefied DME can enter into *U. pinnatifida* cells and goes out together with *U. pinnatifida* components include fucoxanthin. The amount of fucoxanthin could approach to 390 µg/g dry *U. pinnatifida* when the amount of DME used was 286 g. Compared with supercritical CO_2 and ethanol Soxhlet extraction, the result was quite high. Thus, DME extraction process appears to be a good method for fucoxanthin recovery from *U. pinnatifida* with improved yields. On the basis of these results, it is proposed that DME extraction is applicable to isolate carotenoids from other types of biomass.

Acknowledgments

A part of this research was supported by a grant from the Precursory Research for Embryonic Science and Technology Program of Japan Science and Technology Agency.

Author Contributions

Kanda, H.: responsible to research publication, Kamo, Y.: carried out the experiment, Machmudah, S. and Wahyudiono: conducted all data analyses, Goto, M.: supervised this work and provided all experimental and analytical equipment. All authors read and approved the final manuscript.

References

1. Maeda, H.; Hosokawa, M.; Sashima, T.; Funayama, K.; Miyashita, K. Fucoxanthin from edible seaweed, *Undaria pinnatifida*, shows antiobesity effect through UCP1 expression in white adipose tissues. *Biochem. Biophys. Res. Commun.* **2005**, *332*, 392–397.
2. Yan, X.; Chuda, Y.; Suzuki, M.; Nagata, T. Fucoxanthin as the major antioxidant in *Hijikia fusiformis*, a common edible seaweed. *Biosci. Biotechnol. Biochem.* **1999**, *63*, 605–607.
3. Jung, H.A.; Islam, M.N.; Lee, C.M.; Jeong, H.O.; Chung, H.Y.; Woo, H.C.; Choi, J.S. Promising antidiabetic potential of fucoxanthin isolated from the edible brown algae *Eisenia bicyclis* and *Undaria pinnatifida*. *Fish. Sci.* **2012**, *78*, 1321–1329.
4. Maeda, H.; Hosokawa, M.; Sashima, T.; Miyashita, K. Dietary combination of fucoxanthin and fish oil attenuates the weight gain of white adipose tissue and decreases blood glucose in obese/diabetic KK-A^y mice. *J. Agric. Food Chem.* **2007**, *55*, 7701–7706.
5. Beppu, F.; Hosokawa, M.; Niwano, Y.; Miyashita, K. Effects of dietary fucoxanthin on cholesterol metabolism in diabetic/obese KK-A^y mice. *Lipids Health Dis.* **2012**, *11*, 112–119.
6. Hu, X.; Li, Y.; Li, C.; Fu, Y.; Cai, F.; Chen, Q.; Li, D. Combination of fucoxanthin and conjugated linoleic acid attenuates body weight gain and improves lipid metabolism in high-fat diet-induced obese rats. *Arch. Biochem. Biophys.* **2012**, *519*, 59–65.
7. Fung, A.; Hamid, N.; Lu, J. Fucoxanthin content and antioxidant properties of *Undaria pinnatifida*. *Food Chem.* **2013**, *136*, 1055–1062.
8. Sachindra, N.M.; Sato, E.; Maeda, H.; Hosokawa, M.; Niwano, Y.; Kohno, M.; Miyashita, K. Radical scavenging and singlet oxygen quenching activity of marine carotenoid fucoxanthin and its metabolites. *J. Agric. Food Chem.* **2007**, *55*, 8516–8522.
9. Hosokawa, M.; Kudo, M.; Maeda, H.; Kohno, H.; Tanaka, T.; Miyashita, K. Fucoxanthin induces apoptosis and enhances the antiproliferative effect of the PPARγ ligand, troglitazone, on colon cancer cells. *Biochim. Biophys. Acta* **2004**, *1675*, 113–119.
10. Kotake-Nara, E.; Kushiro, M.; Zhang, H.; Sugawara, T.; Miyashita, K.; Nagao, A. Carotenoids affect proliferation of human prostate cancer cells. *J. Nutr.* **2002**, *131*, 3303–3306.
11. Shiratori, K.; Ohgami, K.; Ilieva, I.; Jin, X.-H.; Koyama, Y.; Miyashita, K.; Yoshida, K.; Kase, S.; Ohno, S. Effects of fucoxanthin on lipopolysaccharide-induced inflammation *in vitro* and *in vivo*. *Exp. Eye Res.* **2005**, *81*, 422–428.
12. Plaza, M.; Cifuentes, A.; Ibanez, E. In the search of new functional food ingredients from algae. *Trends Food Sci. Technol.* **2008**, *19*, 31–39.
13. Shang, Y.F.; Kim, S.M.; Lee, W.J.; Um, B.-H. Pressurized liquid method for fucoxanthin extraction from *Eisenia bicyclis* (Kjellman) Setchell. *J. Biosci. Bioeng.* **2011**, *111*, 237–241.

14. Plaza, M.; Santoyo, S.; Jaime, L.; Garcia-Blairsy, R.G.; Herrero, M.; Senorans, F.J.; Ibanez, E. Screening for bioactive compounds from algae. *J. Pharm. Biomed. Anal.* **2010**, *51*, 450–455.

15. Xiao, X.; Si, X.; Yuan, Z.; Xu, X.; Li, G. Isolation of fucoxanthin from edible brown algae by microwave-assisted extraction coupled with high-speed countercurrent chromatography. *J. Sep. Sci.* **2012**, *35*, 2313–2317.

16. Halim, R.; Harun, R.; Danquah, M.K.; Webley, P.A. Microalgal cell disruption for biofuel development. *Appl. Energy* **2012**, *91*, 116–121.

17. Halim, R.; Danquah, M.K.; Webley, P.A. Extraction of oil from microalgae for biodiesel production: A review. *Biotechnol. Adv.* **2012**, *30*, 709–732.

18. Halim, R.; Rupasinghe, T.W.T.; Tull, D.L.; Webley, P.A. Mechanical cell disruption for lipid extraction from microalgal biomass. *Bioresour. Technol.* **2013**, *140*, 53–63.

19. Hii, S.-L.; Choong, P.-Y.; Woo, K.-K.; Wong, C.-L. Stability studies of fucoxanthin from *Sargassum Binderi*. *Aust. J. Basic Appl. Sci.* **2010**, *4*, 4580–4584.

20. Billakanti, J.M.; Catchpole, O.; Fenton, T.; Mitchell, K. Extraction of fucoxanthin from *Undaria pinnatifida* using enzymatic pre-treatment followed by DME and EtOH co-solvent extraction. In Proceedings of the 10th International Symposium on Supercritical Fluids, San Fransisco, CA, USA, 13–16 May 2012; King, J., Ed.; CASSS: Emeryville, CA, USA, 2012.

21. Wu, J.; Zhou, Y.; Lemmon, E.W. An equation of state for the thermodynamic properties of dimethyl ether. *J. Phys. Chem. Ref. Data* **2011**, *40*, 023104.

22. Eltringham, W.; Catchpole, O.J. Relative permittivity measurements of gaseous, liquid, and supercritical dimethyl ether. *J. Chem. Eng. Data* **2007**, *52*, 363–367.

23. Holldorff, H.; Knapp, H. Binary vapor-liquid-liquid equilibrium of dimethyl ether-water and mutual solubilities of methyl chloride and water: Experimental results and data reduction. *Fluid Phase Equilibria* **1988**, *44*, 195–209.

24. European Food Safety Autorithy. Scientific opinion of the panel on food contact materials, enzymes, flavourings and processing aids (CEF) on dimethyl ether as an extraction solvent. *EFSA J.* **2009**, *984*, 1–13.

25. Varlet, V.; Smith, F.; Augsburger, M. New trends in the kitchen: Propellants assessment of edible food aerosol sprays used on food. *Food Chem.* **2014**, *142*, 311–317.

26. Naito, M.; Radcliffe, C.; Wada, Y.; Hoshino, T.; Liu, X.; Arai, M.; Tamura, M. A comparative study on the autoxidation of dimethyl ether (DME) comparison with diethyl ether (DEE) and diisopropyl ether (DIPE). *J Loss Prev. Process Ind.* **2005**, *18*, 469–473.

27. Li, P.; Kanda, H.; Makino, H. Simultaneous production of bio-solid fuel and bio-crude from vegetal biomass using liquefied dimethyl ether. *Fuel* **2014**, *116*, 370–376.

28. Mendes, R.L.; Fernandes, H.L.; Coelho, J.P.; Reis, E.C.; Cabral, J.M.S.; Novais, J.M.; Palavra, A.F. Supercritical CO_2 extraction of carotenoids and other lipids from *Chlorella vulgaris*. *Food Chem.* **1995**, *53*, 99–103.

29. Machmudah, S.; Shotipruk, A.; Goto, M.; Sasaki, M.; Hirose, T. Extraction of Astaxanthin from *Haematococcus pluvialis* using supercritical CO_2 and ethanol as entrainer. *Ind. Eng. Chem. Res.* **2006**, *45*, 3652–3657.

30. Kitada, K.; Machmudah, S.; Sasaki, M.; Goto, M.; Nakashima, Y.; Kumamoto, S.; Hasegawa, T. Supercritical CO_2 extraction of pigment components with pharmaceutical importance from *Chlorella vulgaris*. *J. Chem. Technol. Biotechnol.* **2009**, *84*, 657–661.
31. Roh, M.K.; Uddin, M.S.; Chun, B.S. Extraction of fucoxanthin and polyphenol from *Undaria pinnatifida* using supercritical carbon dioxide with co-solvent. *Biotechnol. Bioprocess Eng.* **2008**, *13*, 724–729.
32. Richter, B.E.; Jones, B.A.; Ezell, J.L.; Avdalovic, N.; Pohl, C. Accelerated solvent extraction: A technique for sample preparation. *Anal. Chem.* **1996**, *68*, 1033–1039.
33. Kim, S.M.; Jung, Y.J.; Kwon, O.N.; Cha, K.H.; Um, B.H.; Chung, D.; Pan, C.H. A potential commercial source of fucoxanthin extracted from the microalga *Phaeodactylum tricornutum*. *Appl. Biochem. Biotechnol.* **2012**, *166*, 1843–1855.
34. Kanazawa, K.; Ozaki, Y.; Hashimoto, T.; Das, S.K.; Matsushita, S.; Hirano, M.; Okada, T.; Komoto, A.; Mori N.; Nakatsuka, M. Commercial-scale preparation of biofunctional fucoxanthin from waste parts of brown sea algae *Laminalia japonica*. *Food Sci. Technol. Res.* **2008**, *14*, 573–582.
35. Bai, S.; Craig, M.V.; Liu, L.F.; Mayne, C.L.; Pugmire, R.J.; Grant, D.M. CO_2 clustering of 1-decanol and methanol in supercritical fluids by ^{13}C nuclear spin-lattice relaxation. *J. Phys. Chem.* **1997**, *101*, 2923–2928.
36. Bulgarevicg, D.S.; Sako, T.; Sujeta, T.; Otake, K.; Takebayashi, Y.; Kamizawa, C.; Horikawa, Y.; Kato, M. The role or general hydrogen-bonding interaction in the solvation process of organic compounds by supercritical CO_2/*n*-alcohol mixtures. *Ind. Eng. Chem. Res.* **2002**, *41*, 2074–2081.
37. Nehari, A.A.; Kim, S.B.; Lee, Y.B.; Lee, H.T.; Chun, B.S. Characterization of oil including astaxanthin extracted from krill (*Euphausia superba*) using supercritical carbon dioxide and organic solvent as comparative method. *Korean J. Chem. Eng.* **2012**, *29*, 329–336.

8

The Red Seaweed *Gracilaria gracilis* as a Multi Products Source

Matteo Francavilla [1,2,]*, Massimo Franchi [2], Massimo Monteleone [1] and Carmela Caroppo [3]

1. STAR-Agroenergy Research Group, University of Foggia, via Gramsci 89/91, Foggia 71121, Italy; E-Mail: massimo.monteleone@unifg.it
2. Institute of Marine Science, National Research Council, via Pola 4, Lesina 71010, Italy; E-Mail: massimo.franchi@fg.ismar.cnr.it
3. Institute for Coastal Marine Environment, National Research Council, via Roma 3, Taranto 74121, Italy; E-Mail: carmela.caroppo@iamc.cnr.it

* Author to whom correspondence should be addressed; E-Mail: matteo.francavilla@unifg.it

---

Abstract: In recent years seaweeds have increasingly attracted interest in the search for new drugs and have been shown to be a primary source of bioactive natural compounds and biomaterials. In the present investigation, the biochemical composition of the red seaweed *Gracilaria gracilis*, collected seasonally in the Lesina Lagoon (Southern Adriatic Sea, Lesina, Italy), was assayed by means of advanced analytical techniques, such as gas-chromatography coupled with mass spectrometry and spectrophotometric tests. In particular, analysis of lipids, fatty acids, sterols, proteins, phycobiliproteins and carbohydrates as well as phenolic content, antioxidant and radical scavenging activity were performed. In winter extracts of *G. gracilis*, a high content of R-phycoerythrin together with other valuable products such as arachidonic acid (PUFA ω-6), proteins and carbohydrates was observed. High antioxidant and radical scavenging activities were also detected in summer extracts of the seaweed together with a high content of total phenols. In conclusion, this study points out the possibility of using *Gracilaria gracilis* as a *multi products source* for biotechnological, nutraceutical and pharmaceutical applications even although more investigations are required for separating, purifying and characterizing these bioactive compounds.

Keywords: seaweed; biochemical composition; biorefinery; *Gracilaria*; Lesina Lagoon

---

1. Introduction

The marine world, due to its phenomenal biodiversity, is a rich natural resource of many biologically active compounds [1,2]. Many marine organisms live in complex habitats exposed to extreme conditions and, in adapting to new environmental surroundings, they produce a wide variety of primary and secondary metabolites which cannot be found in other organisms. Marine-based bioactive compounds can be derived from a vast array of sources, including marine plants, macro- and microalgae, microorganisms, and sponges, all of which contain their own unique set of biomolecules [3].

Macroalgae, known also as seaweeds, produce many biologically active phytochemicals, which include among others, carotenoids, terpenoids, xanthophylls, chlorophylls, phycobilins, polyunsaturated fatty acids, polysaccharides, vitamins, sterols, tocopherol and phycocyanins [4]. Seaweeds represent 23.4% of the tonnage and 9.7% of the value of the global (marine, brackish water, and freshwater) aquaculture production, estimated at 59.4 million tonnes and $ 70.3 billion in 2004 [5,6]. They are used as food, fodder, feed and fertilizer [7] and many of the bioactive compounds produced by the macroalgae are known to have potential beneficial use in healthcare [8,9].

Gracilaria Greville genus (Gracilariales, Rhodophyta) is represented by more than 300 species of which 160 have been accepted taxonomically. The macroalgae belonging to this genus are important for industrial and biotechnological uses and are considered economically valuable resources, because of their ability to achieve high yields of commercially valuable biomass [10]. In fact, they contain, besides other compounds, phycocolloids, the main source of agar-agar, which is a gelatinous non-toxic colloidal carbohydrate present in the cell wall and intercellular spaces of the algae and has wide use in the preparation of food, ice creams, jellies, soups, bacteriological samples and cosmetics [7,11]. These algae are also sources of important bioactive metabolites with antibiotic activity; but also sources of different prostaglandins and other substances that may be toxic to humans by causing gastrointestinal disorders and lethality [4].

During the period 1970–1990, in the Mediterranean Lesina Lagoon (Southern Adriatic Sea, Lesina, Foggia, Italy) local fishermen used to harvest the red seaweed *Gracilaria gracilis* (Stackhouse) Steentoft, Irvine *et* Farnham (as *Gracilaria confervoides* Greville) [12,13]. The biomass, after the sun drying process, was sold to some private companies as raw material for agar extraction. The harvested algal biomass reached 100 t dried weight per year [14]. Unfortunately, the over-exploitation of this natural resource contributed to a drastic reduction in production of *Gracilaria* biomass. As a consequence, other seaweeds not economically valuable, like *Valonia aegagropila*, *Chaetomorpha* sp. grew quickly in the Lesina Lagoon, reaching sometimes bloom conditions [15,16].

After this period, some field experiments aimed at re-establishing stable coverage of *Gracilaria gracilis* were carried out successfully in the Lesina Lagoon. These studies pointed out not only an increase in *Gracilaria* biomass (around 1200 t wet weight per year), but also the nutrient uptake capacity of this species in the area of its re-colonization in the lagoon [17]. This aspect is particularly interesting taking into account that *Gracilaria* could also represent a bio-remediator useful to control the risk of eutrophication in this coastal lagoon.

Recently, Francavilla *et al.* [18] found that Gracilaria harvested in the Lesina Lagoon could be used as an interesting source of natural porous material with several biotechnological applications.

Moreover, Buldarin *et al.* [19] found that microwave (MW)-mediated pyrolysis of this macroalga produced chemical rich bio-oils which are rich in aromatics, sugars and other high value chemicals.

In the light of those results and of worldwide great interest in algae biomass as a source of bioactive compounds (for the nutraceutical, pharmaceutical and cosmetic industry) and new biomaterials, the aim of this work was to screen (for the first time, to the best of our knowledge) the biochemical composition of *Gracilaria gracilis* collected in the Lesina Lagoon, with the idea of taking into account the algal biomass as a hypothetical *multi product source*. In particular, the analysis of lipids, fatty acids, sterols, protein, phycobiliproteins, carbohydrates, as well as phenolic content, antioxidant and radical scavenging activity were performed. Moreover, because biochemical composition depends on many environmental and seasonal factors [20,21], *Gracilaria* was collected seasonally in order to evaluate if there were significant differences in the biochemical composition of this species.

2. Results and Discussion

2.1. Lipids and Fatty Acids (FAMEs)

Lipids are a large group of natural compounds which includes fats, waxes, sterols, fat-soluble vitamins (such as vitamins A, D, E and K), monoglycerides, diglycerides, phospholipids, carotenoids and others [22]. They play many biological functions including energy storage, structural components of cell membranes, and signalling molecules. Although humans and other mammals use various biosynthetic pathways to both break down and synthesize lipids, some essential lipids can be obtained only from diet [22].

Figure 1 shows the concentration of total lipids (TL), unsaponified fraction (UF) and fatty acids (FAMEs) extracted from *Gracilaria* seaweed. The biomass sampled in July showed the highest lipid content which represented 1.98% dry weight The lowest concentration in TL was found in winter (January, 1.12% dry weight), whereas intermediate concentrations were found in *Gracilaria* sampled in October and April (1.38% and 1.40% dry weight, respectively). Statistical analyses confirmed that TL values were significantly different in all the seasons, only with comparable values in fall and spring (Figure 1). The detected concentrations were very similar to those measured by other authors in *Gracilaria* sp. [23,24] and confirmed the low lipid content in red seaweeds. Therefore, its contribution as a food energy source appears to be low.

The unsaponified fraction of total lipids, which includes waxes, sterols, fat-soluble vitamins and carotenoids, generally was low and ranged between 0.35% dry weight in October and 0.15% dry weight in January (Figure 1). In this case Tukey's test showed significant differences only in the samples collected in autumn and winter (Figure 1).

With regard to the total amount of FAMEs in *Gracilaria*, the highest concentration was observed in biomass collected in April (0.67% dry weight) whereas the lowest concentration was in October (0.31% dry weight). Statistical analyses showed significant differences in terms of FAMEs content between samples collected in April and those sampled in the other months (Figure 1). No significant differences were found between samples collected in summer and winter, while samples collected in winter were similar ($p > 0.05$) to those collected in autumn. Fatty acids are precursors in the biosynthesis of eicosanoids, which are important bioregulators of many cellular processes [25–27].

Eicosapentaenoic (EPA) and docosahexaenoic (DHA) polyunsaturated ω-3 fatty acids are known to: (i) have capacities for cardioprotection; (ii) reduce triacylglycerol and cholesterol levels; (iii) have antiinflamatory and anticancer effects [28]. FAMEs composition in *Gracilaria*, a mixture of 31 compounds, showed significant variation between the analysed seasons (Table 1).

Figure 1. Mean concentration (% dry weight) of Total Lipids (TL), Unsaponified fraction (UF) and Fatty Acids Methyl Esters (FAMEs) extracted from *G. gracilis.* Different superscripts (a–d) indicate significant difference ($p < 0.05$).

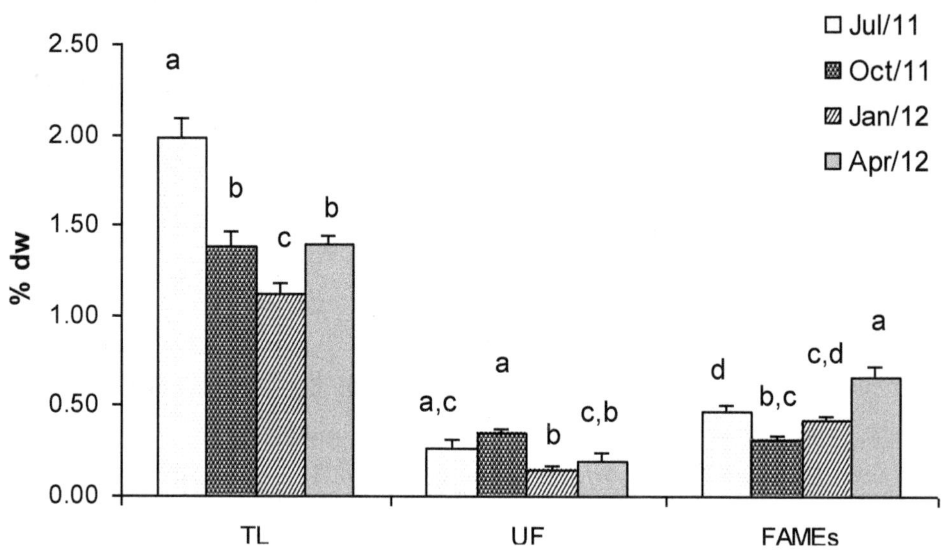

Table 1. Fatty Acids composition (% w/w of total amount of fatty acids) of *G. gracilis* sampled in four different seasons.

Fatty Acids Methyl Esters	Structure	Fatty Acid Content (%)			
		11 July	11 October	12 January	12 April
Methyl decanoate	C10:0	0.05 [a]	0.03 [b]	nd	0.03 [b]
Methyl undecanoate	C11:0	0.01	nd	nd	nd
Methyl dodecanoate	C12:0	0.28 [a]	0.35 [a]	0.06 [b]	0.12 [c]
Methyl tridecanoate	C13:0	0.06 [a]	0.07 [a]	0.02 [b]	0.03 [b]
Methyl 9-tetradecenoate	9-C14:1	0.02 [a]	0.39 [b]	0.74 [c]	0.55 [d]
Methyl tetradecanoate	C14:0	5.48 [a]	5.55 [a]	3.48 [b]	5.13 [a]
cis-10 Pentadecenoic Acid Methyl Ester	10-C15:1	nd	0.15 [a]	0.03 [b]	0.04 [b]
Pentadecanoic Acid Methyl Ester	C15:0	1.03 [a]	1.32 [b]	0.52 [c]	0.53 [c]
Palmitoleic Acid Methyl Ester	9-C16:1	3.84 [a]	8.19 [b]	2.21 [c]	8.81 [d]
Palmitic Acid Methyl Ester	C16:0	25.67 [a]	38.13 [b]	28.55 [c]	29.29 [c]
cis-10 Heptadecenoic Acid Methyl Ester	10-C17:1	0.14 [a]	1.12 [b]	0.23 [c]	0.24 [c]
Heptadecanoic Acid Methyl Ester	C17:0	0.64 [a]	1.14 [b]	0.20 [c]	0.30 [c]
Linolenic Acid Methyl Ester	C18:3*n*-3	0.42 [a]	0.23 [b]	0.09 [c]	3.88 [d]
Linoleic Acid Methyl Ester	C18:2*n*-6c	4.85 [a]	2.49 [b]	1.33 [c]	4.14 [d]
Oleic Acid Methyl Ester	C18:1*n*-9c	10.78 [a]	8.79 [b]	5.76 [c]	9.12 [b]
Elaidic Acid Methyl Ester	C18:1*n*-9t	4.16 [a]	6.15 [b]	3.22 [c]	2.02 [d]
Stearic Acid Methyl Ester	C18:0	3.35 [a]	3.62 [a]	1.87 [b]	2.30 [c]

Table 1. *Cont.*

Arachidonic Acid Methyl Ester	C20:4*n*-6	33.27 [a]	16.05 [b]	47.78 [c]	38.30 [d]
cis-5-8-11-14-17-Eicosapentaenoic Acid Methyl Ester	C20:5*n*-3	1.13 [a]	1.84 [b]	0.24 [c]	3.93 [d]
cis-11,14,17-Eicosatrienoic Acid Methyl Ester	C20:3*n*-3	2.48 [a]	1.23 [b]	2.07 [c]	2.82 [a]
cis-11,14-Eicosdienoic Acid Methyl Ester	11,14-C20:2	0.32 [a]	0.25 [b]	0.23 [b,c]	0.19 [c]
cis-8,11,14-Eicosatrienoic Acid Methyl Ester	8,11,14-C20:3*n*-6	nd	0.32 [a]	0.71 [b]	nd
cis-11 Eicosenoic Acid Methyl Ester	11-C20:1	0.28	nd	nd	nd
Arachidic Acid Methyl Ester	C20:0	0.15 [a]	0.40 [b]	0.08 [c]	0.61 [d]
Heneicosanoic Acid Methyl Ester	C21:1	0.02 [a]	0.02 [a]	0.04 [b]	0.02 [a]
cis-4,7,10,13,16,19-Docosahexaenoic Acid Methyl Ester	C22:6*n*-3	0.23 [a]	0.43 [b]	0.03 [c]	0.12 [d]
Erucidic Acid Methyl Ester	C22:1*n*-9	0.65 [a]	0.34 [b]	0.19 [c]	0.13 [c]
Docosanoic Acid Methyl Ester	C22:0	0.26 [a]	0.38 [b]	0.09 [c]	0.19 [d]
Tricosanoic Acid Methyl Ester	C23:0	0.08 [a]	0.16 [b]	0.02 [c]	0.02 [c]
cis-15-Tetracosenoic Acid Methyl Ester	15-C24:1	0.09 [a]	0.31 [b]	0.15 [c]	0.15 [c]
Tetracosanoic Acid Methyl Ester	C24:0	0.27 [a]	0.56 [b]	0.05 [c]	0.15 [d]
TOTAL (mg g^{-1} dry weight)		4.71 [a]	3.14 [b]	4.18 [a]	6.67 [c]
SFA (% w/w)		37.33 [a]	51.71 [b]	34.94 [c]	34.19 [c]
MUFA (% w/w)		19.98 [a]	25.31 [b]	12.54 [c]	18.60 [a]
PUFA (% w/w)		42.70 [a]	22.52 [b]	51.77 [c]	47.18 [a]
PUFAω6 (% w/w)		38.12 [a]	18.86 [b]	49.82 [c]	37.51 [a]
PUFAω3 (% w/w)		4.26 [a]	3.73 [a,b]	2.43 [b]	9.51 [c]
ω6/ω3		8.96 [a]	5.06 [b]	20.48 [c]	3.95 [b]

SFA: Saturated Fatty Acids; MUFA: Mono-Unsaturated Fatty Acids; PUFA: Poly-Unsaturated Fatty Acids. ω6/ω3: ratio of ω6 and ω3 Fatty Acids; nd: not detected. [a–d] Row wise values with different superscripts of this type indicate significant difference ($p < 0.05$).

In July, the most abundant FAMEs were arachidonic acid (AA-C20:4 *n*-6) and palmitic acid (PA-C16:0). AA reached the highest relative concentration during winter time (January, 48%). The lowest content in AA was found in October (16%) whereas intermediate concentrations were observed in July and April (33% and 38% respectively). Tukey's test revealed significant differences in AA concentration in samples collected in all the seasons (Table 1). Concerning PA, its highest concentration was detected in October (38%) when AA was less concentrated. The lowest value was observed in July (26%). Statistical analyses confirmed significant differences in terms of PA concentration only in these two periods (Table 1). Other abundant fatty acids monitored in *Gracilaria* were: tetradecanoic acid (C14:0), palmitoleic acid (9-C16:1), stearic acid (C18:0), oleic acid (C18:1*n*-9c), elaidic acid (C18:1*n*-9t) and linoleic acid (C18:2*n*-6c). PUFAs (Poly-Unsaturated Fatty Acids) were 52% by total FAMEs in January, when also the lowest content in MUFAs (Mono Unsaturated Fatty Acids) and SFAs (Saturated Fatty Acids) was found (20% and 37%, respectively) (Table 1). In October, concentration of PUFAs decreased to 23% and concentration of MUFAs and SFAs reached their highest values (25% and 52%, respectively). In April, when the highest content in FAMEs was found (0.67% dry weight), PUFAs were 47% of total FAMEs and MUFAs and SFAs represented 19% and 34% respectively. Arachidonic acid and linoleic acid were the only two PUFAs ω-6 present in *Gracilaria*, whereas linolenic acid (C18:3*n*-3), eicosapentanoic acid (C20:5*n*-3), eicosatrienoic acid (C20:3*n*-3) and docosahexanoic acid (C22:6*n*-3) were the PUFAs ω-3 detected.

Statistical analysis showed that in *Gracilaria* biomass the concentrations of palmitoleic acid, elaidic acid, linoleic acid, eicosapentanoic acid and docosahexanoic acid were significantly different in all the seasons. Also the other fatty acids showed significant differences in their concentration but in a reduced number of seasons (Table 1).

Recently, the importance of the ω-6/ω-3 ratio has been discussed in scientific reports. The original value 1 of the ratio ω-6/ω-3 involved the balance of intake of both polyunsaturated ω-6 and ω-3 fatty acids. The high dietary intake of ω-6 fatty acids from a diet rich in vegetable oils causes detrimental turnover of the balance ratio (ω-6/ω-3) to ω-3 fatty acids. This ratio is 50 in Europe and the United States, and 12 in Japan, which is compared to 1 for Greenland Eskimos due to their higher consumption of fish fatty acids [29]. A huge intake of ω-6 fatty acids and an excessively high ω-6/ω-3 ratio promotes the pathogenesis of cardiovascular, inflammatory, and autoimmune diseases and cancer [30]. The WHO currently recommends a ω-6/ω-3 ratio lower than 10. The ω-6/ω-3 ratio in *Gracilaria* extracts was constantly below this limit. The lowest value was found in samples collected in April when the highest amount of FAMEs was found too. Only in January was this ratio higher than 10 because of the high concentration of AA (48%, Table 1). Therefore, taking into account all the results in terms of FAMEs, content of PUFAs and ω-6/ω-3 ratio, *Gracilaria* biomass seems to be an interesting candidate as source of PUFAs and AA if it is harvested in spring, according also to the results reported by MacArtain *et al.* [31].

2.2. Sterols

The composition of the sterols fraction extracted from *Gracilaria* was investigated in order to explore if phytosterols, sterols with C_{28} and C_{29} structure, were present. This is because of the great interest in these compounds for their bioactivity and promising nutraceutical and pharmaceutical use. In fact, they are precursors of some bioactive molecules (e.g., ergosterol is a precursor of Vitamin D2) and have also been shown to lower total and LDL cholesterol levels in humans by inhibiting cholesterol absorption from the intestine [32]. Moreover, phytosterols possess anti-inflammatory activity and may be characterized by anti-cancer and anti-oxidative activities [33].

Total sterols (TS) concentration, analysed in the unsaponified fraction, was 0.45 mg/g dry weight of biomass in July (Figure 2). This value decreased drastically in January when TS was 0.11 mg/g dry weight cholesterol and TS concentrations were statistically different in all the samples of *Gracilaria* (Figure 2). Seven sterols were identified and corresponded respectively to: cholesterol (St_1), (22*Z*)-ergosta-7,22-dien-3β-ol (St_2), (24*R*)-methylcholest-5-en-3β-ol (St_3), ergostan-5α-3β-ol (St_4), (24*S*) stigmasta 5,22-dien-3β-ol (St_5), Stigmast-5-en-3β-ol (St_6) and stigmastan-5α-3β-ol (St_7). Unfortunately, cholesterol was the most abundant, representing about 90% of TS. *St6* and *St7*, two phytosterols, were 4% and 3% of TS respectively, whereas the remaining sterols were only in traces. Therefore, unfortunately, *Gracilaria* biomass cannot be considered as a source of phytosterols for nutraceutical applications because of the high content of cholesterol.

Figure 2. Concentration (mg g^{-1} dry weight of biomass) of Total Sterol (TS) and sterols identified in *G. gracilis*. Cholesterol (*St*$_1$), (22*Z*)-ergosta-7,22-dien-3β-ol (*St*$_2$), (24*R*)-methylcholest-5-en-3β-ol (*St*$_3$), ergostan-5α-3β-ol (*St*$_4$), (24*S*) stigmasta 5,22-dien-3β-ol (*St*$_5$), Stigmast-5-en-3β-ol (*St*$_6$) and stigmastan-5α-3β-ol (*St*$_7$). Different superscripts (a–d) indicate significant difference ($p < 0.05$).

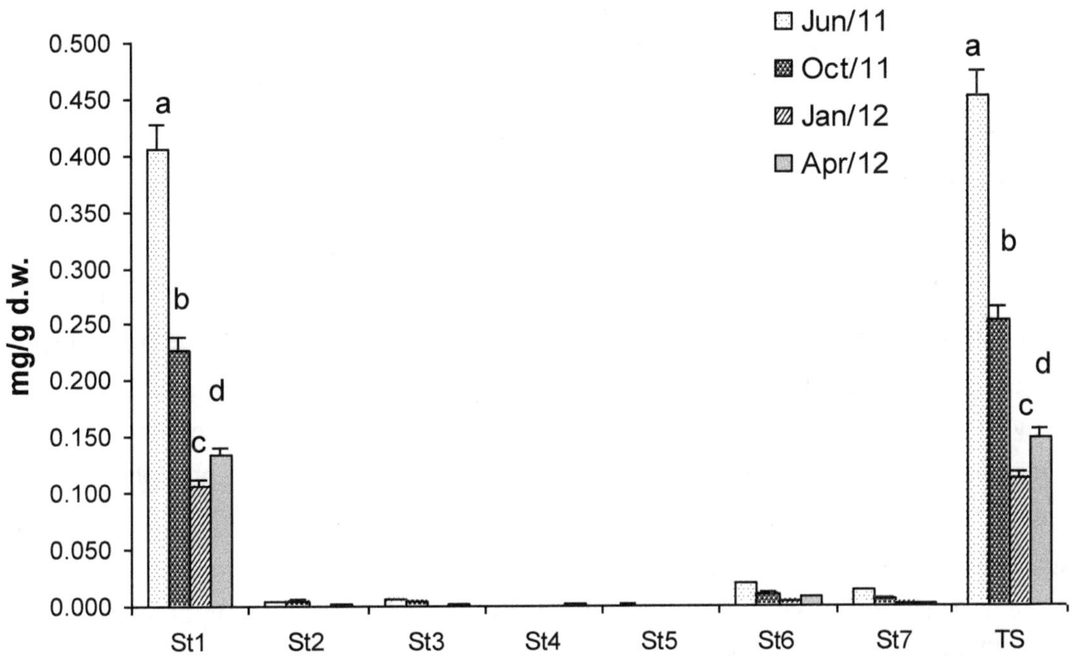

2.3. Proteins

Macroalgae are a unique source of proteins although the content is very variable. Proteins are composed of different amino acids and hence the nutritional quality can be determined basically by the content, proportion and availability of these amino acids. Proteins can have antibacterial, antioxidant, immunostimulating, antithrombotic and anti-inflammatory activities; they can be used for prevention and treatment of hypertension, diabetes and hepatitis among other positive effects in the organism. All these health promoting effects make these compounds of great relevance as nutraceuticals [22]. Analyses of total protein in algae are often done in order to search for new sources of protein supplements. In *Gracilaria* species the protein content found ranged from 5.6% to 30% [34–38].

In this study, *Gracilaria* showed a very interesting protein concentration (Figure 3). In particular the protein content observed in January was 45% dry weight This value decreased in October and April (about 41% dry weight) and reached the lowest concentration in summer (July, 31% dry weight).

Tukey's test revealed that the protein concentrations were significantly different in all the seasons, except for the samples collected in April when the concentrations were significantly different only with respect to the value detected in July (Figure 3). Therefore, macroalgal biomass is a very promising source of protein if it is harvested in winter and spring.

Figure 3. Concentration (% dry weight) of total proteins in *G. gracilis*. Different superscripts (a–d) indicate significant difference ($p < 0.05$).

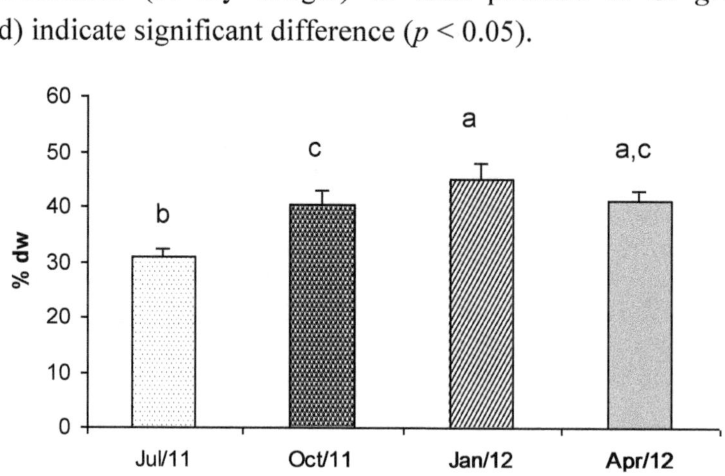

2.4. Phycobiliproteins

The phycobiliproteins are proteins with linear tetrapyrrole prosthetic groups (bilins) that, in their functional state, are covalently linked to specific cysteine residues of the proteins. These proteins are found in cyanobacteria (blue-green algae), in a class of biflagellate unicellular eukaryotic algae (cryptomonads), and in Rhodophyta (red algae). In all of them the phycobiliproteins act as photosynthetic accessory pigments.

Figure 4 shows the concentrations of phycobiliproteins in *Gracilaria* extracts. R-phycoerythrin (R-PE) is the most abundant with a concentration which ranged between 7 mg/g dry weight in January and 3.6 mg/g dry weight in October. Allophycocyanin (APC) showed lower concentration than R-PE which ranged between 3.5 mg/g dry weight and 1.5 mg/g dry weight, in January and October respectively. Phycocyanin (PC) was the less abundant phycobiliprotein showing a concentration that ranged from 3 mg/g dry weight (d.w.) in January to 0.7 mg/g d.w. in October. Generally, the highest concentrations of the three phycobiliproteins were found in biomass sampled in winter time when light radiation and nutrient concentration in the lagoon are favourable for the biosynthesis of these metabolites [17].

Statistical analyses showed significant differences in terms of phycobiliprotein concentration in all the seasons, except for samples collected in April when the concentrations were significantly different only with respect to the values detected in October (Figure 4).

Phycobiliproteins can be used as very useful fluorescent probes because of their excellent spectroscopic properties [39,40], stability, high absorption coefficients, and high quantum yields. They are highly soluble in water and exhibit a large Stokes shift which is very important for detection [41]. Phycobiliproteins are attractive since they are not harmful to humans if they are applied to an external surface or ingested. They are already used as photosensitizers for treatment of tumors and have potential to substitute Photofrin (a kind of light sensitive agent in photodynamic therapy) in common use which is purified from animal blood [42]. Phycobiliproteins are also widely used as natural colorants for food and cosmetics. For their wide usage phycobiliproteins have a great economic potential [43,44]. The concentration of R-PE found in *Gracilaria* harvested in winter is very interesting from a biotechnological point of view. Therefore, the algal biomass could be proposed as a novel industrial source of R-PE. Actually, purified R-PE is a very expensive compound which costs about 50 euro per mg (AnaSpec, Inc., Fremont, CA, USA).

Figure 4. Concentration (mg g^{-1} dry weight) of phycobiliproteins in *G. gracilis*. Allophycocyanin (APC), phycocyanin (PC) and R-phycoerythrin (R-PE). Different superscripts (a–d) indicate significant difference ($p < 0.05$).

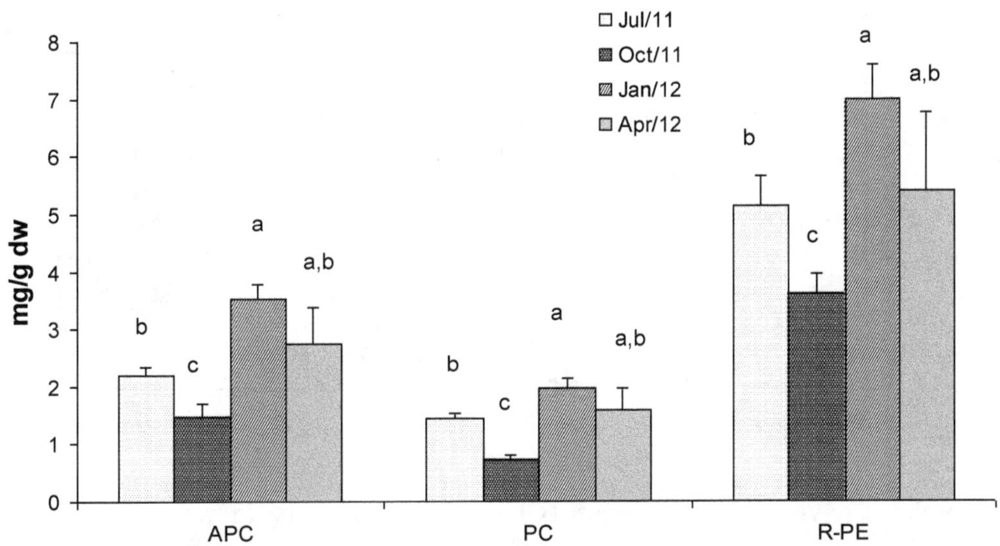

2.5. Fractionated Extraction of Algal Biomass for Antioxidant Assays and Analysis of Total Phenolic Content

Figure 5 shows concentrations of *Gracilaria* extracts obtained by using four solvents with increasing polarity: *n*-hexane, ethyl acetate, methanol and water (at 80 °C). Results evidenced that the higher the solvent polarity the higher the concentration of extract. Therefore, the highest concentration was found using water (21% dry weight), whereas the lowest one was found using *n*-hexane (0.16% d.w.). Statistically significant differences were detected mainly in the water extracts during all the seasons, except in April when values were comparable with those observed in October (Figure 5). Methanol and ethyl acetate showed intermediate concentrations (4.2% and 0.43% dry weight in April respectively). Generally, extracts of biomass in winter were more concentrated than extracts of biomass in other seasons. Exceptions were represented by the *n*-hexane and methanol extracts that showed the highest concentration in April (0.40% dry weight). Interestingly, in the same sample the highest concentrations of FAMEs were also found, which might have been extracted by *n*-hexane.

The dried extracts, dissolved in methanol, were used for antioxidant assays and analysis of phenolic content.

2.6. Antioxidant Activity Assays

Reactive oxygen species such as hydroxyl, super oxide and peroxyl radicals produced in human tissue cells are responsible for extensive oxidative damage that leads to age related degenerative conditions, cancer and a wide range of other human diseases [45,46].

Figure 5. Concentration (% dry weight) of extracts of *G. gracilis* obtained by using four solvents with increasing polarity: *n*-hexane, ethyl acetate, methanol and deionized water (at 80 °C). Different superscripts (a–d) indicate significant difference ($p < 0.05$).

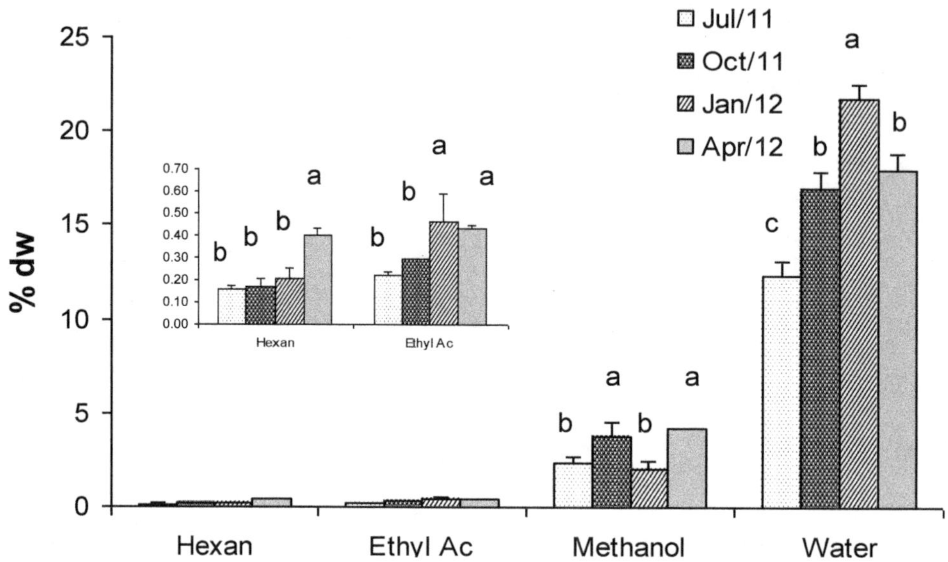

Many researchers have found different types of antioxidants in various species of higher plants [47,48]. However, seaweeds also are considered to be an interesting source of antioxidant compounds. In fact, several researchers have reported the antioxidant properties of both brown and red seaweeds from across the globe [49–52]. Further, the evidence available in the literature suggests the potential protective effects of seaweeds against oxidative stress in target tissues and lipid oxidation in food [53]. To date no reports have identified *Gracilaria gracilis* as a source of antioxidant compounds. While there are two reports of *Gracilaria edulis* showing antioxidant activity it is not possible to compare the antioxidant results reported here and those reported in *Gracilaria edulis* as different extractions methods were used [51,54].

Many authors have stressed the need to carry out more than one type of antioxidant activity measurement to take into account the various mechanisms of antioxidant action [52,55], as no single assay will accurately reflect all of the radical sources or all antioxidants in a mixed or complex system [56].

Three different antioxidant assays were carried out in this study to test the activity of *Gracilaria* extracts: FRAP, ABTS and DPPH assays.

2.6.1. Ferric-Reducing Antioxidant Power (FRAP) Assay

The four seasonal extracts showed a high variability in antioxidant activity (as μmol Trolox g^{-1} extract) (Table 2). Analysing the biomass sampled in July, the highest activity (809 μmol Trolox g^{-1}) was found in the ethyl acetate extract. The n-hexane extract showed an intermediate value (370 μmol Trolox g^{-1}) whereas in the methanol and water fraction the lowest antioxidant activities were found (100 and 13 μmol Trolox g^{-1}, respectively). Interestingly, these values decreased to about 60%–70% in extracts of biomass sampled in October. In fact, the antioxidant activity was 312 μmol Trolox g^{-1} in the ethyl acetate extract, 114 μmol Trolox g^{-1} in hexane, 27 μmol Trolox g^{-1} in methanol and 9 μmol Trolox g^{-1} in the water extract.

Table 2. Ferric-reducing antioxidant power (FRAP) (μmol Trolox g^{-1} extract) of extracts obtained from *G. gracilis* sampled in four different seasons ($n = 3$).

Date	Extracts				CTR (BHT)
	HE	EA	ME	WT	
July/11	370.74 ± 18.5 [a]	808.90 ± 40.4 [a]	99.90 ± 5.0 [a]	13.32 ± 0.7 [a]	1917.93 ± 95.9
October/11	114.28 ± 5.7 [b]	312.47 ± 15.6 [b]	26.90 ± 1.3 [b]	9.06 ± 0.5 [b]	
January/12	189.23 ± 9.5 [c]	62.02 ± 3.1 [c]	53.05 ± 2.7 [c]	9.19 ± 0.5 [b]	
April/12	117.24 ± 5.9 [b]	29.82 ± 1.5 [d]	50.14 ± 2.5 [c]	14.61 ± 0.7 [a]	

All the values are mean ± SD; SD: standard deviation.HE: n-Hexan; EA: Ethyl Acetate; ME: methanol; WT: Water (80 °C); CTR: Control (butylated hydroxytoluene). [a–d] Column wise values with different superscripts of this type indicate significant difference ($p < 0.05$).

Statistical analyses evidenced significant differences in all the seasons with regard to the ethyl acetate extracts. In the *n*-hexane and methanol extracts differences were evidenced in all the seasons except in April when values were comparable with those detected in October and January, respectively. In water extracts values showed a lower seasonality (Table 2).

Comparing these results with a known and commercial antioxidant, butylated hydroxytoluene (BHT), we found that the ethyl acetate extract (of July) had 42% antioxidant activity by BHT (1918 μmol Trolox g^{-1}) (Table 2).

2.6.2. ABTS Assay

Results of ABTS antioxidant assay are reported in Table 3 (as mmol Trolox g^{-1} extract). Similar to that found using FRAP assay, the highest activity was found in the ethyl acetate extract of biomass sampled in July (0.43 mmol Trolox g^{-1}). However, this activity decreased only marginally in April (0.34 mmol Trolox g^{-1}). The hexane extract showed the highest activity in April (0.26 mmol Trolox g^{-1}) and the lowest in October (0.07 mmol Trolox g^{-1}). Interestingly, in the methanol and water extracts of *Gracilaria* seaweed the lowest antioxidant activities were found. They ranged between 0.06 and 0.02 mmol Trolox g^{-1} for methanol extracts and between 0.15 and 0.05 mmol Trolox g^{-1} for water extracts. As already observed for the FRAP assay, Tukey's test showed significant differences in all the seasons with regard to the ethyl acetate extracts. In the *n*-hexane extracts differences were evidenced in all the seasons except in October when values were comparable with those detected in July. In the methanol and water extracts values showed a lower seasonality (Table 3).

Surprisingly, antioxidant activity of ethyl acetate extracts for July and April were higher than the activity of gallic acid (0.28 mmol Trolox g^{-1}) used as control.

2.6.3. DPPH Assay

Table 4 shows the scavenging activity of extracts of *Gracilaria* biomass at different concentrations. EC_{50} value (mg extract mL^{-1}), the effective concentration at which 50% of the DPPH radicals were scavenged, is used to compare the scavenging activity of extracts.

Table 3. Antioxidant activity (mmol Trolox g^{-1} extract) of extracts obtained from *G. gracilis* sampled in four different seasons measured using ABTS assay ($n = 3$).

	Extracts				
Date	**HE**	**EA**	**ME**	**WT**	**CTR (GA)**
July/11	0.09 ± 0.02 ^{a}	0.43 ± 0.04 ^{a}	0.06 ± 0.01 ^{a}	0.07 ± 0.02 ^{a}	0.28 ± 0.04
October/11	0.07 ± 0.01 ^{a}	0.18 ± 0.02 ^{b}	0.02 ± 0.01 ^{b}	0.05 ± 0.01 ^{a}	
January/12	0.16 ± 0.03 ^{b}	0.26 ± 0.02 ^{c}	0.06 ± 0.02 ^{a}	0.10 ± 0.03 ^{a,b}	
April/12	0.26 ± 0.03 ^{c}	0.34 ± 0.03 ^{d}	0.03 ± 0.01 ^{b}	0.15 ± 0.03 ^{b}	

All the values are mean ± SD; SD: standard deviation; HE: *n*-Hexan; EA: Ethyl Acetate; ME: methanol; WT: Water (80 °C); CTR: Control (gallic acid). $^{a–d}$ Column wise values with different superscripts of this type indicate significant difference ($p < 0.05$).

The ethyl acetate extract of seaweed sampled in July showed the highest scavenging effect, corresponding to the lowest EC_{50} (0.82 mg mL^{-1}) (Table 4). The hexane extract of the same sample of biomass (July), showed intermediate EC_{50} value (1.1 mg mL^{-1}) and methanol showed the highest value (2.94 mg mL^{-1}). At the concentrations tested, it was not possible to measure the Ec_{50} of the water extract. Therefore, we reported EC_{40}, the concentration at which 40% of the DPPH radicals were scavenged. The EC_{40} of the water extract of July was 10 mg mL^{-1}.

Table 4. DPPH radical scavenging activity (EC_{50} value, mg extract mL^{-1} methanol) in extracts of *G. gracilis* sampled in four different seasons ($n = 3$).

	Extracts				
Date	**HE**	**EA**	**ME**	**WT**	**CTR (BHT)**
11 July	1.1 ± 0.06 ^{a}	0.82 ± 0.04 ^{a}	2.94 ± 0.15 ^{a}	10 ± 0.50 *	0.67 ± 0.03
11 October	3.43 ± 0.17 ^{b}	2.55 ± 0.13 ^{b}	5.73 ± 0.29 ^{b}	30.4 ± 1.52 *	
12 January	3.32 ± 0.17 ^{b}	3.47 ± 0.17 ^{c}	9.72 ± 0.49 ^{c}	33.17 ± 1.66 ^{a}	
12 April	4.29 ± 0.21 ^{c}	7.03 ± 0.35 ^{d}	8.72 ± 0.44 ^{c}	35.03 ± 1.75 ^{a}	

All the values are mean ± SD; SD: standard deviation; HE: *n*-Hexan; EA: Ethyl Acetate; ME: methanol; WT: Water (80 °C); CTR: Control (gallic acid). $^{a–d}$ Column wise values with different superscripts of this type indicate significant difference ($p <0.05$). * EC_{40} value (mg extract mL^{-1} methanol).

The extracts of *Gracilaria* sampled in other months showed higher EC_{50} values than the sample for July (Table 4). In fact, the EC_{50} of the ethyl acetate extract of biomass sampled in October was 2.55 mg mL^{-1}, hexane extract was 3.43 mg mL^{-1} and methanol extract was 5.73 mg mL^{-1}. The EC_{40} of water extract was 30.4 mg mL^{-1}. These values increased in the samples for January and April.

Also the DPPH assay confirmed what already had been observed for the ABTS and FRAP assays. In fact statistical analysis revealed a marked seasonality regarding ethyl acetate extracts. With regard to the *n*-hexane extracts, samples collected in October and January showed comparable values, but significantly different with respect to those collected in the other seasons. Also methanol extracts were characterized by significant differences except for samples collected in spring which were found to be similar to samples collected in winter (Table 4).

Finally, similar to that which we found for ABTS assay, the EC_{50} radical scavenging activity of the ethyl acetate extract of *Gracilaria* sampled in July was very similar to the scavenging activity of BHT used as control (Table 4).

2.7. Total Phenolic Content

Phenols are an important group of natural products with antioxidant and other biological activities [57]. These compounds play an important role in algal cell defence against abiotic and biotic stress. Many of them possess antioxidant, antimicrobial and antiviral activities that are important for the protection of algal cells against stress conditions.

Several authors have recently published results regarding the total phenol content and antioxidant activity of algae [51,52].

The phenolic content in extracts of *Gracilaria* was very variable even if a marked seasonality was not so evident, as statistical analyses evidenced (Figure 6). The highest concentration was found in the ethyl acetate extract (65 mg GAE g^{-1} of extract in July and October), whereas the lowest one was found in the methanol extract (2.3 mg GAE g^{-1} of extract in October). Hexane and water extracts showed intermediate values of phenolic content that ranged from 11.56 to 31.8 mg GAE g^{-1} and from 4.7 to 8.8 mg GAE g^{-1} respectively. Interestingly, the phenolic content that we found in ethyl acetate extracts of *Gracilaria gracilis* in July and October was higher than the value reported by Ganesan *et al.* [51] for *Gracilaria edulis* (16.26 mg GAE g^{-1} of extract).

Figure 6. Total Phenolic content (mg GAE g^{-1} dry weight extract) in extracts of *G. gracilis*. Different superscripts (a–d) indicate significant difference ($p < 0.05$).

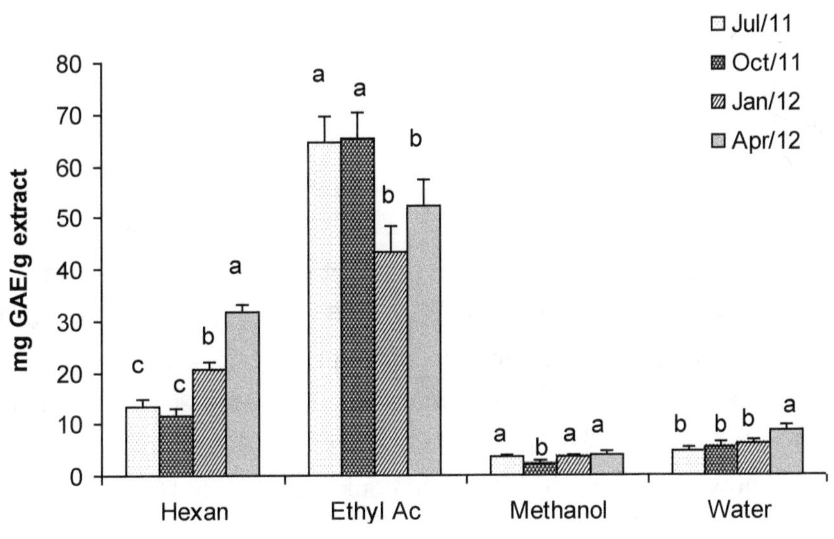

The overlapping of FRAP, TEAC and DPPH assays with the Total Phenols content highlights a positive correlation. In other words, it seems that the antioxidant and scavenging activity is related to phenolic compounds soluble in *n*-hexane and ethyl acetate.

2.8. Total Carbohydrates

Carbohydrates perform numerous essential roles in living beings [22]. Monosaccharides are the major source of energy for metabolism while polysaccharides serve for storage of energy and can act as structural components. Furthermore, they have been shown to have other beneficial health effects, including their prebiotic effect and antioxidant or anti-inflammatory activity [58].

In Figure 7 the concentration of total carbohydrates detected in *Gracilaria* is represented. In April, the average carbohydrates content was 34.1% dry weight, while in October it was 24.8% dry weight. In July and January intermediate concentrations were found (27.5% and 31.1% dry weight respectively). Values showed significant differences in all the tested samples (Figure 7). The seasonal variation and the carbohydrate contents that were found in *Gracilaria* biomass were very similar to those reported by Marinho-Soriano and Bourret [59] for the same species collected in the Thau lagoon (France).

Figure 7. Total carbohydrate content (% dry weight biomass) in *G. gracilis* extract. Different superscripts (a–d) indicate significant difference ($p < 0.05$).

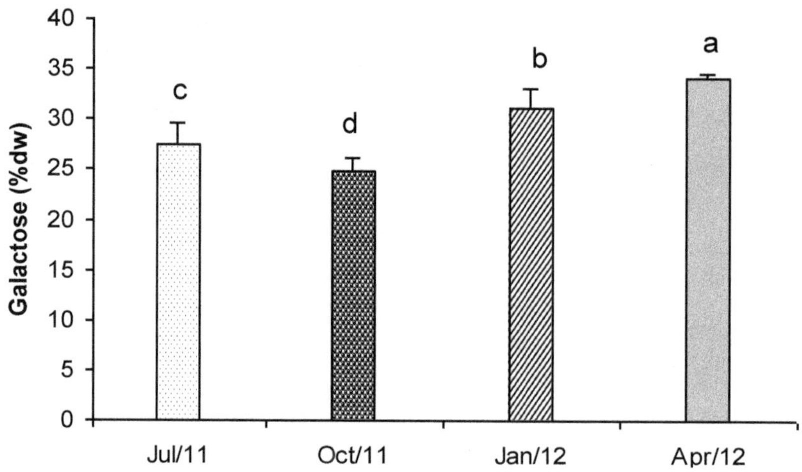

3. Experimental Section

3.1. Biomass Sampling

Gracilaria gracilis was collected from the western area of the Lesina lagoon, where a stable coverage of this seaweed was found (41.866470° N, 15.363350° E). About 3 Kg of wet biomass were sampled in July and October 2011 and January and April 2012, respectively. Algal biomass was washed with distilled water and their epiphytes removed. The fresh seaweed was placed in a freezer (−20 °C) immediately after collection. The cleaned seaweed was freeze-dried at −110 °C for 3 days and then ground to fine powder and stored in airtight containers at −20 °C.

3.2. Lipid Extraction

Lipids from the macroalgal pellets were extracted according to Bligh and Dyer [60]. Briefly, 0.5 g of freeze-dried algal biomass were homogenized for 2 min at 12,000 rpm in a mixture of chloroform and methanol (2:1, v/v, 25 mL) using a Kinematica GmgH homogenizer. The mixture was centrifuged and the solid residue re-suspended in a chloroform/methanol mixture (25 mL). Homogenization continued for a further 2 min and the solution was centrifuged again and washed with fresh solvent (25 mL). The combined supernatants were cleaned using a saturated NaCl water solution and the chloroform phase was recovered using a separatory funnel. The water phase was extracted several times with chloroform until the organic solvent was colourless. The chloroform phase was combined and dried with sodium sulphate overnight. The solvent was removed on a rotary evaporator (Büchi Rotavapor) and the purified lipids were weighed.

3.3. Saponification of Lipids for Sterols Analysis

Lipids (50 mg under N_2) were saponified by refluxing in 20 mL of a 5% (w/v) KOH methanol/water (4:1, v/v) solution for 2 h. The refluxed mixture was then transferred into a separatory funnel, and the reflux bottle was washed with 40 mL of Milli-Q water. The unsaponified matter in the combined solution was then extracted four times with 20 mL of *n*-hexane. The hexane phases were then combined, dried with sodium sulphate overnight, filtered and evaporated. Sterols content in the unsaponified fraction was analysed by means of GC-MS.

3.4. Sterols Analysis by Gaschromatography-Mass Spectrometry (GC-MS)

Purified sterols fraction of *Gracilaria* biomass was analysed by gas chromatography-mass spectrometry as previously described by Francavilla *et al.* [61]. A Varian Saturn 2200 GC/MS/MS ion trap (Varian Analytical Instruments, Walnut Creek, CA, USA) was used. The GC/MS was equipped with a Varian 3800 CP gas chromatograph (GC). A CP-Sil-8 (30 m × 250 μm × 0.25 μm) fused silica column (Varian Analytical Instruments, Walnut Creek, CA, USA) was installed in the GC and helium was used as carrier gas at a constant flow rate of 1.0 mL min^{-1}.

3.5. Fatty Acids Methyl Esters (FAMEs) Analysis

Hydroalcolic residue of extraction process of unsaponified fraction was acidified until pH 1 by adding HCl 2 M and extracted 3 times with 25 mL of *n*-hexane. The hexane phases were then combined, dried with sodium sulphate anhydrous, filtered and evaporated. The yellowish viscous liquid obtained (Fatty Acids) was treated with 10 mL of freshly prepared 5% methanolic HCl in a pyrex screw-topped tube. This mixture was carefully mixed and the tube was closed under nitrogen and then heated for 2 h in an oven at 60 °C. After cooling to room temperature, 5 mL of 6% aqueous K_2CO_3 were added and the mixture was extracted 4 times with 20 mL of *n*-hexane. The hexane phases were then combined, dried with sodium sulphate anhydrous, filtered and evaporated. The process yielded a yellowish viscous liquid (Fatty Acids Methyl Esters) that was weighed and stored at −20 °C until GC-MS analysis.

FAMEs were analysed by means of the same GC-MS/MS equipment described above for sterols analysis, but with a different column temperature regime: 140 °C for 5 min, followed by a 2 °C/min ramp up to 240 °C, followed by 5 min at 240 °C. The injector temperature was 250 °C and the injected volume was 0.2 μL.

FAME peaks were identified by comparison of their retention times and mass spectra with those of a standard mixture (PUFA C4-C24, Supelco, Bellafonte, PA, USA) whereas they were quantified by using calibration curves made with PUFA C4-C24 standard (Supelco, Bellafonte, PA, USA) and using C15:0 FAME (Supelco, Bellafonte, PA, USA) as internal standard.

3.6. Protein Extraction and Analysis

An amount of 50 mg of freeze dried algal sample mixed with 50 mg of alumina was ground manually with pestle and mortar. The mixture was suspended in 5 mL of lysis buffer (0.1 N NaOH, 0.05 M EDTA, 2% SDS, 2% β-mercaptoethanol) [62] for 20 min. After the incubation with buffer, samples were ground for 5 min using a Potter homogeniser (Marconi, model MA099) and kept at room temperature. The mixture was transferred in a centrifuge glass tube and 5 mL of lysis buffer were added to rinse the homogeniser and recover all water-ground material. After this step, samples were centrifuged at 21 °C, 15,000× *g* for 20 min. Supernatants were collected for protein assay. The final volume of the extract was 10.0 mL. The extract was diluted (1:100) in MilliQ water for spectrophotometric analysis according to the Lowry method [63]. The spectrophotometer was blanked with MilliQ water to which was added lysis buffer in the same dilution rate of the samples. Calibration curves were prepared using bovine serum albumin (BSA) (Sigma Co., St. Louis, MO, USA) at the maximum concentration of 100 μg mL^{-1}.

3.7. Phycobiliprotein Extraction and Analysis

An amount of 0.5 g of freeze dried algal sample mixed with 0.5 g of alumina was ground manually with pestle and mortar. The mixture was suspended in 10 mL of 1 M acetic acid–sodium acetate buffer (pH 5.5) with 0.01% of sodium azide for 30 min in the dark. After the incubation with buffer, samples were ground for 5 min using a Potter homogeniser (Marconi, model MA099). The mixture was transferred to a centrifuge glass tube centrifuged at 5 °C, 15,000× *g* for 20 min. Supernatant was collected and the pellet was extracted three times again with buffer as described. Supernatants were combined and the final volume of the extract was about 40.0 mL. Phycobiliproteins (identified as R-phycoerythrin R-PE, phycocyanin PC and allophycocyanin APC) were quantified by spectrophotometry according to Kursar *et al.* [64].

The absorbance of the aqueous supernatants was determined at 498.5, 614.0, and 651.0 nm and their APC, PC, and R-PE contents were calculated as μg mL^{-1} using the following Equations (1):

$$\begin{aligned} \text{APC} &= 181.3A_{651} - 22.3A_{614} \\ \text{PC} &= 151.1A_{614} - 99.1A_{651} \\ \text{R-PE} &= 155.8A_{498.5} - 40.0A_{614} - 10.5A_{651} \end{aligned} \tag{1}$$

3.8. Fractionated Extraction of Algal Biomass

The extraction procedure of Hajimahmoodi *et al.* [65] was performed. Four solvents with different polarity were used to perform the extractions. They were conveniently selected in order to perform selective extraction of compounds (or better, group of compounds) with different polarities, according to the principle "*similia solunt similibus*". Briefly, a precisely weighed amount (2 g) of ground dried algal biomass was homogenized at 17,000 rpm, using a Kinematica GmgH homogenizer, twice each time with 20 mL of *n*-hexane at room temperature for 1 min followed by centrifugation. Further extractions in a similar manner were performed sequentially on the pellet using ethyl acetate, methanol and then deionized water (at 80 °C). The isolated supernatants for each solvent were combined. Hexane, ethyl acetate and methanol extracts were evaporated under vacuum and the dried samples were weighed, dissolved in 1 mL of methanol and stored at −20 °C until analysis. The water extract was freeze dried, weighed and dissolved in 5 mL of methanol. The four extracts were used for antioxidant activity assay and analysis of the total phenolic content.

3.9. Antioxidant Activity Assays

3.9.1. Reducing Power: The FRAP Assay

The FRAP (ferric-reducing antioxidant power) assay is based on the reduction of a ferric-tripyridyl triazine complex to its ferrous-blue-coloured form in the presence of antioxidants [66]. Briefly, the FRAP reagent contained 5 mL of a TPTZ (2,4,6-tripyridyl-*S*-triazine) solution (10 mmol L^{-1}) in HCl (40 mmol L^{-1}) plus 5 mL of $FeCl_3$ (20 mmol L^{-1}) and 50 mL of acetate buffer (0.3 mol L^{-1}, pH 3.6). It was freshly prepared and warmed to 37 °C. A 100 μL sample of each extract (hexane, ethyl acetate, methanol and water) was mixed with 3 mL of FRAP reagent and the absorbance of the reaction mixture was measured at 593 nm after incubation at 37 °C for 10 min. The results could be expressed in micromole of Fe (II), (+) catechin, vitamin C, Trolox or BHT equivalent [67]. For the present study, the standard curve was constructed using Trolox solution (1–30 μM) and the results were expressed as μmol Trolox g^{-1} dry weight of extract. Butylated hydroxytoluene (BHT) whose concentration was 0.1 mg mL^{-1} was used as positive control.

3.9.2. ABTS Assay

The antioxidant activities of seaweed extracts were also measured by using the ABTS [2-2′-azino-bis (3-ethylbenz-thiazoline-6-sulfonic acid)] assay as described by Matanjun *et al.* [52]. ABTS is a chromogen (colourless) that would be converted to blue-green coloured $ABTS^+$ radical cation by an oxidative reagent. $ABTS^+$ could also be reduced to its colourless form by antioxidant. The absorbance was measured spectrophotometrically at 645 nm as a function of concentration and the scavenging percentage of $ABTS^+$ was calculated relative to Trolox, a water-soluble analogue of vitamin E used as an antioxidant standard.

Briefly, $ABTS^+$ radical cation was generated by a reaction of 7 mM sample with 2.45 mM potassium persulphate. The reaction mixture was allowed to stand in the dark for 16 h at room temperature. A working solution was prepared by diluting 3 mL of ABTS stock solution with 57 mL phosphate buffered saline (PBS, 5 mM, pH 7.4). A volume of 2 mL of this working solution was dispensed to test tubes. Addition of 200 μL diluted methanolic extracts (10 μL of extract was diluted with 190 μL of methanol, dilution 1:20) initiated the reaction and absorbance was read after exactly 6 min. Gallic acid (GA) at the concentration of 0.1 mg mL^{-1} was used as positive control. The results were expressed as mmol Trolox g^{-1} dry weight of extract.

3.9.3. DPPH Radical Scavenging Activity

This activity was detected by the method of Hu *et al.* [68]. An aliquot of each sample (30 μL, 0.5–30 mg/mL) in acetone/MeOH (1/1, v/v) was mixed with 200 μL of 100 μM DPPH (2,2-diphenyl-2-picrylhydrazyl hydrate) prepared with methanol. The mixture was shaken vigorously and then left to stand at room temperature for 60 min in the dark. The absorbance was measured spectrophotometrically at 520 nm against an acetone/MeOH (1/1, v/v) blank. The lower absorbance indicated the stronger scavenging activity. EC_{50} value (mg extract mL^{-1}), the effective concentration at which 50% of the DPPH radicals were scavenged, was obtained from the plot of scavenging activity against the concentration of extract. The scavenging activity was calculated based on the percentage of DPPH radical scavenged using the follow Equation (2) [50]:

$$\text{Scavenging effect (\%)} = [1 - (A_{\text{sample}} - A_{\text{sample blank}})/A_{\text{control}}] \times 100 \quad (2)$$

where the A_{control} is the absorbance of the control (DPPH solution without extract), the A_{sample} is the absorbance of the test extract (DPPH solution plus test extract), and the $A_{\text{sample blank}}$ is the absorbance of the extract only (extract without DPPH solution). Butylated hydroxytoluene (BHT), a synthetic antioxidant, was used as positive control.

3.10. Determination of Total Phenolic Content

Total phenolics were colorimetrically determined using Folin-Ciocalteu reagent as described by Velioglu *et al.* [69] with slight modifications. The extract (200 μL) was mixed with 1.5 mL of Folin-Ciocalteu reagent (previously diluted tenfold with distilled water) and allowed to stand at room temperature for 5 min. A 1.5 mL sodium bicarbonate solution (60 g L^{-1}) was added to the mixture. After incubation for 90 min at room temperature, the absorbance was measured at 750 nm. Total phenolics were quantified by a calibration curve obtained from measuring the absorbance of known concentrations of gallic acid standard solutions (25–150 μg mL^{-1} in 50% methanol). The results were calculated as gallic acid equivalent (GAE) g^{-1} dry weight of extract.

3.11. Total Carbohydrates

Total carbohydrate content was assayed by the phenolsulphuric acid method [70] after extraction with 2.5 N HCl for 3 h at 100 °C. The results were calculated from a galactose standard curve and reported as % dry weight of algal biomass.

3.12. Statistical Analyses

All the experiments were repeated three times. Unless otherwise stated, all data were expressed as mean ± standard deviation (SD). The means of all the parameters were examined for significance by analysis of variance (ANOVA) using the software JMP version 9 (SAS Institute Inc., Cary, NC, USA). When F values showed significance, individual means were compared using Tukey's honest significant difference (HSD). Significant differences were considered when $p < 0.05$.

4. Conclusions

Gracilaria gracilis of Lesina Lagoon is a natural marine biomass which has already been found interesting for several applications including agar, as mesoporous material and for bio-oils production. The results of this study highlight that this marine biomass seems to be also a promising source of R-phycoerythrin if it is harvested in winter time. In the same season, the red seaweed also contains a high amount of other valuable products including arachidonic acid (PUFA ω-6), proteins and carbohydrates. High antioxidant and radical scavenging activity, comparable to that of commercial antioxidants compounds, was found in biomass harvested in summer time when the highest concentration of total phenols was also detected. However, more investigations are required for separating, purifying and characterizing these bioactive compounds. Furthermore, it will be necessary to deepen the knowledge of the biology of this macroalga in the Lesina Lagoon, by considering the life cycle and the reproductive peaks to optimize the exploitation of the species and to safeguard its presence in the lagoon.

In conclusion, this study pointed out the possibility of using *Gracilaria gracilis* as a *multi products source* for biotechnological, nutraceutical and pharmaceutical applications according to the new concept of "Biorefinery". Nevertheless, it will be very important to investigate the economic and environmental affordability and sustainability of *Gracilaria* natural resource by means of Life Cycle Assessment.

Acknowledgments

Matteo Francavilla gratefully acknowledges the European Commission, Directorate-General for Research & Innovation, for funding the project "STAR AgroEnergy" (FP7 Regpot 2011-1, Grant Agreement No. 286269) as well as Fondazione Cassa di Risparmio di Puglia for funding the project MADLENA.

References

1. Molinski, T.F.; Dalisay, D.S.; Lievens, S.L.; Saludes, J.P. Drug development from marine natural products. *Nat. Rev. Drug Discov.* **2009**, *8*, 69–85.
2. Lordan, S.; Ross, R.P.; Stanton, C. Marine bioactives as functional food ingredients: Potential to reduce the incidence of chronic diseases. *Mar. Drugs* **2011**, *9*, 1056–1100.
3. Rasmussen, R.S.; Morrissey, M.T. Marine biotechnology for production of food ingredients. *Adv. Food Nutr. Res.* **2007**, *52*, 237–292.
4. De Almeida, C.L.F.; de S. Falcão, H.; de M. Lima, G.R.; de A. Montenegro, C.; Lira, N.S.; de Athayde-Filho, P.F.; Rodrigues, L.C.; de Souza, M.F.V.; Barbosa-Filho, J.M.; Batista, L.M. Bioactivities from marine algae of the genus *Gracilaria*. *Int. J. Mol. Sci.* **2011**, *12*, 4550–4573.
5. Food and Agriculture Organization of the United Nations. In *Year Book of Fishery and Aquaculture Statistics (2006)*; Food and Agricultural Organisation of the United Nations: Rome, Italy, 2008; Volumn 98, p. 57.
6. Bourgougnon, N.; Stiger-Pouvreau, V. Red and Brown Macroalgae Along the French Coasts, Metropole and Overseas Departements and Territories. In *Handbook of Marine Macroalgae: Biotechnology and Applied Phycology*, 1st ed.; Se-Kwon, K., Ed.; JohnWiley & Sons, Ltd.: New Delhi, India, 2012; pp. 58–105.
7. Ghosh, R.; Banerjee, K.; Mitra, A. Eco-Biochemical Studies of Common Seaweeds in the Lower Gangetic Delta. In *Handbook of Marine Macroalgae: Biotechnology and Applied Phycology*, 1st ed.; Se-Kwon, K., Ed.; JohnWiley & Sons, Ltd.: New Delhi, India, 2012; pp. 45–57.

8. Smit, A.J. Medicinal and pharmaceutical uses of seaweed natural products: A review. *J. Appl. Phycol.* **2004**, *16*, 245–262.
9. Kadam, S.; Prabhasankar, P. Marine foods as functional ingredients in bakery and pasta products. *Food Res. Int.* **2010**, *43*, 1975–1980.
10. Capo, T.R.; Jaramillo, J.C.; Boyd, A.E.; Lapointe, B.E.; Serafy, J.E. Sustained high yields of *Gracilaria* (Rodophyta) grown in intensive large-scale culture. *J. Appl. Phycol.* **1999**, *11*, 143–147.
11. Kerton, F.M.; Liu, Y.; Omaria, K.W.; Hawboldt, K. Green chemistry and the ocean-based biorefinery. *Green Chem.* **2013**, *15*, 860–871.
12. Trotta, P. On the Rhodophyta *Gracilaria confervoides* Grev. in Lesina lagoon: Field Survey and *in Vitro* Culture. In Proceedings of the International Congress Phytodepuration and Biomass Utilization, Parma, Italy, 15–16 May 1981; pp. 91–96.
13. D'Adamo, R.; Cecere, E.; Fabbrocini, A.; Petrocelli, A.; Sfriso, A. The Lagoons of Lesina and Varano. In *Flora and Vegetation of the Italian Transitional Water Systems*; Cecere, E., Petrocelli, A., Izzo, G., Sfriso, A., Eds.; Spinea, CORILA, Multigraph: Venezia, Italy, 2009; pp. 159–171.
14. Trotta, P. La pesca, la qualità delle acque, gli scarichi in laguna. In Proceedings of the Conference La laguna di Lesina: Quali prospettive, Quali soluzioni, Lesina (FG), Italy, 11 December 1994; pp. 11–21.
15. Manini, E.; Fiordelmondo, C.; Gambi, C.; Pusceddu, A.; Danovaro, R. Benthic microbial loop functioning in coastal lagoons: A comparative approach. *Oceanol. Acta* **2003**, *26*, 27–38.
16. Nonnis Marzano, C.; Scalera Liaci, L.; Franchini, A.; Gravina, F.; Mercurio, M.; Corriero, G. Distribution, persistence and change in the macrobenthos of the lagoon of Lesina (Apulia, southern Adriatic Sea). *Oceanol. Acta* **2003**, *26*, 57–66.
17. Francavilla, M. Riabilitazione delle acque lagunari salmastre attraverso la gestione di praterie di macroalghe di valore commerciale. Ph.D. Thesis, Università di Foggia, Facoltà di Agraria, Foggia, Italy, 2007.
18. Francavilla, M.; Pineda, A.; Lin, C.S.K.; Franchi, M.; Trotta, P.; Romero, A.A.; Luque, R. Natural porous agar materials from macroalgae. *Carbohydr. Polym.* **2013**, *92*, 1555–1560.
19. Budarin, V.L.; Zhao, Y.; Gronnow, M.J.; Shuttleworth, P.S.; Breeden, S.W.; Macquarrie, D.J.; Clark, J.H. Microwave-mediated pyrolysis of macro-algae. *Green Chem.* **2011**, *13*, 2330–2333.
20. Dawczynski, C.; Schubert, R.; Jahreis, G. Amino acids, fatty acids, and dietary fibre in edible seaweed products. *Food Chem.* **2007**, *103*, 891–899.
21. Bocanegra, A.; Bastida, S.; Benedí, J.; Ródenas, S.; Sánchez-Muniz, F.J. Characteristics and nutritional and cardiovascular-health properties of seaweeds. *J. Med. Food* **2009**, *12*, 236–258.
22. Bernal, J.; Mendiola, J.A.; Ibañez, E.; Cifuentes, A. Advanced analysis of nutraceuticals. *J. Pharm. Biomed. Anal.* **2011**, *55*, 758–774.
23. Burtin, P. Nutritional value of seaweeds. *Electron. J. Environ. Agric. Food Chem.* **2003**, *2*, 498–503.
24. Mišurcová, L. Chemical Composition of Seaweeds. In *Handbook of Marine Macroalgae: Biotechnology and Applied Phycology*, 1st ed.; Se-Kwon, K., Ed.; JohnWiley & Sons, Ltd.: New Delhi, India, 2012; pp. 181–182.

25. Gressler, V.; Yokoya, N.Y.; Fujii, M.T.; Colepicolo, P.; Mancini Filho, J.; Pavan Torres, R.; Pinto, E. Lipid, fatty acid, protein, amino acid and ash contents in four Brazilian red algae species. *Food Chem.* **2010**, *120*, 585–590.
26. Benatti, P.; Peluso, G.; Nicolai, R.; Calvani, M. Polyunsaturated fatty acids: Biochemical, nutritional and epigenetic properties. *J. Am. Coll. Nutr.* **2004**, *23*, 281–302.
27. Von Schacky, C. The role of Omega-3 fatty acids in cardiovascular disease. *Curr. Atheroscler. Rep.* **2003**, *5*, 139–145.
28. Maeda, H.; Tsukui, T.; Sashima, T.; Hosokawa, M.; Miyashita, K. Seaweed carotenoid, fucoxanthin, as multi-functional nutrient. *Asia Pac. J. Clin. Nutr.* **2008**, *17*, 196–199.
29. Simopoulos, A.P. Omega-3 fatty acids in inflammation and autoimmune diseases. *J. Am. Coll. Nutr.* **2002**, *21*, 495–505.
30. Simopoulos, A.P. The importance of the omega-6/omega-3 fatty acid ratio in cardiovascular diseases and other chronic diseases. *Exp. Biol. Med.* **2008**, *233*, 674–688.
31. MacArtain, P.; Gill, C.I.R.; Brooks, M.; Campbell, R.; Rowland, I.R. Nutritional value of edible seaweeds. *Nutr. Rev.* **2007**, *65*, 535–543.
32. Fernandes, P.; Cabral, J.M.S. Phytosterols: Applications and recovery. *Bioresour. Technol.* **2007**, *98*, 2335–2350.
33. Platt, D.; Pelled, D.; Shulman, A. Oils Enriched with Diacylglycerols and Phytosterolesters for Use in the Reduction of Cholesterol and Tryclycerides. PCT Patent WO2004069150, 23 September 2004.
34. McDermid, K.J.; Stuercke, B. Nutritional composition of edible Hawaiian seaweeds. *J. Appl. Phycol.* **2003**, *15*, 513–524.
35. Marrion, O.; Fleurence, J.; Schwertz, A.; Guéant, J.-L.; Mamelouk, L.; Ksouri, J.; Villaume, C. Evaluation of protein *in vitro* digestibility of *Palmaria palmata* and *Gracilaria verrucosa*. *J. Appl. Phycol.* **2005**, *17*, 99–102.
36. Renaud, S.M.; Luong-Van, J.T. Seasonal variation in the chemical composition of tropical Australian marine macroalgae. *J. Appl. Phycol.* **2006**, *18*, 381–387.
37. Wen, X.; Peng, C.; Zhou, H.; Lin, Z.; Lin, G.; Chen, S.; Li, P. Nutritional composition and assessment of *Gracilaria lemaneiformis* Bory. *J. Integr. Plant Biol.* **2006**, *48*, 1047–1053.
38. Marinho-Soriano, E.; Câmara, M.R.; Cabral, T.M.; Carneiro, M.A.A. Preliminary evaluation of the seaweed *Gracilaria cervicornis* (Rhodophyta) as a partial substitute for the industrial feeds used in shrimp (*Litopenaeus vannamei*) farming. *Aquac. Res.* **2007**, *38*, 182–187.
39. Vernon, T.O.; Alexander, N.; Glazer, A.N.; Stryer, L. Fluorescent phycobiliprotein conjugates for analyses of cells and molecules. *J. Cell Biol.* **1982**, *93*, 981–986.
40. Ayyagari, M.S.; Pande, R.; Kamtekar, S.; Marx, K.A.; Tripathy, S.K.; Gao, H.; Kumar, J.; Akkara, J.A.; Kaplan, D.L. Molecular assembly of proteins and conjugated polymers: Toward development of biosensors. *Biotechnol. Bioeng.* **1995**, *45*, 116–125.
41. Glazer, A.N. Light harvesting by phycobilisome. *Annu. Rev. Biophys. Biophys. Chem.* **1985**, *14*, 47–77.
42. Li, G.W.; Wang, G.C.; Li, Z.G.; Tseng, C.K. Biological effect of R-phycoerythrin-mediated photosensitization on DNA. *Prog. Biochem. Biolophys.* **2000**, *27*, 621–624.

43. Niu, J.F.; Wang, G.C.; Tseng, C.K. Method for large-scale isolation and purification of R-phycoerythrin from red alga *Polysiphonia urceolata* Grev. *Protein Expr. Purif.* **2006**, *49*, 23–31.
44. Bermejo, R.; Ruiz, E.; Acien, F.G. Recovery of B-phycoerythrin using expanded bed adsorption chromatography: Scale-up of the process. *Enzym. Microb. Technol.* **2007**, *40*, 927–933.
45. Reaven, P.D.; Witzum, J.L. Oxidised LDL in atherogenesis. Role of dietary modification. *Annu. Rev. Nutr.* **1996**, *16*, 51–71.
46. Aruoma, I.O. Antioxidant action of plant foods. Use of oxidative DNA damage, as a tool for studying antioxidant efficacy. *Free Radic. Res.* **1999**, *30*, 419–427.
47. Shon, M.Y.; Kim, T.H.; Sung, N.J. Antioxidants and free radical scavenging activity of *Phellinus baumii* (Phellius of Hymenochaetaceae) extracts. *Food Chem.* **2003**, *82*, 593–597.
48. Kumaran, A.; Karunakaran, R.J. *In vitro* antioxidant properties of methanol extracts of five *Phillanthus* species from India. *Food Sci. Technol.* **2007**, *40*, 344–352.
49. Kuda, T.; Tsunekawa, M.; Goto, H.; Araki, Y. Antioxidant properties of four edible algae harvested in the Noto Peninsula, Japan. *J. Food Compos. Anal.* **2005**, *18*, 625–633.
50. Duan, X.J.; Zhang, W.W.; Li, X.M.; Wang, B.G. Evaluation of antioxidant property of extract and fractions obtained from a red alga, *Polysiphonia urceolata. Food Chem.* **2006**, *95*, 37–43.
51. Ganesan, P.; Chandini, S.K.; Kumar, C.S.; Bhaskar, N. Antioxidant properties of methanol extract and its solvent fractions obtained from selected Indian red seaweeds. *Bioresour. Technol.* **2008**, *99*, 2717–2723.
52. Matanjun, P.; Mohamed, S.; Mustapha, N.M.; Muhammad, K.; Ming, C.H. Antioxidant activities and phenolics content of eight species of seaweeds from north Borneo. *J. Appl. Phycol.* **2008**, *20*, 367–373.
53. Yuan, Y.V.; Bone, D.E.; Carrington, M.F. Antioxidant activity of dulse (*Palmaria palmate*) extract evaluated *in vitro*. *Food Chem.* **2005**, *91*, 485–494.
54. Murugan, K.; Iyer, V.V. Antioxidant and Antiproliferative Activities of Marine Algae, *Gracilaria edulis* and *Enteromorpha lingulata*, from Chennai Coast. *Int. J. Cancer Res.* **2012**, *8*, 15–26.
55. Frankel, E.N.; Meyer, A.S. The problems of using one-dimensional methods to evaluate multifunctional food and biological antioxidants. *J. Sci. Food Agric.* **2000**, *80*, 1925–1941.
56. Prior, R.; Wu, X.; Schaich, K. Standardized methods for the determination of antioxidant capacity and phenolics in foods and dietary supplements. *J. Agric. Food Chem.* **2005**, *53*, 4290–4302.
57. Onofrejová, L.; Vašičková, J.; Klejdusa, B.; Stratil, P.; Mišurcovà, L.; Kràčmar, S.; Kopecký, J.; Vaceka, J. Bioactive phenols in algae: The application of pressurized-liquid and solid-phase extraction techniques. *J. Pharm. Biomed. Anal.* **2010**, *51*, 464–470.
58. Wijesekara, I.; Pangestuti, R.; Kima, S.-K. Biological activities and potential health benefits of sulfated polysaccharides derived from marine algae. *Carbohydr. Polym.* **2011**, *84*, 14–21.
59. Marinho-Soriano, E.; Bourret, E. Effects of season on the yield and quality of agar from *Gracilaria* species (Gracilariaceae, Rhodophyta). *Bioresour. Technol.* **2003**, *90*, 329–333.
60. Bligh, E.G.; Dyer, W.J. A rapid method of total lipid extraction and purification. *Can. J. Biochem. Physiol.* **1959**, *37*, 911–917.
61. Francavilla, M.; Trotta, P.; Luque, R. Phytosterols from *Dunaliella tertiolecta* and *Dunaliella salina*: A potentially novel industrial application. *Bioresour. Technol.* **2010**, *101*, 4144–4150.

62. Von der Haar, T. Optimized protein extraction for quantitative proteomics of yeasts. *PLoS One* **2007**, *2*, e1078.
63. Lowry, O.H.; Rosebrough, N.J.; Farr, A.L.; Randall, R.J. Protein measurement with the Folin phenol reagent. *J. Biol. Chem.* **1951**, *193*, 265–275.
64. Kursar, T.A.; van der Meer, J.; Alberte, R.S. Light harvesting system of red alga *Gracilaria tikvahiae*. I. biochemical analyses of pigment mutations. *Plant Physiol.* **1983**, *73*, 353–360.
65. Hajimahmoodi, M.; Faramarzi, M.A.; Mohammadi, N.; Soltani, N.; Oveisi, M.R.; Nafissi-Varcheh, N. Evaluation of antioxidant properties and total phenolic contents of some strains of microalgae. *J. Appl. Phycol.* **2010**, *22*, 43–50.
66. Benzie, I.F.F.; Strain, J.J. The ferric reducing ability of plasma as a power: The FRAP assay. *Anal. Biochem.* **1976**, *239*, 70–76.
67. Katalinic, V.; Milos, M.; Kulisic, T.; Jukic, M. Screening of 70 medicinal plant extracts for antioxidant capacity and total phenols. *Food Chem.* **2006**, *94*, 550–557.
68. Hu, C.C.; Lin, J.T.; Lu, F.J.; Chou, F.P.; Yang, D.J. Determination of carotenoids in *Dunaliella salina* cultivated in Taiwan and antioxidant capacity of the algal carotenoid extract. *Food Chem.* **2008**, *109*, 439–446.
69. Velioglu, Y.S.; Mazza, G.; Gao, L.; Oomah, B.D. Antioxidant activity and total phenolics in selected fruits, vegetables and grain products. *J. Agric. Food Chem.* **1998**, *46*, 4113–4117.
70. Dubois, M.; Gilles, K.A.; Hamilton, J.K.; Rebers, P.A.; Smith, F. Colorimetric method for determination of sugars and related substances. *Anal. Chem.* **1956**, *28*, 350–356.

9

Identification of Marine Neuroactive Molecules in Behaviour-Based Screens in the Larval Zebrafish

Si-Mei Long [1,†], Feng-Yin Liang [1,†], Qi Wu [1], Xi-Lin Lu [1], Xiao-Li Yao [1], Shi-Chang Li [2], Jing Li [2], Huanxing Su [3], Ji-Yan Pang [2,*] and Zhong Pei [1,*]

[1] Department of Neurology, National Key Clinical Department and Key Discipline of Neurology, Guangdong Key Laboratory for Diagnosis and Treatment of Major Neurological Diseases, The First Affiliated Hospital, Sun Yat-sen University, Guangzhou 510080, China; E-Mails: longsm@mail.sysu.edu.cn (S.-M.L.); liangfy@mail2.sysu.edu.cn (F.-Y.L.); wuqi@medmail.com.cn (Q.W.); goodxilin@163.com (X.-L.L.); liliyao71@163.com (X.-L.Y.)

[2] School of Chemistry & Chemical Engineering, Sun Yat-sen University, Guangzhou 510275, China; E-Mails: lishich@mail2.sysu.edu.cn (S.-C.L.); zsulijing@163.com (J.L.)

[3] State Key Laboratory of Quality Research in Chinese Medicine, Institute of Chinese Medical Sciences, University of Macau, Macao 999078, China; E-Mail: Huanxingsu@umac.mo

† These authors contributed equally to this work.

* Authors to whom correspondence should be addressed; E-Mails: peizhong@mail.sysu.edu.cn (Z.P.); cespjy@mail.sysu.edu.cn (J.-Y.P.)

Abstract: High-throughput behavior-based screen in zebrafish is a powerful approach for the discovery of novel neuroactive small molecules for treatment of nervous system diseases such as epilepsy. To identify neuroactive small molecules, we first screened 36 compounds (**1–36**) derived from marine natural products xyloketals and marine isoprenyl phenyl ether obtained from the mangrove fungus. Compound **1** demonstrated the most potent inhibition on the locomotor activity in larval zebrafish. Compounds **37–42** were further synthesized and their potential anti-epilepsy action was then examined in a PTZ-induced epilepsy model in zebrafish. Compound **1** and compounds **39**, **40** and **41** could significantly attenuate PTZ-induced locomotor hyperactivity and elevation of *c-fos* mRNA in larval zebrafish. Compound **40** showed the most potent inhibitory action against PTZ-induced hyperactivity. The structure-activity analysis showed that the OH group at

12-position played a critical role and the substituents at the 13-position were well tolerated in the inhibitory activity of xyloketal derivatives. Thus, these derivatives may provide some novel drug candidates for the treatment of epilepsy.

Keywords: behavior-based screen; zebrafish; PTZ; *c-fos*

Abbreviations

HT, high-throughput; PTZ, pentylenetetrazol; GABA, γ-Aminobutyric acid; 5-HT, 5-hydroxy-tryptamine; NMDA, *N*-Methyl-D-aspartic acid; CBZ, carbamazepine; DMSO, dimethyl sulfoxide; dpf, days post-fertilization; hpf, hours post fertilization; CNS, central nervous system; IEGs, immediate early genes; EEG, electroencephalography.

1. Introduction

Disorders of the central nervous system (CNS) are very common and devastating. However, CNS diseases are usually poorly treated due to of the limited availability of selective neuroactive drugs. Thus, the development of novel neuroactive drugs is of high priority. It has been very difficult to discover novel neuroactive drugs in the past years [1]. Many current behavior-altering drugs were discovered by chance in the 1940s and 1950s. A major obstacle to the discovery of novel neuroactive drugs is the lack of available relevant model systems for screening large numbers of active compounds. Modeling the brain activity *in vitro* is problematic because of the complex networks of the brain. In addition, the screens in mice and rats are low-throughput due to the expense and ethical issues [2]. Recently, zebrafish has become a powerful model system for whole organism small molecule screening. Zebrafish are small, cheap to keep, fast to develop, and easy to breed. Similar to mammals, zebrafish larvae can display diverse behaviors including the optokinetic response [3], the optomotor response [4], prepulse inhibition [5] and sleep [6,7]. Combined with the video track system, several high-throughput behavior-based assays have been successfully applied to identify novel neuroactive small molecules in the zebrafish [8,9].

The marine habitat is a rich resource for the discovery of new drugs because of its vast chemical and biological diversity. However, most marine-derived lead compounds are stereochemically complex or have low activity. Thus, the appropriate structural modifications of lead compounds are important to develop chemically simple and active drug candidates [10]. In this paper, we conducted a behavior-based screen for neuroactive small molecules on 12 benzopynan compounds derived from natural xyloketals from marine mangrove fungus (NO. 2508) [11] and 24 isoprenyl phenyl ether derivatives modified from marine isoprenyl phenyl ether from Mangrove fungus (NO. B60) [12] (Chart 1). We further modified compound **1** (Chart 2) to study structure-activity relationships and optimize the biological activity of compound **1** derived compounds. Finally, we explored the potential of compound **40** as a new antiepileptic candidate in pentylenetetrazol (PTZ)-induced epilepsy model in zebrafish.

Chart 1. Structures selected for neuroactive screening.

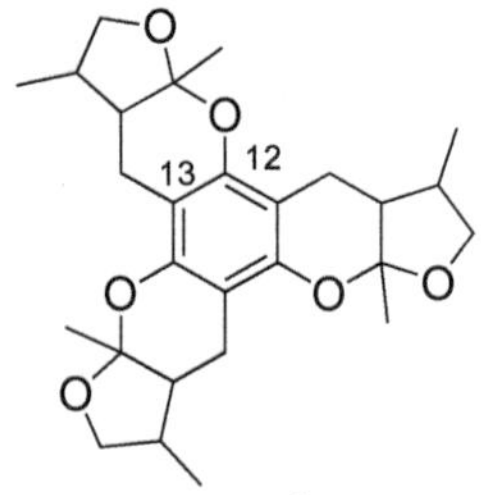
1

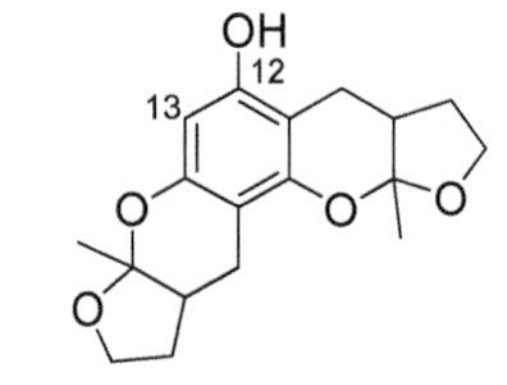
2

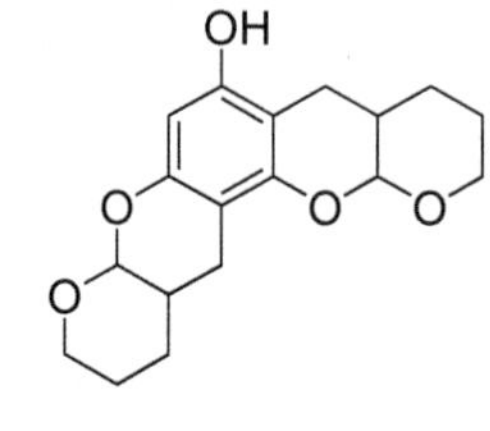
3

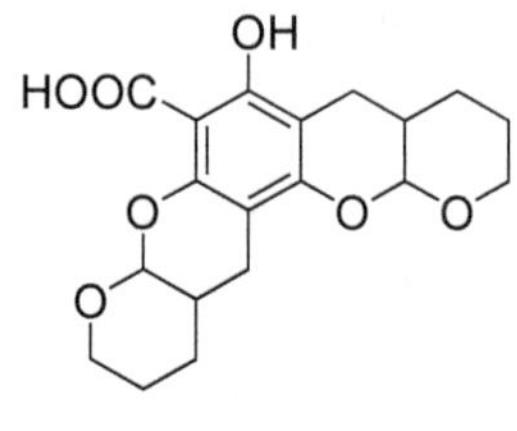
4

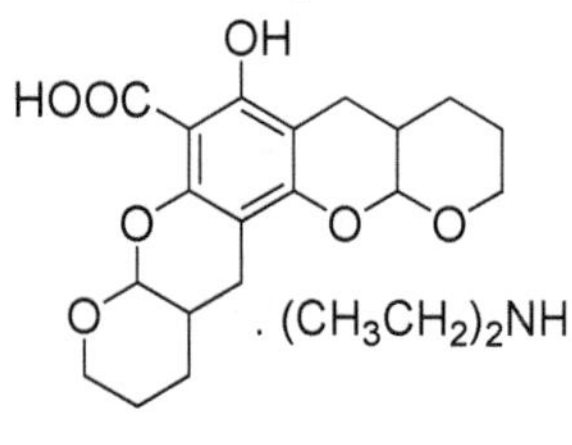
5

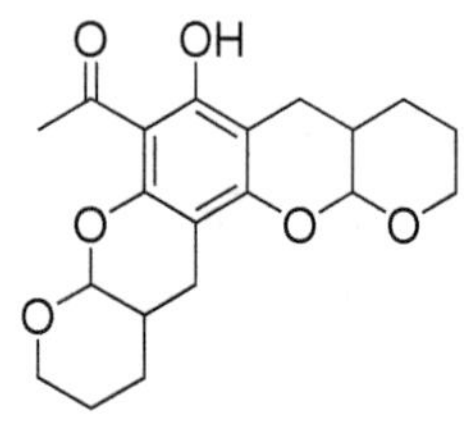
6

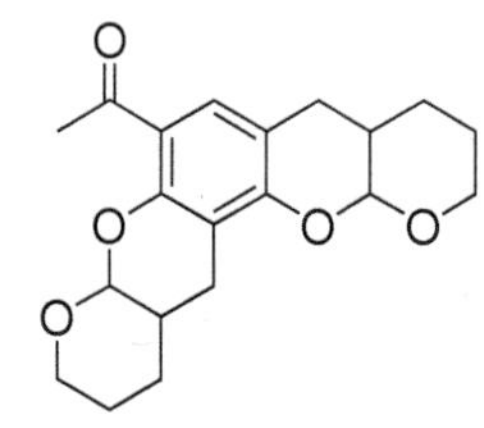
7

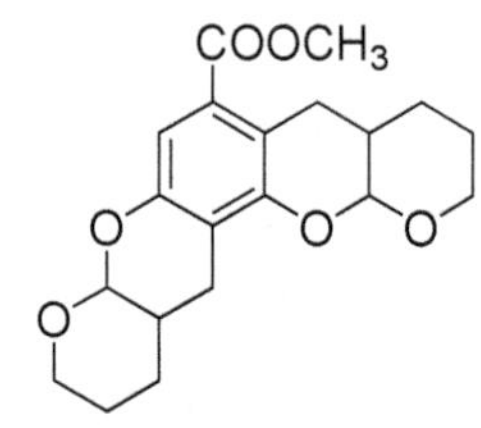
8

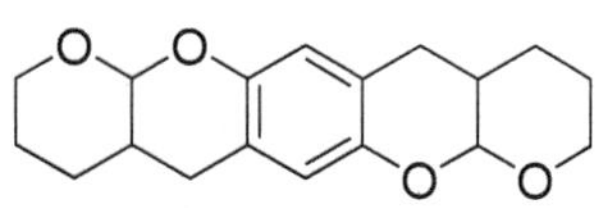
9

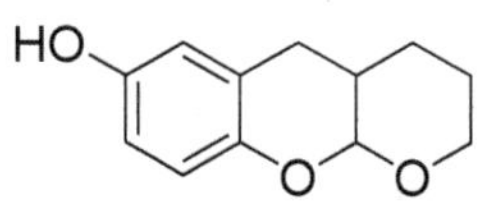
10

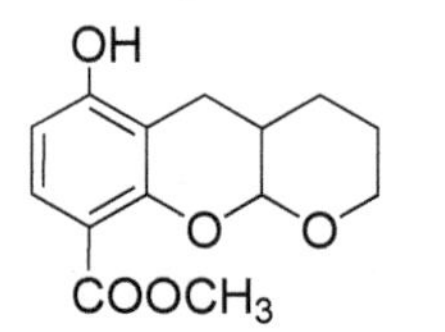
11

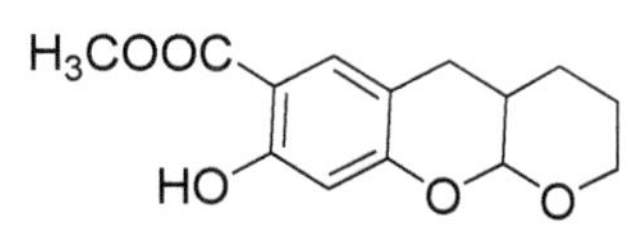
12

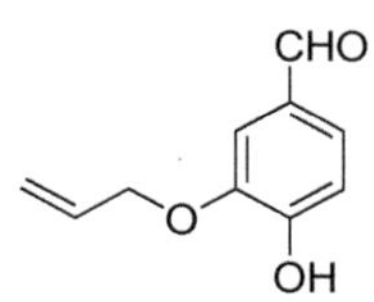
13

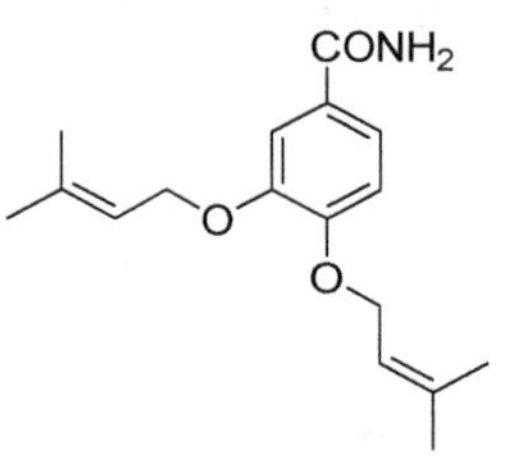
14

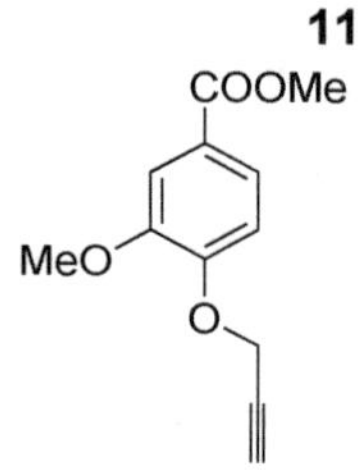
15

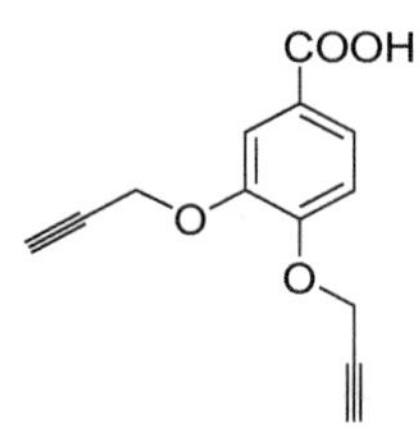
16

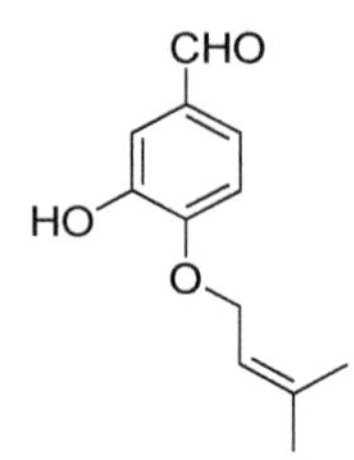
17

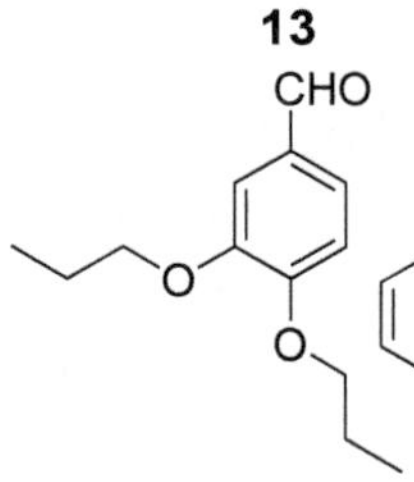
18

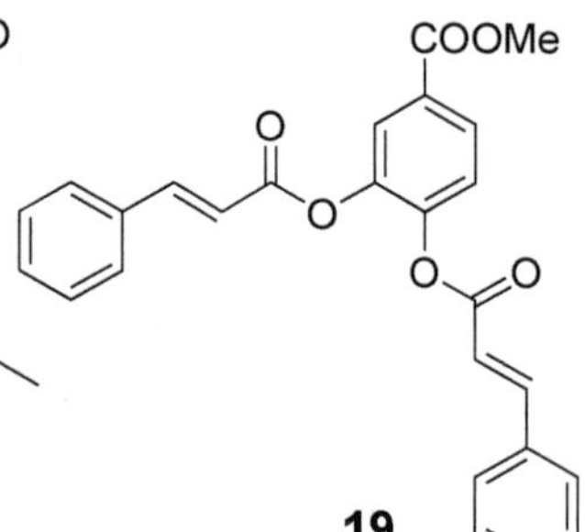
19

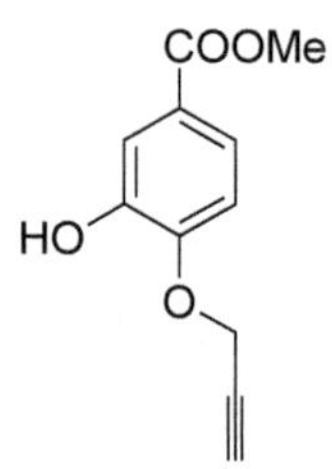
20

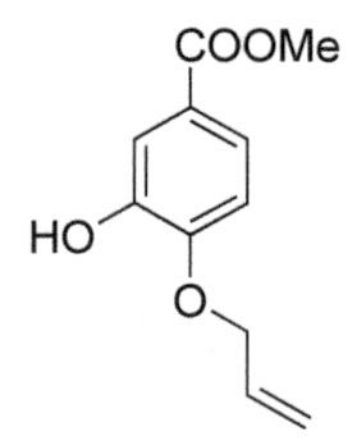
21

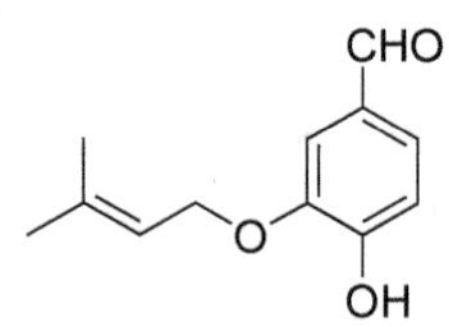
22

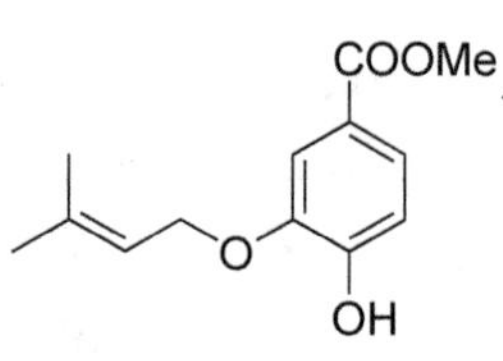
23

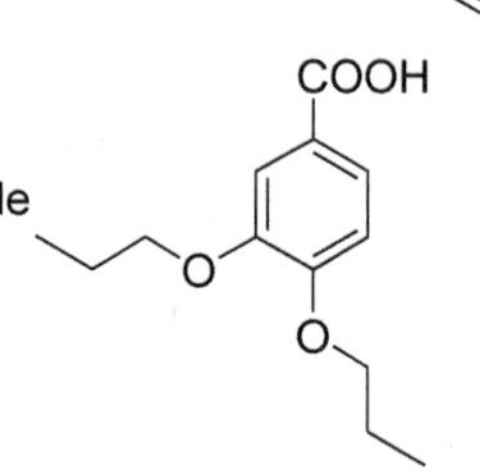
24

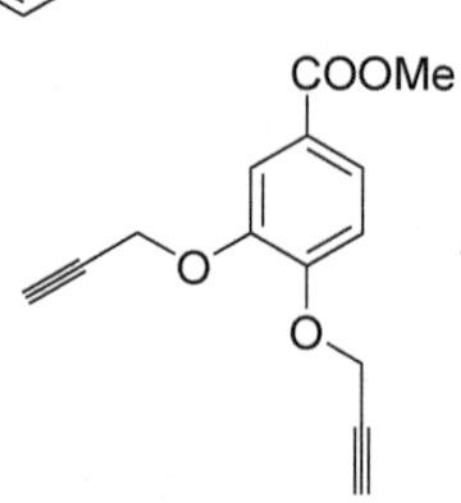
25

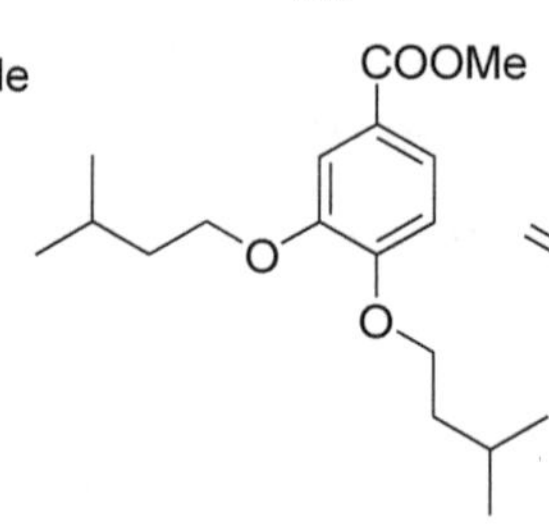
26

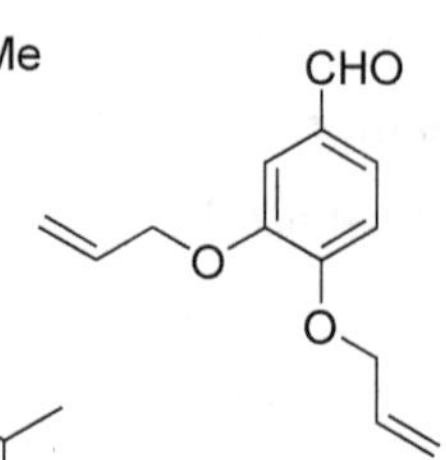
27

Chart 1. *Cont.*

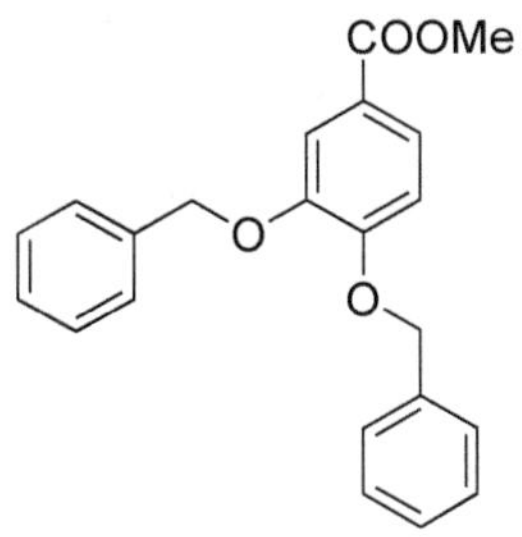

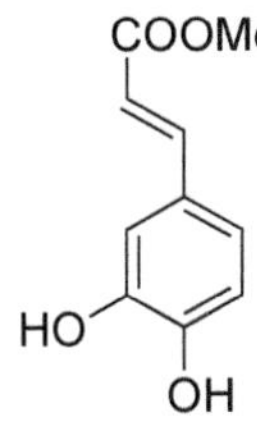

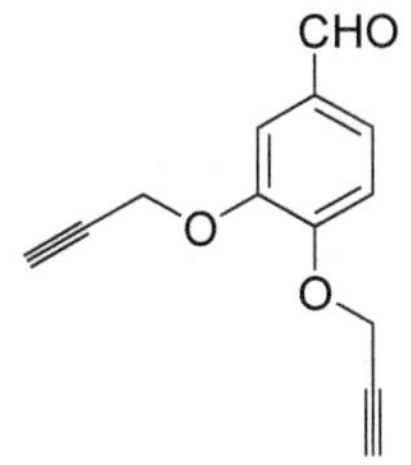

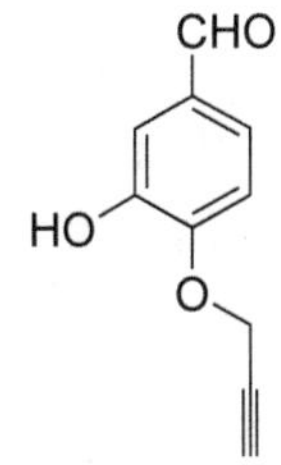

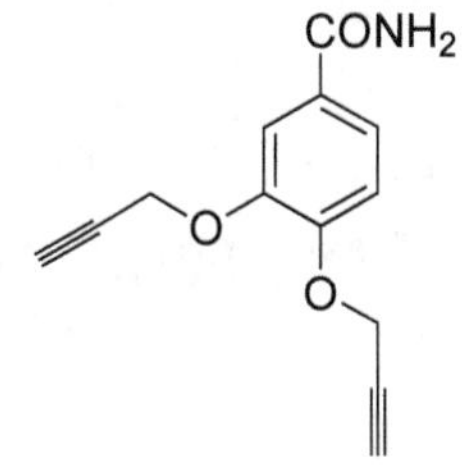

28 **29** **30** **31** **32**

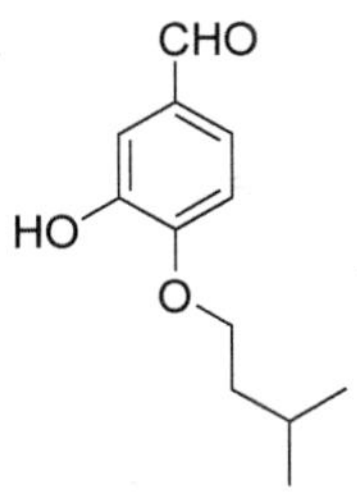

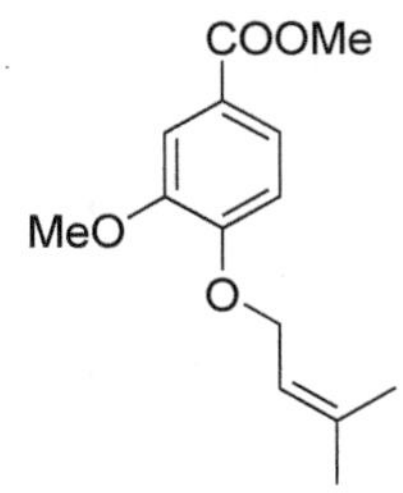

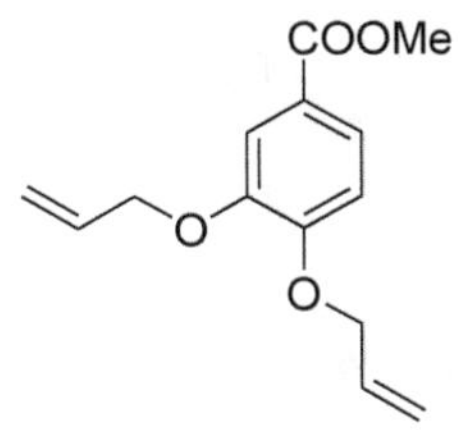

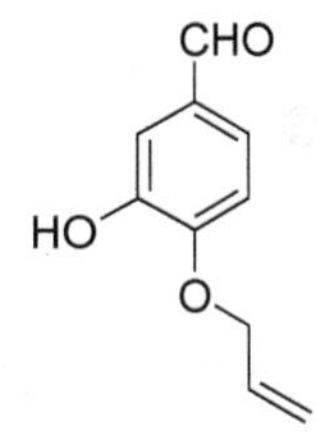

33 **34** **35** **36**

Chart 2. Modification of compound **1.**

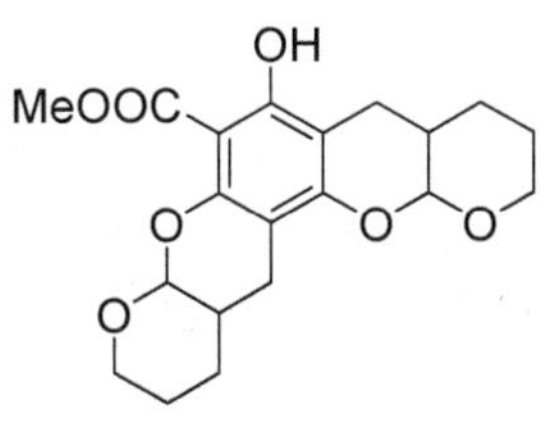

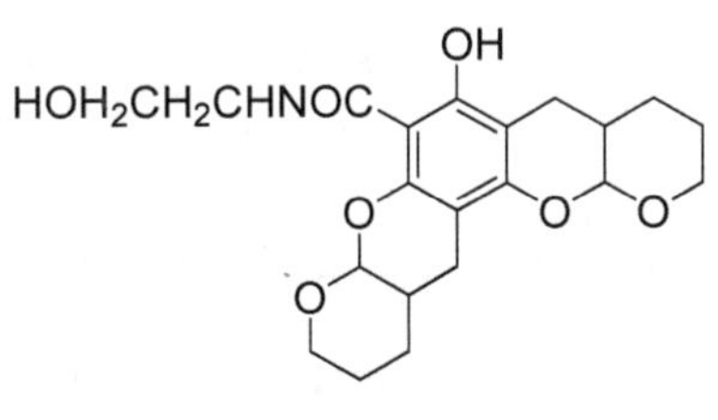

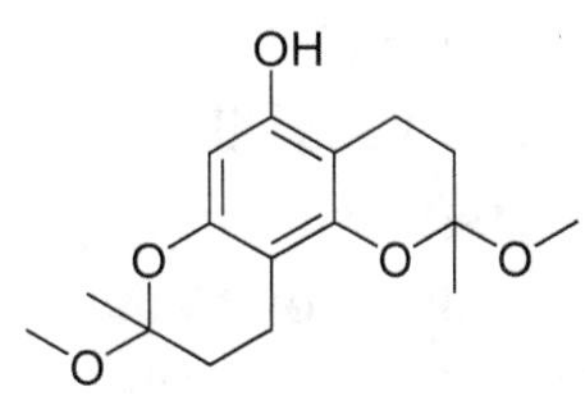

37 **38** **39**

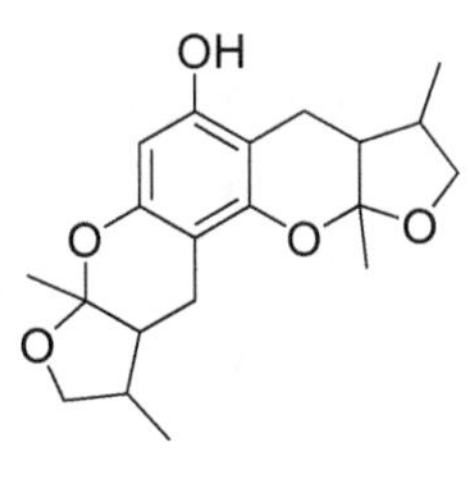

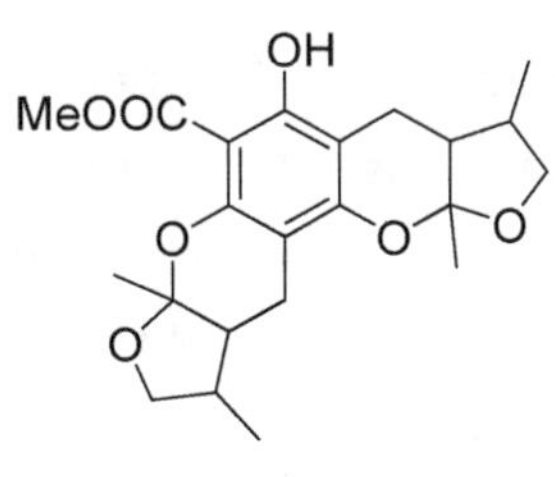

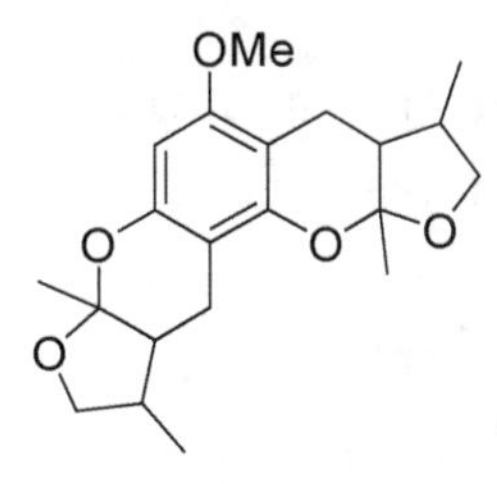

40 **41** **42**

2. Results and Discussion

2.1. Chemistry

Forty-two analogues of the natural xyloketals and isoprenyl phenyl ether were obtained, and the general synthetic routes of compounds **1–42** have been described previously [10,12–14]. Compounds **9–12** were new compounds synthesized by reduction and electrophilic aromatic substitution reactions of 3,4-dihydro-2*H*-pyran-5-carboxylate with different phenols with 29%–52% yield (Scheme 1). The reduction product of 3,4-dihydro-2*H*-pyran-5-carboxylate was unstable and was immediately used for the subsequent experiment. The title compounds were isolated as their bis- and mono- adducts in different phenol proportions. Different substituted compounds of methyl 2,4-dihydroxybenzoate **11** and **12** could be obtained in one pot reaction and detected using TLC, followed by purifying, easily done by flash chromatography. However, the final products were obtained in low yields and further optimizations were required. Compounds **9–12** were fully characterized by HRMS and NMR.

Scheme 1. Synthesis of compounds **9–12**.

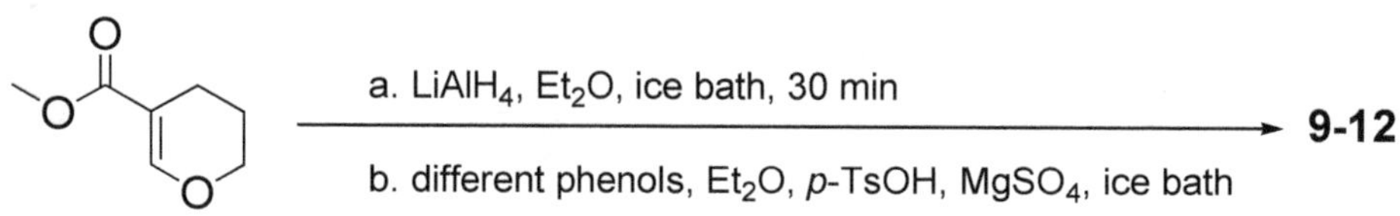

2.2. Neuroactive Activity of 36 Compounds were Evaluated in Zebrafish Behavioral Assay

Zebrafish have neural networks similar to mammals and their locomotor activity is a measurable complex behavior. Similar to higher vertebrates, locomotor activity of zebrafish is also regulated by light. In zebrafish behavioral assays, the total swimming distance traveled is often measured to determine the changes in the locomotor activity. Therefore, we examine the effect of different compounds on the total distance during the initial screen. All marine-derived compounds (**1–36**) were dissolved in DMSO and diluted with E3 buffer (5 mM NaCl, 0.17 mM KCL, 0.33 mM $CaCl_2{\cdot}2H_2O$, 0.33 mM $MgSO_4$) to a final concentration of 20 μM. The results (Figure 1) showed that three compounds including compounds **1**, **2** and **14** significantly inhibited locomotor activity ($p < 0.01$ *vs.* DMSO) to 33%, 40% and 38%, respectively. Meanwhile, several compounds exhibited a hyperactive effect on locomotor activity ($p < 0.01$ *vs.* DMSO). For example, compounds **4**, **5** and **35** could significantly increase locomotor activity by 91%, 84% and 64%, respectively.

Figure 1. The larval zebrafish behavioral assay was performed on 120-hpf zebrafish dosed with compounds at 20 μM concentrations in DMSO. Each group had 24 replicates and three independent experiments were performed. The data of total distance are normalized as the percentage of control and representative of three independent experiments. Data was analyzed using One-way ANOVA followed by Post Hoc test (Bonferroni's Multiple Comparison Test). * $p < 0.05$ *vs.* DMSO, ** $p < 0.01$ *vs.* DMSO, *** $p < 0.001$ *vs.* DMSO.

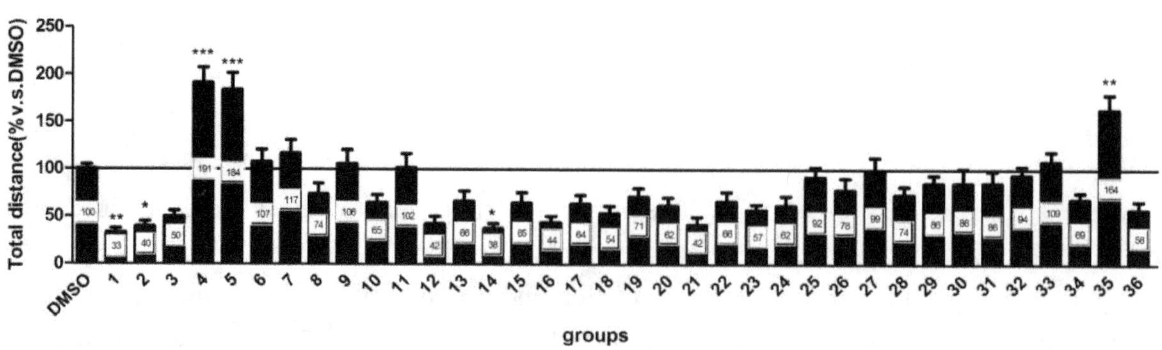

2.3. Neuroactive Compounds Exhibited Different Behavioral Patterns

The behavioral assay used here has been well-characterized. During this assay, zebrafish typically exhibited robust but transient behavioral activity in response to sudden transitions from light to dark [15]. In the present study, we used a modified version of this test consisting of a single transition from light to dark. The basal swimming activity was recorded during 10 min with lights on. Immediately following the basal activity recording, the lights were suddenly turned off for 10 min. Consistent with previous reports, the control animals displayed a normal pattern of locomotor activity, *i.e.*, the activity of zebrafish decreased when the visible light was on (light) whereas the activity of zebrafish rapidly and markedly increased when the light was off (dark). We further analyzed these six neuroactive (compounds **1**, **2**, **4**, **5**, **14**, **35**) compounds obtained from the initial screen. We found that neuroactive compounds exhibited distinct patterns and some neuroactive compounds altered the orderly normal activity pattern (Figure 2). Based on the behavior assay, marine-derived compounds tested could be simply categorized as two major types: hyperactive and hypoactive compounds. The behaviors also varied among different compounds within the same types. For example, among these hyperactive compounds, compounds **4** and **5** showed hyperactive activities which were more prevalent in darkness than in light (Figure 2A,E) whereas compound **35** induced a constant hyperactivity all the time regardless of dark or light conditions (Figure 2B,F). Among hypoactive compounds, compound **2** decreased activities during the first dark and light cycle and activities were then returned to normal in the subsequent cycles whereas compound **14** decreased activity during all the periods (Figure 2C,G). In addition, animals treated with compound **1** decreased activity in every dark and light period (Figure 2D,H).

Figure 2. The behavioral assay of neuroactive compounds. The larvae were placed into the compound solutions (final concentration is 20 μM) and recording began 20 min later in alternating periods of darkness and light for a total duration of 70 min. (**A**–**D**) Total activity (distance moved, cm) in two-minute intervals for total duration of 70 min. (**E**–**H**) Total activity (distance moved, cm) in each 10-minute light and dark period. Values are reported as mean (n = 21–24 larvae/concentration/plate, for 2 plates) ±SEM. * $p < 0.05$ *vs.* DMSO, ** $p < 0.01$ *vs.* DMSO.

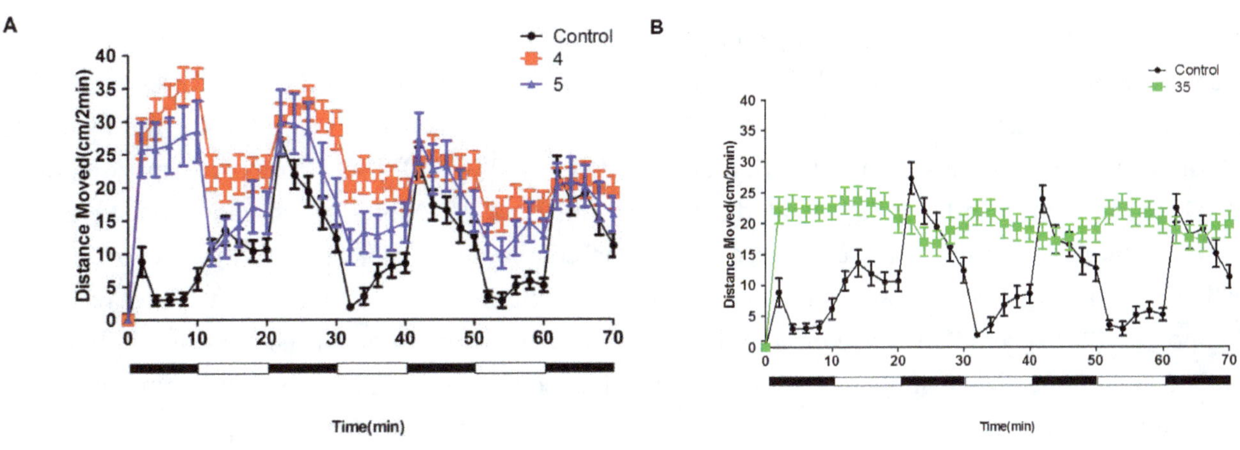

Figure 2. *Cont.*

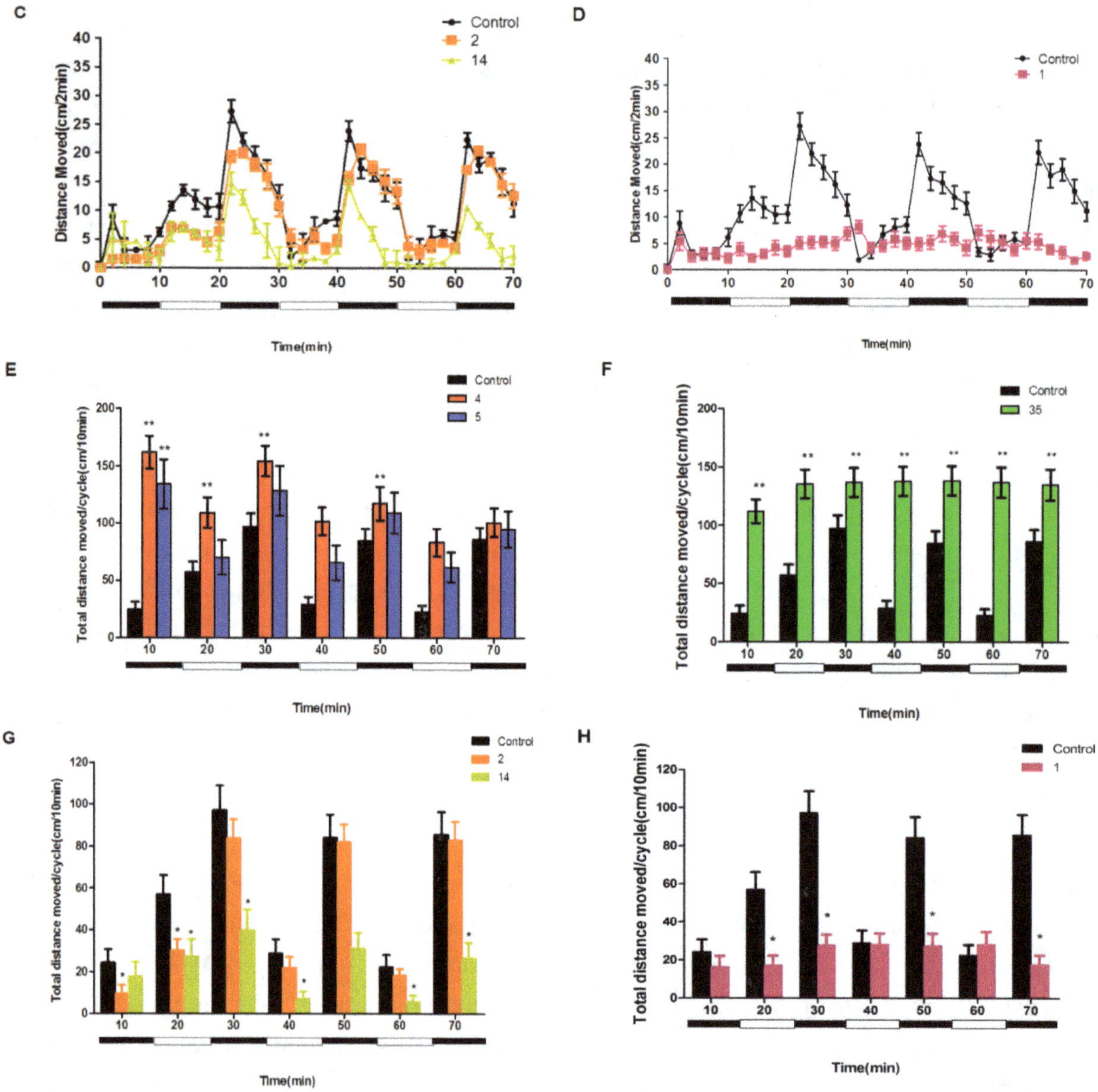

*2.4. Compounds **1** and **37–41** Suppressed Locomotor Activity in Larval Zebrafish*

To further study the structure-activity relationships and optimize biological activity of compound **1**, three compounds (**40–42**) derived from **1** and three more compounds(**37–39**) were synthesized (Scheme 1). The results (Figure 3A) showed that the majority of compounds inhibited the locomotor activity (total distance) at the concentration of 20 μM ($p < 0.05$ *vs.* DMSO). Among them, compounds **37–41** could significantly reduce total distance by up to 57%. Moreover, compounds **40** and **41** displayed more potent inhibition compared to the lead compound **1** (reduced activity level to 27% and 28%, respectively). Six derivatives displayed different patterns of locomotor activity (Figure 3B–D). Compound **42** displayed a reverse effect with a higher activity during all of the dark and light cycles. Similar to compound **1**, compounds **37**, **40** and **41** induced a constant inhibition on locomotor activity throughout every dark and light period. Furthermore, compounds **40** and **41** displayed much lower activity compare to compound **1**. Although compounds **38** and **39** suppressed the locomotor activity, they did not disturb the normal orderly pattern of activity. Altogether, we found that compound **40** showed the most potent inhibitory action on locomotor activity.

Structure–activity investigation was conducted to identify the active components within xyloketal derivatives. We found that xyloketal derivatives possessing the benzopyrano furan skeleton exhibited more potent inhibitory action compared to that possessing benzopyrano pyran. However, methyl ether derivative **42** did not show any suppression of activity compared to controls. Thus, an unsubstituted hydroxyl group or easily cleaved hydroxyl prodrug, such as a phenolic ester at 12-position of the benzopyran scaffold, may be important for the inhibitory action. Furthermore, the substituents at the 13-position may be tolerated but not necessary for the inhibitory activity. However, compound **1** does not possess an OH group at this position perhaps hinting at different binding modes between **1** and **40/41**.

Figure 3. Compounds **1** and **37–41** suppressed larval zebrafish locomotor activity. (**A**) The behavior assay was performed with compounds **37–42**. The relative distance of DMSO group was set as control and other groups were shown relative to DMSO group. (**B–D**) Different activity patterns of compound **1** and compounds **37–42**. * $p < 0.05$ *vs.* DMSO, *** $p < 0.001$ *vs.* DMSO.

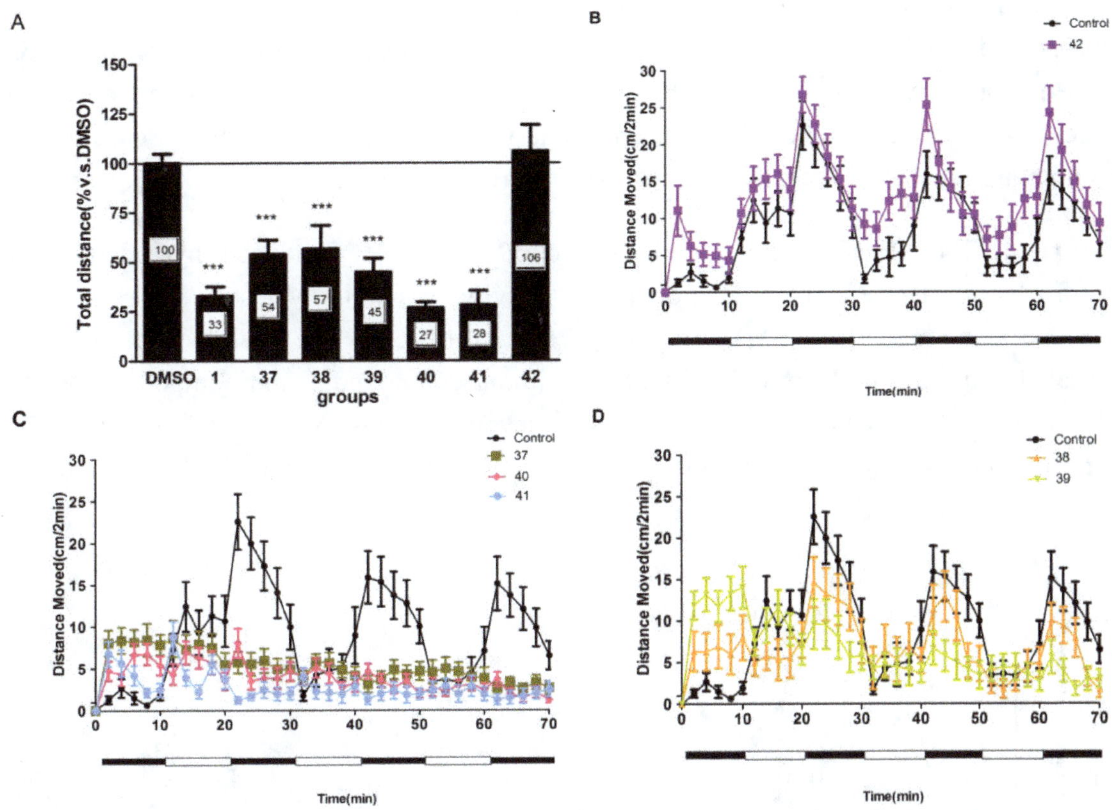

2.5. The Potential of Selected Compounds for New Antiepileptic Drug Development in PTZ-Induced Seizure Model

Epilepsy is one of the most common CNS disorders and affects about 50 million people worldwide [16]. A variety of pharmacological and genetic models of epilepsy have been developed to study seizure mechanisms and to identify new anti-epileptic drugs. PTZ is a noncompetitive antagonist of the GABA receptor complex and its epileptogenic properties have been widely used for anti-epileptic drug discovery. Similarly, PTZ can induce measurable seizure-related behaviors in zebrafish. These behaviors can be reversed by known anti-epileptic drugs. Thus, zebrafish are emerging as a useful model system for anti-epileptic drug discovery.

The potential of compounds **1** and **39–41** as anti-epileptic drugs was further explored in a PTZ-induced epilepsy model in zebrafish. Consistent with previous reports [17], PTZ at 10 mM induced a robust locomotor activity (hyperactivity) in zebrafish larvae (Figure 4A). In contrast, all three compounds (**39–41**) demonstrated significant inhibitory action (hypoactivity) in this PTZ model. Among compounds tested, compound **40** showed the most potent inhibitory activity (Figure 4B).

PTZ induced changes not only in activity but also in gene expression. The expression levels of IEGs are well correlated with seizure activity. The expression of *c-fos* has been used as an indicator of neuronal activity in zebrafish which is often measured by using the electroencephalography (EEG) [18], To investigate whether compounds **1** and **39–41** affects PTZ-induced expression of the gene *c-fos*, quantitative real time PCR of *c-fos* was performed on larval zebrafish (Figure 4C). The results indicated that the PTZ treatment (10 mM for 60 min) induced about 43-fold expression of *c-fos* mRNA in five dpf larval zebrafish whereas compounds **1** and **39–41** significantly attenuated the PTZ-induced increase in *c-fos* expression. The inhibitory action on *c-fos* expression by compounds tested corresponded to their suppression of PTZ-induced movement in the larval zebrafish. Consistent with the behavior assay, compound **40** exhibited one of the most potent inhibitory actions. Given that PTZ induces *c-fos* expression exclusively in central nervous system in zebrafish, our data also suggests that the compounds are able to cross the blood brain barrier [19].

Figure 4. The effect of compounds **1** and **39–41** on PTZ-induced hyperactivity in zebrafish. (**A**) The effect of PTZ on larval activity was initially tested across a broad concentration range (0–40 mM). *** $p < 0.001$ *vs.* DMSO. (**B**) The total distances in larval zebrafish exposure to 10 mM PTZ with or without compound **1** and compounds **39–41**. *** $p < 0.001$ *vs.* PTZ. (**C**) The expression levels of *c-fos* mRNA in larval zebrafish exposure to PTZ (10 mM for 60 min) with or without tested compounds (** $p < 0.01$ *vs.* PTZ, ***, $p < 0.0001$).

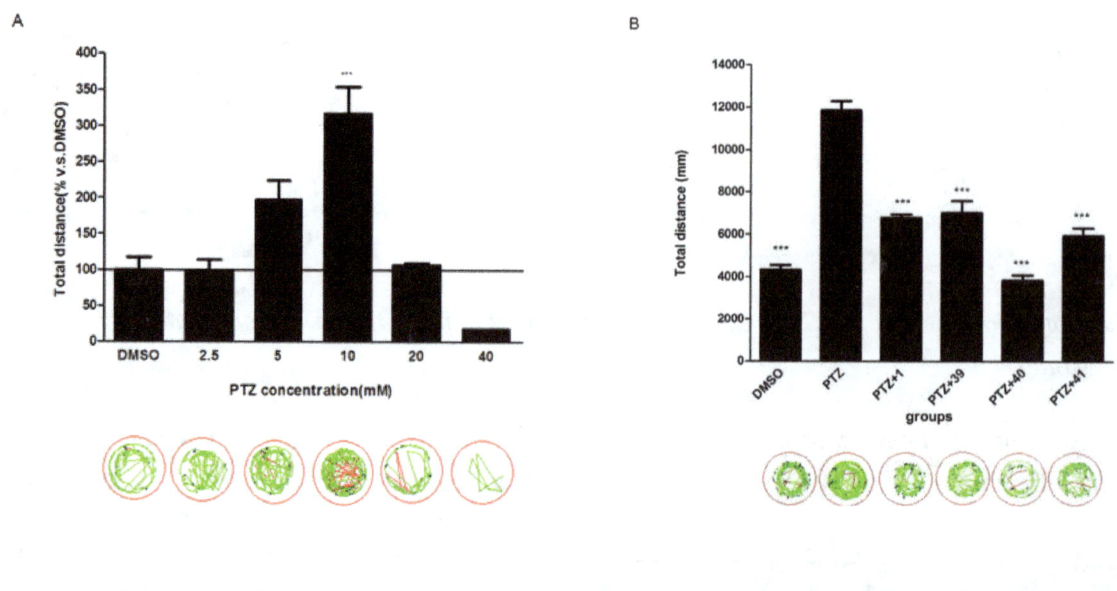

Figure 4. *Cont.*

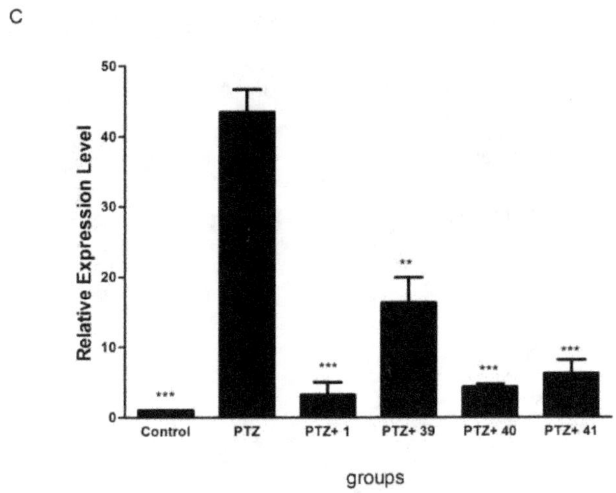

*2.6. The Dose-Response Study of Compound **40** in PTZ Model*

We then conducted a dose-response study on compound **40**, the most potent inhibitory compound tested, to explore its efficacy and toxicity. We found that compound **40** at doses between 10 μm and 1 mM significantly reduced PTZ-induced hyperactivity in larval zebrafish (Figure 5F). To investigate whether compound **40** causes potential toxicity to zebrafish, we examined the morphology of compound **40**-treated larval zebrafish. The results (Figure 5A–E) showed normal morphology in larval zebrafish receiving different doses of **40**, indicating that compound **40** alone at doses between 1 μm and 1 mM did not cause toxic effects on larval zebrafish. Images of larval zebrafish were taken using an OLYMPUS IX71 inverted microscope at 4× magnification.

Figure 5. The effect of compound **40** on PTZ-induced morphological changes. (**A**) The morphology of larval zebrafish at five dpf. (**B–E**) The morphology of five dpf larval zebrafish exposed to compound **40** at different concentrations (100 nM, 1, 10, 100 μm and 1 mM). Scale bar represents 100 μm. (**F**) The total distances in larval zebrafish exposure to PTZ with or without different concentrations of compound **40**. * $p < 0.05$ *vs.* PTZ, ** $p < 0.01$ *vs.* PTZ.

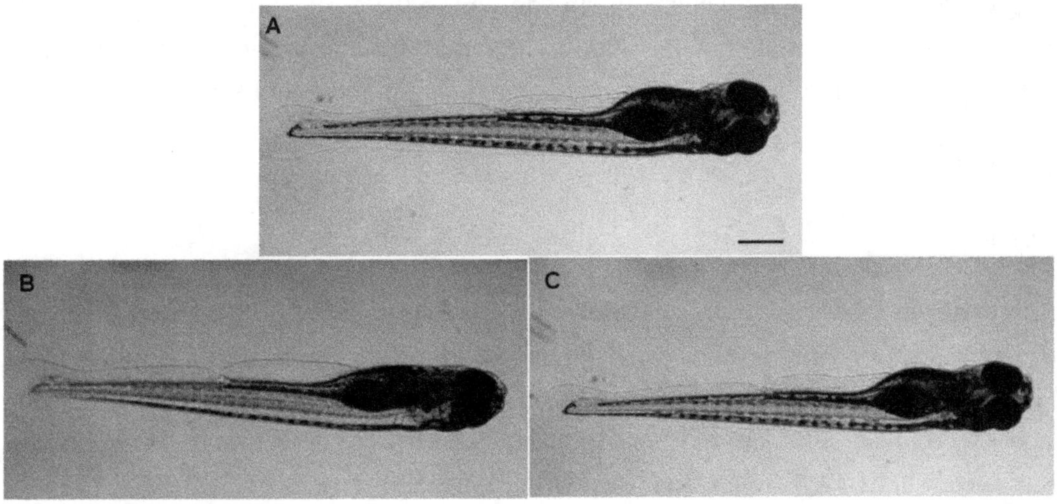

Figure 5. *Cont.*

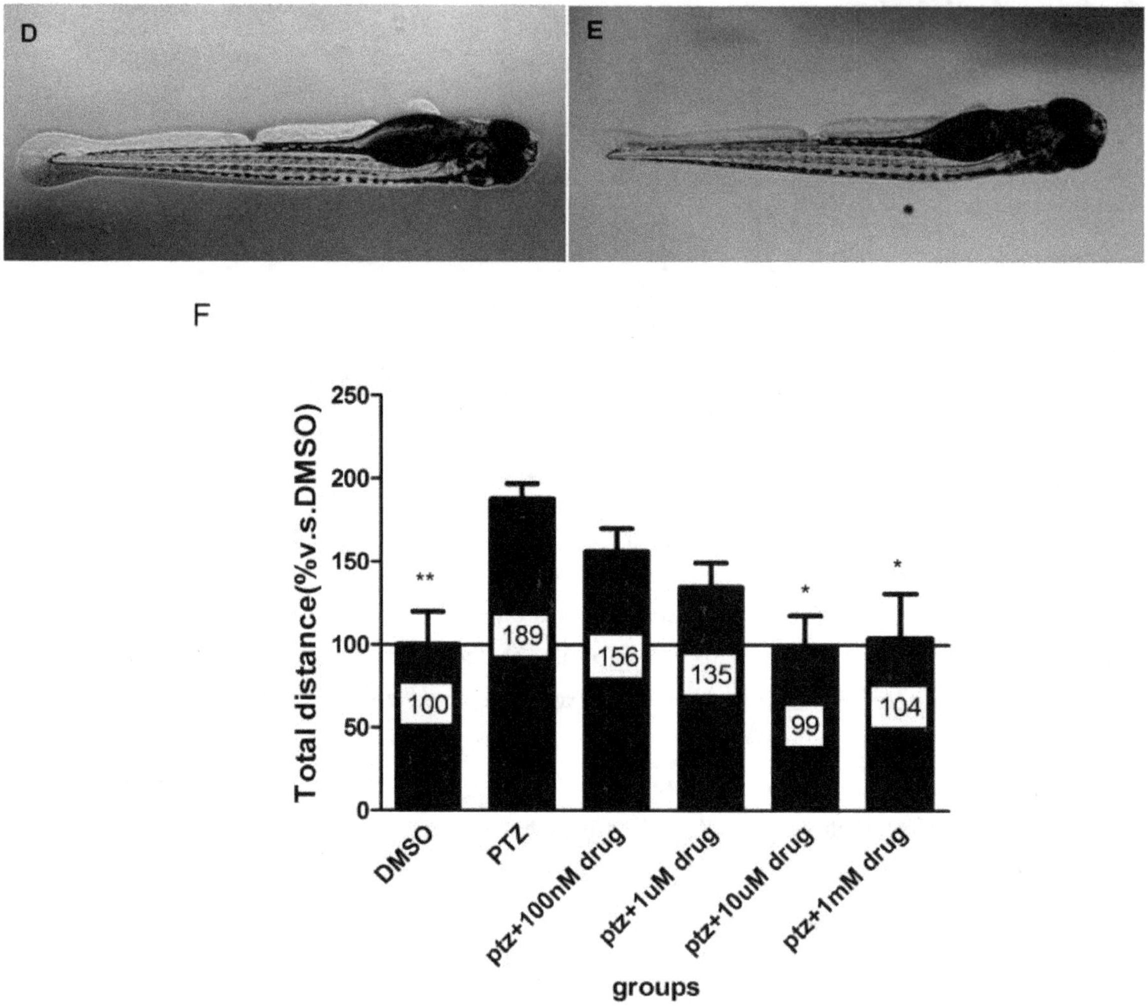

Calcium has been implicated in the pathophysiology of seizure. It has been reported that calcium influx mediates seizure induction whereas blockage of calcium channels reduces epileptic activity [20–22]. Xyloketal derivatives have multiple-properties and their inhibitory action on L-calcium channel activity has been reported [23]. Therefore, xyloketal derivatives may exert their anti-convulsive effect by inhibiting the activity of calcium channels. In addition to the direct anti-convulsive action, calcium channel antagonists can potentiate the anticonvulsant action of other anti-epileptics such as diazepam. Xyloketal derivatives deserve further investigation as a potential anticonvulsant drug candidate.

3. Experimental Section

3.1. Chemistry

All commercial available reagents and solvents were used directly without further purification unless otherwise stated. ^{1}H and ^{13}C NMR data were recorded on a 300 MB NMR spectrometer operating at 400 MHz and 100 MHz for ^{1}H and ^{13}C, respectively. Tetramethylsilane (TMS) as internal standard and all chemical shifts are in ppm (δ). Mass spectra were detected on DSQ (Low Resolution Mass Spectrometer) and MAT95XP (High Resolution Mass Spectrometer).

*3.2. General Procedure of Synthesizing Compounds **9–12**; Take **9** for Example*

3.2.1. Compound **9**

To a suspension of lithium aluminum hydride (0.76 g, 20 mmol) in ether (30 mL) with an ice water bath was added a solution of 3,4-dihydro-2*H*-pyran-5-carboxylate (1.88 g, 13.2 mmol, 3.2 molar equivalent to the subsequent reagent hydroquinone) in ether (20 mL) dropwise. The resultant mixture was allowed to warm to room temperature and stirred for additional 1 hour. Then, aqueous solution of 2 M sodium hydroxide (1.2 g) was added at 0 °C. The resultant mixture was filtered, washed with ether (3 × 50 mL), and the combined filtrates were concentrated in vacuo below 10 °C to afford the reduction product as a colorless liquid. This material was immediately diluted with ether (20 mL) and used for the subsequent experiment without storage because of its instability.

To a suspension of hydroquinone (0.52 g, 4.1 mmol) and anhydrous magnesium sulfate (2 g) in 40 mL ether at 0 °C was added the prepared above ether solution as soon as possible. After stirring for 1 min, *p*-toluene sulfonic acid (0.1 g) was added. The resultant mixture was allowed to warm to room temperature and stirred for 30 min. The reaction mixture was then filtered and the filter-cake was washed with ether (3× 50 mL). The combined filtrates were washed with water (100 mL) and saturated brine (100 mL) followed by drying over anhydrous magnesium sulfate and concentrated in vacuo. Purification by flash chromatography using ether: dichloromethane (2:25) afforded 0.71 g title compound **9** as a colorless crystal in 50% yield. ^{1}H NMR (400 MHz, $CDCl_3$) δ 6.68 (s, 1H), 6.52 (s, 1H), 5.28–5.14 (m, 2H), 4.06–3.93 (m, 2H), 3.72–3.63 (m, 2H), 2.93–2.39 (m, 4H), 2.25–2.05 (m, 2H), 1.71–1.59 (m, 8H). EI-MS *m/z* 302 (M). EI-HR-MS *m/z* Found: 302.1510, Calcd. for $C_{18}H_{22}O_4$: 302.1513.

3.2.2. Compound **10**

The title compound **10** was obtained 0.48 g from hydroquinone as a colorless crystal in 52% yield. ^{1}H NMR (400 MHz, $CDCl_3$) δ 6.69 (d, J = 8.7 Hz, 1H), 6.58 (dd, J = 8.7, 3.0 Hz, 1H), 6.49 (d, J = 2.9 Hz, 1H), 6.13 (s, 1H), 5.25 (d, J = 2.6 Hz, 1H), 4.05–3.96 (m, 1H), 3.73–3.66 (m, 1H), 2.81 (dd, J = 16.7, 6.1 Hz, 1H), 2.57 (dd, J = 16.7, 5.3 Hz, 1H), 2.17–2.06 (m, 1H), 1.69–1.59 (m, 4H). ^{13}C NMR (100 MHz, $CDCl_3$) δ 149.73, 146.07, 120.57, 116.82, 115.35, 114.46, 96.45, 62.76, 31.51, 28.50, 24.11, 23.12. EI-MS *m/z* 206 (M). EI-HR-MS *m/z* Found: 206.0940, Calcd. for $C_{12}H_{14}O_3$: 206.0937.

3.2.3. Compound **11**

The title compound was obtained 0.18 g from methyl 2,4-dihydroxybenzoate as a colorless crystal in 29% yield. ^{1}H NMR (400 MHz, $CDCl_3$) δ 11.18 (s, 1H), 7.62 (d, J = 8.9 Hz, 1H), 6.43 (d, J = 8.9 Hz, 1H), 5.34 (d, J = 2.5 Hz, 1H), 4.06–3.96 (m, 1H), 3.90 (s, 3H), 3.79–3.68 (m, 1H), 2.85–2.61 (m, 2H), 2.29–2.12 (m, 1H), 1.76–1.58 (m, 4H). ^{13}C NMR (100 MHz, $CDCl_3$) δ 170.77, 161.21, 158.86, 128.67, 108.42, 107.88, 105.09, 96.96, 62.57, 51.87, 30.92, 24.09, 23.63, 23.02. EI-MS *m/z* 264 (M). EI-HR-MS *m/z* Found: 264.0998, Calcd. for $C_{14}H_{16}O_5$: 264.0992.

3.2.4. Compound **12**

The title compound **12** was obtained 0.21 g from methyl 2,4-dihydroxybenzoate as a colorless crystal in 34% yield. ^{1}H NMR (400 MHz, $CDCl_3$) δ 10.61 (s, 1H), 7.52 (s, 1H), 6.45 (s, 1H), 5.38 (s, 1H), 4.04–3.93 (m, 1H), 3.90 (s, 3H), 3.79–3.63 (m, 1H), 2.97–2.74 (m, 1H), 2.65–2.49 (m, 1H), 2.27–2.07 (m, 1H), 1.73–1.53 (m, 4H). ^{13}C NMR (100 MHz, $CDCl_3$) δ 170.16, 161.81, 159.56, 130.88, 111.86, 106.15, 103.82, 97.13, 62.22, 51.84, 31.77, 28.41, 23.63, 23.57. EI-MS *m/z* 264 (M).

3.3. Zebrafish Behavior-Based Activity Screen

3.3.1. Zebrafish Maintenance

Zebrafish (Danio rerio) were maintained according to standard animal care protocols [24]. AB strain zebrafish were bred to yield embryos. Embryos and larvae were maintained on a re-circulating Tecniplast aquatic system at 28 ± 1 °C and between pH 7.0 and 7.5 on a 14/10 h light/dark (L/D) cycle. Embryos were collected from multiple AB/Tubingen breeding. Embryos were collected, rinsed in E3 buffer (5 mM NaCl, 0.17 mM KCl, 0.33 mM $CaCl_2 \cdot 2H_2O$, 0.33 mM $MgSO_4$), washed three times, transferred to Petri dishes and incubated at 28 °C until 4 dpf. They were washed twice every day to remove dead or unfertilized embryos.

3.3.2. Zebrafish Behavioral Assay

AB strain zebrafish were bred to yield embryos. At 4 dpf, larval zebrafish were transferred to 96-well microplates and acclimated at 28 °C overnight. Twenty-four h later, compounds were added and incubated with zebrafish for 30 min at 28 °C. The 96-well microplates were then put into the zebrafish tracking box (Viewpoint Life Sciences Inc., Montreal, Canada) and the activity of zebrafish was monitored using automated video-tracking (the Viewpoint video tracking system and software).

After zebrafish acclimated at the tracking box for 20 min, their behaviors were recorded in infrared light (which will hereafter be referred to as "dark"), and the infrared light remained on throughout the recording session. In the behavior assay, the visible light (or "light") was switched on for 10 min and then switched off for 10 min (dark period). This light/dark transition was continued for a total duration of 70 min. Each drug challenge was conducted on at least two separate plates with n = 12 larvae/plate in the compounds challenges, and 24 larvae/plate in the PTZ challenge (for controls, n = 12 larvae/plate in the compounds challenge and 24 larvae/plate in the PTZ challenge). All experiments consisted of 20 min of acclimation in the dark followed by seven 10 min-cycles containing a darkness and light phase (90 min total). To capture the different types of activity, a threshold was set at 25 mm/s in order to separate the fast/darting activity and 4 mm/s to separate the slow activity.

3.3.3. Morphology Assay

Briefly, 5 dpf zebrafish with or without compounds were anaesthetized by treatment with 0.4% tricaine and mounted onto glass slide. The morphology was visualized at 4× magnification under an OLYMPUS IX71 inverted microscope.

3.3.4. Quantitative RT Real Time PCR

Total RNA from zebrafish was extracted using E.Z.N.A. total RNA extraction kit (OMEGA biotek, Inc., Norcross, GA, USA). Intact RNA was checked by running a 1.0% agarose/formaldehyde gel and quantified spectrometrically (Beckman Coulter DU 800) before proceeding to subsequent steps. Five hundred ng of total RNA was reverse-transcribed using PrimeScript™ RT Master Mix (Perfect Real Time) Kit (Takara Inc., Otsu, Japan) according to manufacturer's instructions. Real-Time PCR was performed on an Opticon MONITORTM Software (MJ Research Inc., Quebec, Canada) using SYBR® Premix Ex Taq™ II (Tli RNaseH Plus) (Takara Inc.). Expression levels for each target gene were calculated by the 2~DDCT method [25]. All analyses were performed in triplicates. Primers used for the RT real-time PCR are as follow:

c-fos Forward primer: 5′-GCTCCATCTCAGTCCCAGAG-3′
c-fos Reverse primer: 5′-AGAGTGGGCTCCAGATCAGA-3′
β-actin Forward primer: 5′-CCGTTGCCCCGAGGCTCTCT-3′
β-actin Reverse primer: 5′-CGCATCCTGAGTCAATGCGCCA-3′

3.3.5. Data Analysis

Dead larvae or larvae with physical abnormalities were not included in any data analyses or figures. Any observed death or abnormalities were not due to drug exposures, and the total amount of dead and abnormal animals on each plate represented less than 5% of the total population. Data were analyzed using Graph Pad Prism software. Replicate experiments were run on two separate days (n = 12/day) for both the therapeutic dilution series test and the PTZ dilution series test, with a carrier control group in each plate. The data for each drug were first assessed using a repeated measures analysis of variance (ANOVA) with time and dose as the independent variables and locomotor activity (distance moved/time) as the dependent variable. All data are presented as mean ± standard error of the mean (SEM).

4. Conclusions

In the present study, we have evaluated the neuroactive activity of 36 natural compounds and six designed novel derivatives in zebrafish model. Compound **1** and compounds **39**, **40** and **41** could significantly attenuate PTZ-induced locomotor hyperactivity and elevation of *c-fos* mRNA in larval zebrafish. Compound **40** showed the most potent inhibitory action against PTZ-induced changes. The structure-activity analysis showed that the OH group at the 12-position had a critical role and the substituents at the 13-position had a favorable role in the inhibitory activity of xyloketal derivatives. Thus, these derivatives may provide some novel drug candidates for the treatment of epilepsy.

Acknowledgments

This study was supported by grants from the National Key Clinical Department, National Key Discipline, Guangdong Key Laboratory For Diagnosis and Treatment of Major Neurological Diseases, the National Natural Science Foundation of China (No. 81371255, 81100936, 21172271), Doctoral Program of Higher Education of China (No. 20110171110058), Guangdong Technological grant

(No. 2010B050700024, 2011B050400031, 2012B031800107), and the Natural Science Foundation of Guangdong Province (No. S2011010004860). This study was also supported by the multi-year research grant, university of Macau, MYRG122 (Y1-L3)-ICMS12-SHX.

Author Contributions

Conceived and designed the experiments: S.-M. Long, F.-Y. Liang, J.-Y. Pang, Z. Pei; Performed the experiments: S.-M. Long, F.-Y. Liang, Q. Wu, X.-L. Liu, X.-L. Yao, S.-C. Li, J. Li; Analyzed the data: S.-M. Long, H. Su, J.-Y. Pang; Wrote the paper: S.-M. Long, J.-Y. Pang, Z. Pei.

References

1. Pangalos, M.N.; Schechter, L.E.; Hurko, O. Drug development for CNS disorders: Strategies for balancing risk and reducing attrition. *Nat. Rev. Drug Discov.* **2007**, *6*, 521–532.
2. Kokel, D.; Peterson, R.T. Chemobehavioural phenomics and behaviour-based psychiatric drug discovery in the zebrafish. *Brief. Funct. Genomics Proteomics* **2008**, *7*, 483–490.
3. Brockerhoff, S.E.; Hurley, J.B.; Janssen-Bienhold, U.; Neuhauss, S.C.; Driever, W.; Dowling, J.E. A behavioral screen for isolating zebrafish mutants with visual system defects. *Proc. Natl. Acad. Sci. USA* **1995**, *92*, 10545–10549.
4. Neuhauss, S.C.; Biehlmaier, O.; Seeliger, M.W.; Das, T.; Kohler, K.; Harris, W.A.; Baier, H. Genetic disorders of vision revealed by a behavioral screen of 400 essential loci in zebrafish. *J. Neurosci.* **1999**, *19*, 8603–8615.
5. Burgess, H.A.; Granato, M. Sensorimotor gating in larval zebrafish. *J. Neurosci.* **2007**, *27*, 4984–4894.
6. Prober, D.A.; Rihel, J.; Onah, A.A.; Sung, R.J.; Schier, A.F. Hypocretin/orexin overexpression induces an insomnia-like phenotype in zebrafish. *J. Neurosci.* **2006**, *26*, 13400–13410.
7. Zhdanova, I.V.; Wang, S.Y.; Leclair, O.U.; Danilova, N.P. Melatonin promotes sleep-like state in zebrafish. *Brain Res.* **2001**, *903*, 263–268.
8. Kokel, D.; Bryan, J.; Laggner, C.; White, R.; Cheung, C.Y.; Mateus, R.; Healey, D.; Kim, S.; Werdich, A.A.; Haggarty, S.J.; *et al.* Rapid behavior-based identification of neuroactive small molecules in the zebrafish. *Nat. Chem. Biol.* **2010**, *6*, 231–237.
9. Rihel, J.; Prober, D.A.; Arvanites, A.; Lam, K.; Zimmerman, S.; Jang, S.; Haggarty, S.J.; Kokel, D.; Rubin, L.L.; Peterson, R.T.; *et al.* Zebrafish behavioral profiling links drugs to biological targets and rest/wake regulation. *Science* **2010**, *327*, 348–351.
10. Li, S.; Shen, C.; Guo, W.; Zhang, X.; Liu, S.; Liang, F.; Xu, Z.; Pei, Z.; Song, H.; Qiu, L.; Lin, Y.; Pang, J. Synthesis and neuroprotective action of xyloketal derivatives in Parkinson's disease models. *Mar. Drugs* **2013**, *11*, 5159–5189.
11. Liu, R.; Zhou, Z.Y.; Jiang, M.Y.; Wang, F.; Liu, J.K. A new isoprenyl phenyl ether riboside from the culture of basidiomycete Laccaria amethystea. *J. Asian Nat. Prod. Res.* **2010**, *12*, 723–726.
12. Li, J.; Zhang, D.; Zhu, X.; He, Z.; Liu, S.; Li, M.; Pang, J.; Lin, Y. Studies on synthesis and structure-activity relationship (SAR) of derivatives of a new natural product from marine fungi as inhibitors of influenza virus neuraminidase. *Mar. Drugs* **2011**, 9, 1887–1901.

13. Pettigrew, J.D.; Wilson, P.D. Synthesis of xyloketal A, B, C, D, and G analogues. *J. Organ. Chem.* **2006**, *71*, 1620–1625.
14. Xu, Z.; Li, Y.; Xiang, Q.; Pei, Z.; Liu, X.; Lu, B.; Chen, L.; Wang, G.; Pang, J.; Lin, Y. Design and synthesis of novel xyloketal derivatives and their vasorelaxing activities in rat thoracic aorta and angiogenic activities in zebrafish angiogenesis screen. *J. Med. Chem.* **2010**, *53*, 4642–4653.
15. Guo, S. Linking genes to brain, behavior and neurological diseases: What can we learn from zebrafish? *Genes Brain Behav.* **2004**, *3*, 63–74.
16. Epilepsy Fact Sheet, World Health Organization. October 2012. Available online: http://www.who.int/mediacentre/factsheets/fs999/en/ (accessed on 10 May 2014).
17. Mussulini, B.H.; Leite, C.E.; Zenki, K.C.; Moro, L.; Baggio, S.; Rico, E.P.; Rosemberg, D.B.; Dias, R.D.; Souza, T.M.; Calcagnotto, M.E.; *et al.* Seizures induced by pentylenetetrazole in the adult zebrafish: a detailed behavioral characterization. *PLoS One* **2013**, *8*, e54515.
18. Scallet, A.C.; Kowalke, P.K.; Rountree, R.L.; Thorn, B.T.; Binienda, Z.K. Electroencephalographic, behavioral, and c-*fos* responses to acute domoic acidexposure. *Neurotoxicol. Teratol.* **2004**, *26*,331–342.
19. Fleming, A.; Diekmann, H.; Goldsmith, P. Functional characterisation of the maturation of the blood-brain barrier in larval zebrafish. *PLoS One* **2013**, *8*, e77548.
20. Ghasemi, M.; Shafaroodi, H.; Nazarbeiki, S.; Meskar, H.; Heydarpour, P.; Ghasemi, A.; Talab, S.S.; Ziai, P.; Bahremand, A.; Dehpour, A.R. Voltage-dependent calcium channel and NMDA receptor antagonists augment anticonvulsant effects of lithium chloride on pentylenetetrazole-induced clonic seizures in mice. *Epilepsy Behav.* **2010**, *18*, 171–178.
21. Meyer, F.B.; Anderson, R.E.; Sundt, T.M., Jr.; Yaksh, T.L.; Sharbrough, F.W. Suppression of pentylenetetrazole seizures by oral administration of a dihydropyridine Ca^{2+} antagonist. *Epilepsia* **1987**, *28*, 409–414.
22. Moron, M.A.; Stevens, C.W.; Yaksh, T.L. The antiseizure activity of dihydropyridine calcium channel antagonists in the conscious rat. *J. Pharmacol. Exp. Ther.* **1990**, *252*, 1150–1155.
23. Wu, X.Y.; Liu, X.H.; Lin, Y.C.; Luo, J.H.; She, Z.G.; Jin, L.H.; Chan, W.L.; Antus, S.; Kurtan, T.; Elsässer, B.; *et al.* Xyloketal F: A strong L-calcium channel blocker from the mangrove fungus *xylaria* sp. (#2508) from the South China Sea Coast. *Eur. J. Org. Chem.* **2005**, *2005*, 4061–4064.
24. Westerfield, M. *The Zebrafish Book: A Guide for Laboratory Use of Zebrafish (Danio rerio)*; University of Oregon Press: Eugene, OR, USA, 1995.
25. Livak, K.J.; Schmittgen, T.D. Analysis of relative gene expression data using real-time quantitative PCR and the 2(−Delta Delta C(T)) method. *Methods* **2001**, 25, 402–408.

10

Pigment Cell Differentiation in Sea Urchin Blastula-Derived Primary Cell Cultures

Natalya V. Ageenko [1,2,*], Konstantin V. Kiselev [3], Pavel S. Dmitrenok [4] and Nelly A. Odintsova [1,2]

[1] Cytotechnology Laboratory, A.V. Zhirmunsky Institute of Marine Biology, FEB RAS, Vladivostok 690041, Russia
[2] Far Eastern Federal University, Sukhanova Str. 8, Vladivostok 690950, Russia; E-Mail: nelodin54@yahoo.com
[3] Laboratory of Biotechnology, Institute of Biology and Soil Sciences, FEB RAS, Vladivostok 690022, Russia; E-Mail: kkv5@mail.ru
[4] Laboratory of Instrumental and Radioisotope Methods of Analysis, G.B. Elyakov Pacific Institute of Bioorganic Chemistry, FEB RAS, Vladivostok 690022, Russia; E-Mail: paveldmt@piboc.dvo.ru

* Author to whom correspondence should be addressed; E-Mail: natkuprina@mail.ru

---

Abstract: The quinone pigments of sea urchins, specifically echinochrome and spinochromes, are known for their effective antioxidant, antibacterial, antifungal, and antitumor activities. We developed *in vitro* technology for inducing pigment differentiation in cell culture. The intensification of the pigment differentiation was accompanied by a simultaneous decrease in cell proliferation. The number of pigment cells was two-fold higher in the cells cultivated in the coelomic fluids of injured sea urchins than in those intact. The possible roles of the specific components of the coelomic fluids in the pigment differentiation process and the quantitative measurement of the production of naphthoquinone pigments during cultivation were examined by MALDI and electrospray ionization mass spectrometry. Echinochrome A and spinochrome E were produced by the cultivated cells of the sand dollar *Scaphechinus mirabilis* in all tested media, while only spinochromes were found in the cultivated cells of another sea urchin, *Strongylocentrotus intermedius*. The expression of genes associated with the induction of pigment differentiation was increased in cells cultivated in the presence of shikimic acid, a precursor of naphthoquinone pigments. Our results should contribute to the development of new techniques in marine biotechnology, including the generation of cell cultures producing complex bioactive compounds with therapeutic potential.

Keywords: cell culture; echinochrome; gene expression; MALDI TOF MS; ESI MS; marine biotechnology; naphthoquinone pigments; pigment differentiation; proliferation; sea urchin

1. Introduction

Marine inhabitants are the most phylogenetically diverse organisms, demonstrating a significant potential for biodiscovery research [1] and as possible sources of valuable biologically active substances for the pharmaceutical and food industries [2]. For example, sea urchins are a source of pharmacologically important quinone pigments—specifically echinochrome and the spinochromes—that constitute a group of polyketide compounds. Like many marine secondary metabolites, polyketide compounds are known for their highly effective antioxidant, antibacterial, antifungal, and antitumor activities. In addition, these compounds may play decisive roles in the regulation of lipid peroxidation and in immune defense [3–6]. They are generated via a series of enzymatic, oxidative and photochemical reactions from shikimic acid (ShA)—a precursor of naphthoquinone pigments (Figure 1) [7]. Echinochrome is one of these pigments and is synthesized in the sea urchin pigment cells demonstrating a strong bactericidal effect during embryonic and larval development [8,9]. The morphology and behavior of pigment cells are similar to those of macrophages, confirming the involvement of sea urchin pigment cells in the larval immune system [10]. A drug (Histochrome®, Moscow, Russia) with cardiological and ophthalmological activity based on the echinochrome structure has been developed to correct metabolic processes and act as an oxygen transporter [11].

Figure 1. The structures of shikimic acid, a precursor of naphthoquinone pigments, and echinochrome and spinochromes in accordance with a previous report [7].

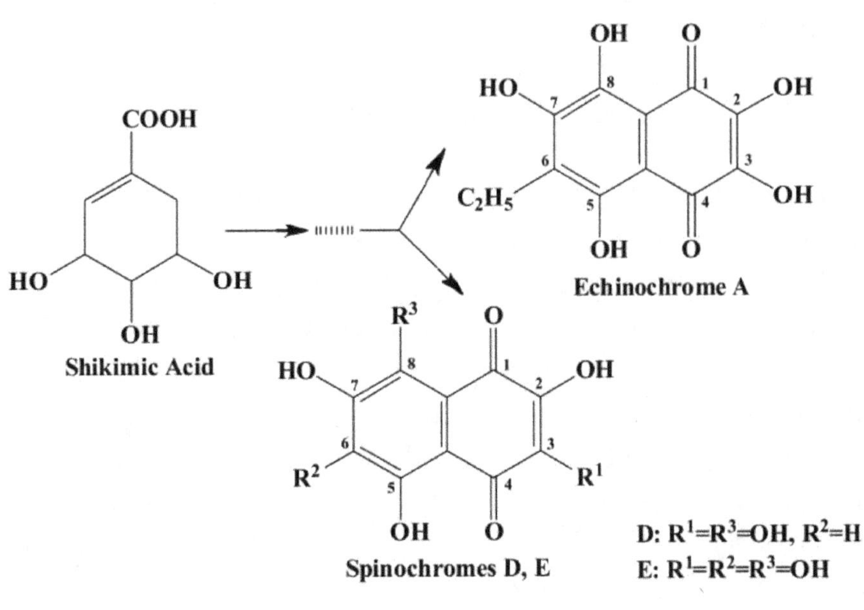

An industrial-scale procurement of echinochrome could lead to the extinction of the organisms that produce this substance. Therefore, cultured sea urchin pigment cells might provide a source of pharmacologically important quinone pigments, as well as help reduce the impact on the adult sea urchin population and pave the way for solving several pharmacological problems.

The aim of our work is to develop a technology for directed-pigment differentiation in sea urchin culture for solving practical tasks in marine biotechnology. Two sea urchin species differing in their number of embryonic pigment cells are chosen: the sea urchin *Strongylocentrotus intermedius* and the sand dollar *Scaphechinus mirabilis.* As shown in [12], the embryos of the sand dollar contain a lot of pigment cells. Previously, Calestani *et al.* [13] revealed three groups of genes expressed in sea urchin pigment cells that play key roles in the biosynthesis of naphthoquinone pigments [14]. The central parts of these genes in *S. intermedius* have been reported to be similar to those of the same genes in the closely related sea urchin *S. purpuratus* [15]. In this study, we evaluated the gene expression associated with the induction of pigment differentiation in cell cultures.

As previously shown, sand dollar embryos transfected with the foreign gene (the yeast *gal4* gene encoding the transcription activator) were dissociated into single cells that produced pigments after one month of cultivation [16]. We then developed an *in vitro* technology for inducing pigment differentiation without transfecting foreign genes into the sea urchin embryos but using the coelomic fluids from sea urchins [15]. Here, to estimate the contributions of the specific components of the coelomic fluids from sea urchins to the pigment differentiation process and the production of naphthoquinone pigments during cultivation, we used matrix assisted laser desorption/ionization (MALDI) time-of-flight (TOF) mass spectrometry and electrospray ionization (ESI) mass spectrometry.

This study has three main results. First, we developed an *in vitro* technology for inducing pigment differentiation in cell culture. Second, our data support the hypothesis that specific components of sea urchin coelomic fluids might act as inductive signals in pigment differentiation through the regulation of genes implicated in naphthoquinone synthesis. Third, echinochrome was produced only in the sand dollar cells, and its maximum level was found in cells cultured in coelomic fluids rather than seawater.

2. Results and Discussion

2.1. Differentiation of Pigment Cells in a Blastula-Derived Cell Culture

The growth patterns and morphology of cultured cells is often determined by peculiarities of the culture medium. To evaluate the effect of different culture media on the development of pigment differentiation in the cell cultures of both sea urchin species, we tested three types of media (Figures 2 and 3): seawater, the coelomic fluid preparations of control sea urchins and injured sea urchins. Injured sea urchins were obtained by needle pricks in the area of Aristotle's lantern (see the Experimental Section). The first distinctions in the appearance of pigment cells and their pigmentation became obvious after 3–7 days of cultivation. The cells cultivated in seawater were faintly pigmented and not numerous, whereas the pigment cells were more abundant in the coelomic fluid-cultivated cells. This picture was observed both in a short-time culture of the sea urchin *S. mirabilis* and in a long-time culture of the sea urchin *S. intermedius*.

Figure 2. Embryonic pigment cells in a blastula-derived cell culture of the sea urchin *Scaphechinus mirabilis* cultivated for 3 days. The cells were cultivated in seawater (**A**); the coelomic fluid of intact sea urchins (**B**); or the coelomic fluid of injured sea urchins (**C**). All culture media were supplemented with 2% fetal bovine serum. Insets in **B** and **C**: higher magnifications. Note the change in morphology of the pigment cells depending on the medium tested. Scale bar, 10 μm.

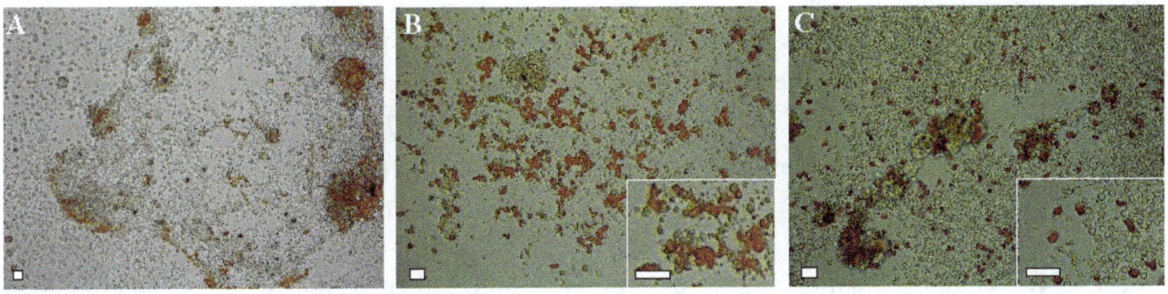

Figure 3. Embryonic pigment cells in a blastula-derived cell culture of the sea urchin *Strongylocentrotus intermedius* cultivated for 5 (**A,B**); 17 (**C,D**) or 41 days (**E,F**). **A**, **C**, **E**—cells cultivated in the coelomic fluid of intact sea urchins; **B**, **D**, **F**—cells cultivated in the coelomic fluid of injured sea urchins. The coelomic fluid was supplemented with 2% fetal bovine serum. Note the change in morphology of the pigment cells during cultivation. Scale bar, 10 μm.

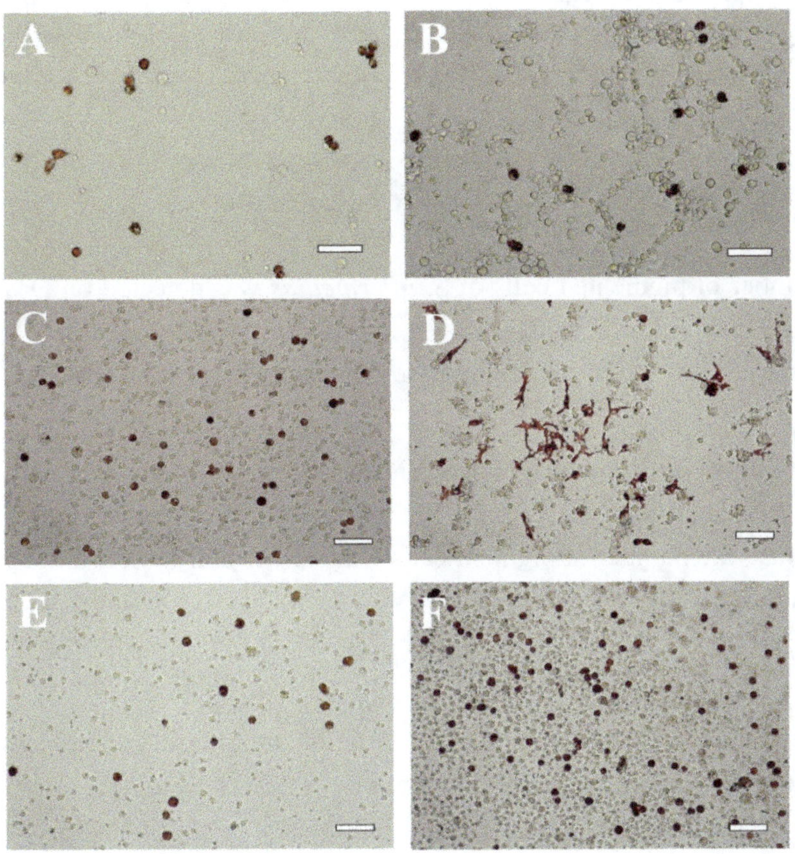

If the coelomic fluid of intact sea urchins was used as the medium, the pigment cells of *S. mirabilis* were well attached and spread during three days in culture (Figure 2B, insert). However, when the injured sea urchin coelomic fluid was used, most of the pigment cells were rounded and failed to spread (Figure 2C, insert).

The medium-dependent differences in cell morphology were also seen in the sea urchin *S. intermedius*. The pigment differentiation of the cells in the seawater occurred slowly and less extensively (data not shown) than that of the cells in the coelomic fluids. Furthermore, the pigment cells in this species changed their morphology during prolonged cultivation for 20 days in the same medium (Figure 3A–D). When the cells were cultivated in injured sea urchin coelomic fluid, the maximal spreading of the pigment cells was observed on Day 17 of cultivation (Figure 3D). Upon further culture, the cells became rounded (Figure 3E,F) and increased in number up to Day 41 of cultivation (Figure 4). Signs of cell degeneration were still detectable after two months of culturing.

Figure 4. Cellular dynamics of pigment cells cultivated in coelomic fluids of intact (**blue** line) and injured (**red** line) sea urchins for 2.5 months. At least 500 cells were counted in each tested culture medium.

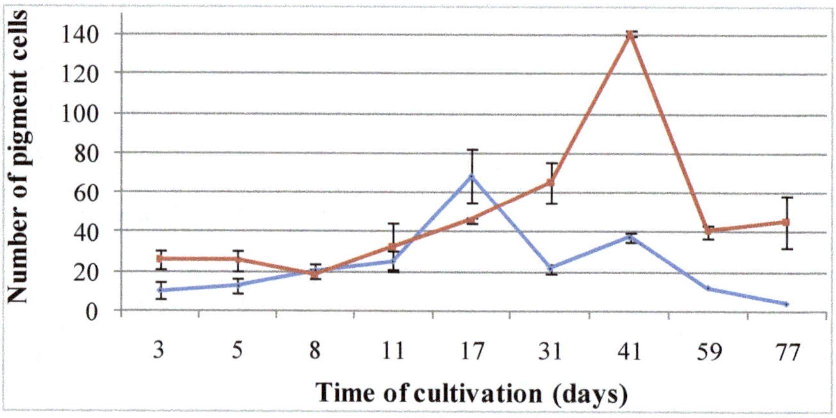

The maximal number of pigmented cells of *S. intermedius* was detected on Day 17, when the cells were cultivated in the coelomic fluid of intact sea urchins (Figure 4, blue line). When the cells were cultivated in the injured sea urchin coelomic fluid, that maximal number was recorded on Day 41 (Figure 4, red line).

The number of pigment cells was dependent on the coelomic fluid used: twice as many pigment cells were detected in the cells cultivated in the injured sea urchin coelomic fluid as in that from intact sea urchins. In contrast, the number of pigment cells was significantly lower among cells cultured in in seawater (data not shown).

2.2. Cell Proliferation in Culture

It is well known that cell proliferation and differentiation are interrelated processes. To compare the contribution of these processes—pigment differentiation (as described above) and cell proliferation—in sea urchin culture, we detected dividing cells with antibodies to phospho-H3-histone, a well-known cell proliferation marker for many organisms [17–19]. At least 300–500 DAPI-stained cells were examined for each experiment with the culture media tested, and the proportion of cells positive for phospho-H3-histone relative to the total number of cells was determined. The number of dividing cells was higher in cells cultured in seawater than in those cultured in the coelomic fluids, where cell division was weak (Figure 5).

Figure 5. The number of dividing cells in a blastula-derived culture of the sea urchin *S. intermedius* depending upon culture medium tested. The cells were cultivated for 24–72 h. The proportion of phospho-H3-histone-positive cells among the total number of examined cells (determined by DAPI staining of nuclei) was counted. SW—seawater; CFn—coelomic fluid obtained from intact sea urchins; CFreg—coelomic fluid obtained from injured sea urchins. * $p < 0.05$; ** $p < 0.01$ *versus* values of dividing cells cultivated in the sea water (24 h).

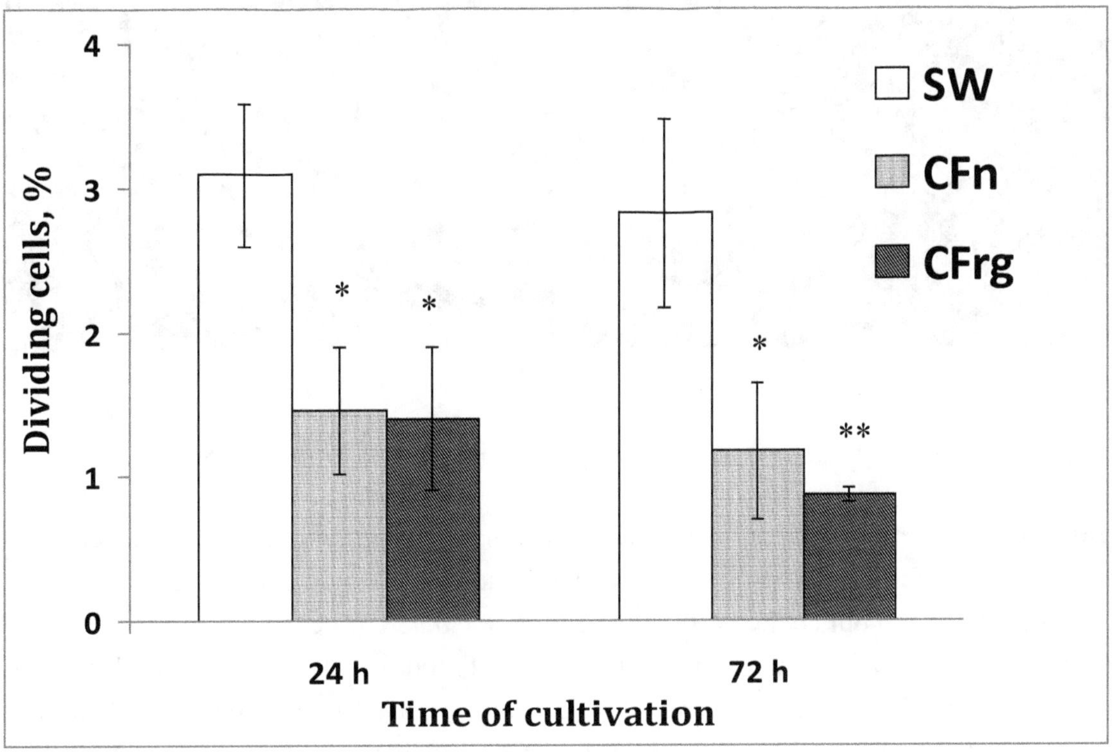

A significant difference in the number of dividing cells in the different media after one day of cultivation was found, and this difference increased for three days, when a general decrease in mitotic activity was then detected. In most cases, the cultures survived well, but their proliferation rate was low.

Dividing (phospho-H3-histone+) cells were detected from 6 to 72 h in all media tested, but the maximal number of dividing cells was found in seawater. The decrease in cell proliferation in the coelomic fluids was accompanied by a simultaneous intensification of the pigment cell differentiation. As already noted, the composition of the culture medium affected the rate of appearance of the pigment cells during cultivation. Figure 6A,B arrows, shows dividing cells in both of the coelomic fluid and seawater after 12 h of cultivation (see details of the immunocytochemical manipulations in the Experimental Section).

Figure 6. Immunofluorescence detection of dividing (phospho-H3-histone-positive) cells in a blastula-derived culture. The preparations were imaged via confocal microscopy. The cells were cultivated on fibronectin-coated coverslips in coelomic fluid obtained from injured sea urchins (**A**) and in seawater (**B**) for 12 h and then labeled with Abs against phospho-H3-histone Abs for the detection of dividing cells (**red**) and tubulin (**green**) for the detection of microtubules. The nuclei were stained with DAPI (**blue**). Arrows show ph-H3-histone-positive cells. Scale bar, 10 μm.

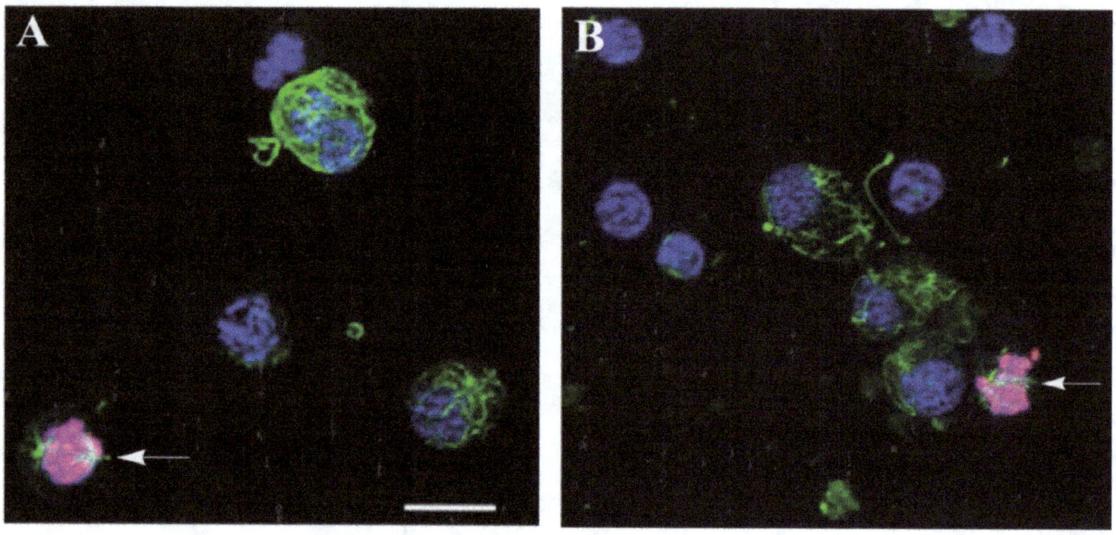

2.3. MALDI MS Analysis of Coelomic Fluids Obtained from Intact and Injured Sea Urchins

To estimate the contribution of specific components in intact and injured sea urchin coelomic fluids to the cell pigment differentiation process during cultivation and to understand its origin, we conducted MALDI MS analyses of both coelomic fluids of *S. intermedius*. Figure 7 illustrates the MALDI mass spectra obtained for proteins with a molecular mass of 4000–9000 Da. Two well-defined peaks corresponding to proteins of 8205 Da and 4119 Da were detected in the spectrum of the intact sea urchin coelomic fluid (Figure 7A). The spectrum of coelomic fluid obtained from the injured sea urchins exhibited three peaks: 8327 Da, 4511 Da, and 4164 Da (Figure 7B). A comparison of the coelomic fluid profiles revealed a new peak in the coelomic fluid of the injured sea urchins corresponding to a protein with a molecular weight of around 4500 Da and a shift of approximately 40–120 Da from the basic components in the intact sea urchin coelomic fluid, which may be due to a protein modification such as phosphorylation, sulfation, *etc.* The molecular weights of the proteins were measured using MALDI MS with standard precision (0.1%) based on the external calibration programs included (Protein Calibration Mixture II, Bruker Daltonik, Germany). The new peak in the spectrum (4511 Da) indicates a change in the composition of the proteins in the coelomic fluid after injury in comparison to the norm. Although we could not identify the molecular nature of the coelomic fluid proteins, our data clearly demonstrated a qualitative difference in the composition of normal coelomic-fluid proteins and that of coelomic-fluid proteins following injury.

Figure 7. The mass spectra of coelomic fluids obtained from intact (**A**) and injured (**B**) sea urchins by MALDI TOF MS using a mass spectrometer (Ultraflex-III TOF/TOF, Bruker Daltonics, Germany). Note a new peak in the coelomic fluid of the injured sea urchins, corresponding to a protein with MW near 4500 Da (marked by an arrow), as well as the shift of the basic components of the intact coelomic fluid to approximately 40–120 Da. The coelomic fluids from three independent experiments were analyzed.

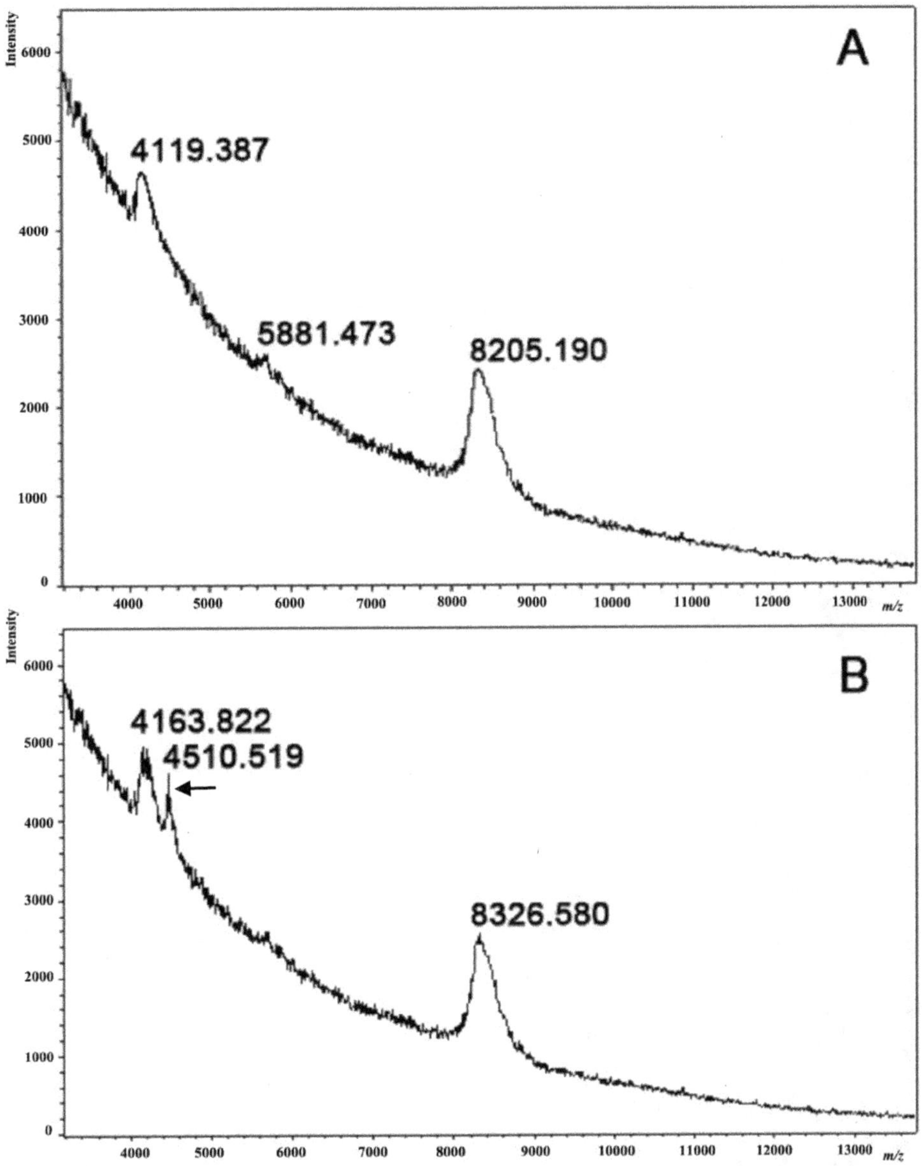

2.4. *Naphthoquinone Pigment Production in Cultivated Sea Urchin Cells*

To elucidate the naphthoquinone pigment profile in the cultivated sea urchin cells and the production of pigments during cultivation, we performed ESI MS. As shown in Figure 8A, a significant peak of echinochrome A and a small peak corresponding to spinochrome E were detected in the pigment extracts from the sand dollar *S. mirabilis*, while only the peaks corresponding to spinochromes D and E were recorded in the pigment cell extracts from *S. intermedius* (Figure 8B). Thus, a distinct difference in pigment expression was demonstrated in the cells of the two sea urchin species.

Figure 8. ESI MS profiles of naphthoquinone pigments from sea urchin cell cultures: (**A**) the pigment extracts from the cells of the sea urchin *S. mirabilis* after 3-day cultivation in seawater; (**B**) the pigment extracts from the cells of the sea urchin *S. intermedius* after 6-day cultivation in seawater.

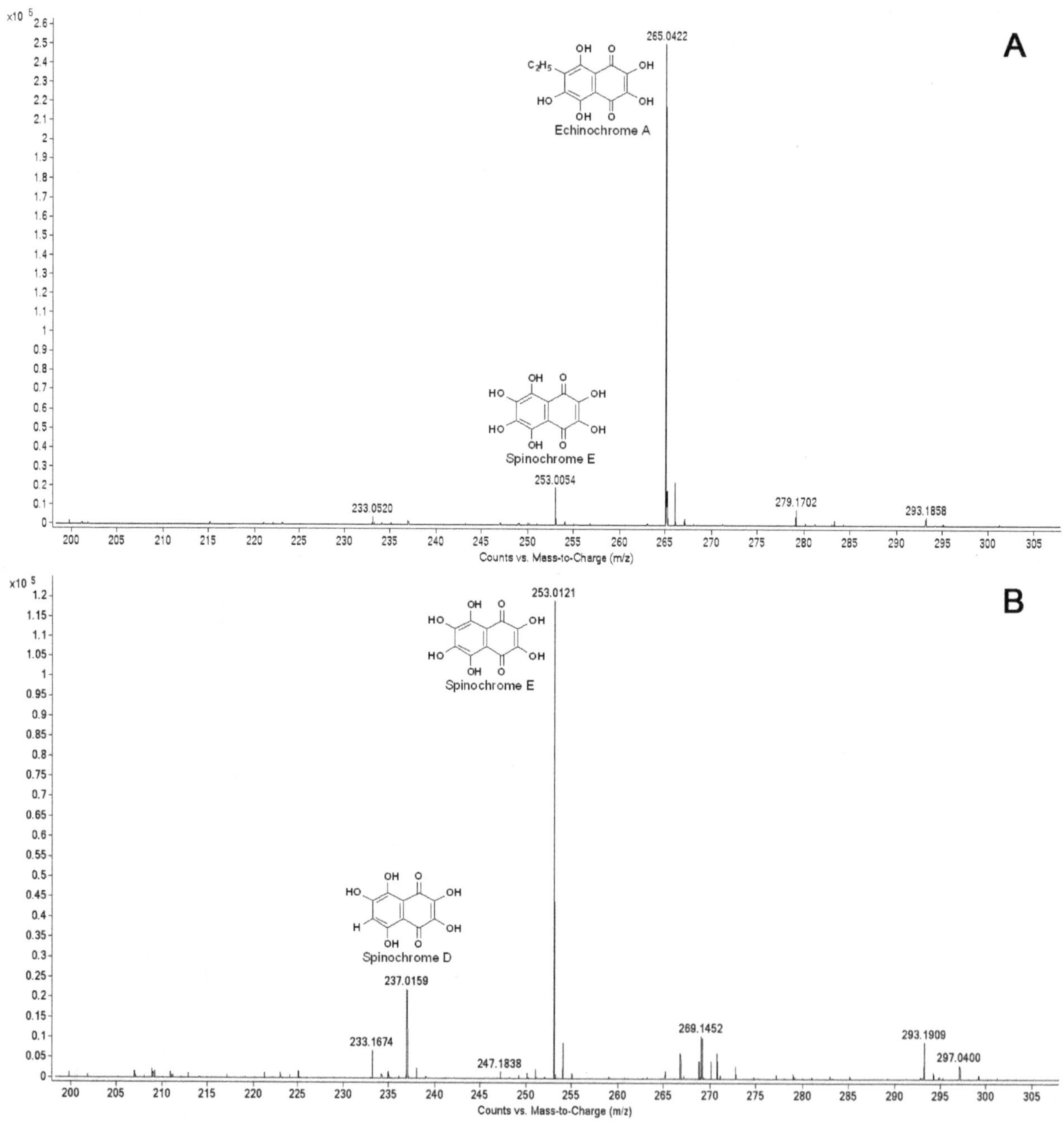

The results of a semi-quantitative measurement (using ESI MS) of the echinochrome and spinochrome production in cells cultivated in the different media are presented in Table 1. Echinochrome A was produced by cultivated *S. mirabilis* cells in all of the tested media, and it was well defined in all of the spectra. The highest level of echinochrome A was detected in the cells cultured in the coelomic fluids as compared to those cultured in seawater. In contrast, no significant differences were seen in the spinochrome D and E levels in the tested media for the cultivated cells of the other sea urchin, *S. intermedius* (Table 1).

Table 1. Naphthoquinone pigment production in sea urchin cultivated cells. Data are presented as the mean ± standard error from two independent experiments (ESI MS). SW—seawater; CFn—coelomic fluid obtained from intact sea urchins; CFreg—coelomic fluid obtained from injured sea urchins.

Pigments, mg/g of Fresh Biomass of Cells	M.W.	Cell Culture Medium		
		SW	CFn	CFreg
	Scaphechinus mirabilis			
Spinochrome E 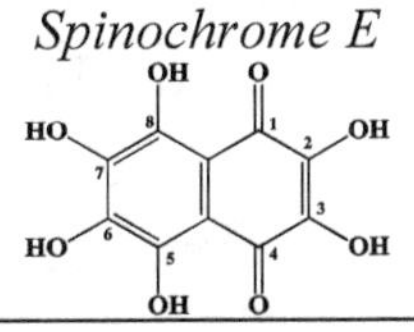	254.00	0.021 ± 0.003	0.062 ± 0.007	0.054 ± 0.006
Echinochrome A 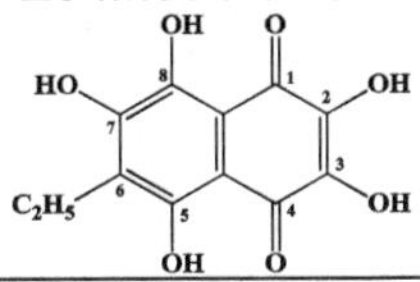	266.04	0.250 ± 0.026	0.640± 0.061	0.540 ± 0.055
	Strongylocentrotus intermedius			
Spinochrome E 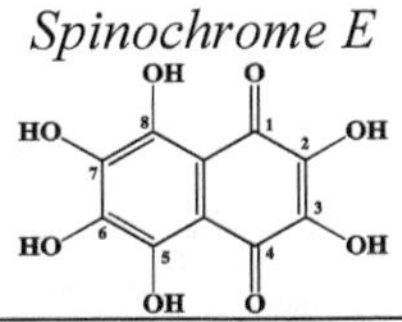	254.00	0.024 ± 0.004	0.013 ± 0.002	0.015 ± 0.002
Spinochrome D 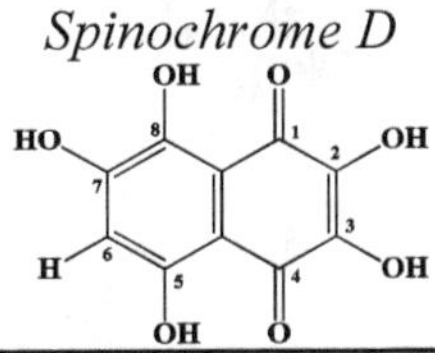	238.02	0.052 ± 0.006	0.044 ± 0.005	0.062 ± 0.007

2.5. Expression of the pks Genes in Cultivated Sea Urchin Cells. Experiments with Shikimic Acid, a Precursor of Naphthoquinone Pigments

To understand the molecular mechanisms underlying sea urchin pigment specialization in culture, we evaluated the gene expression associated with the induction of pigment differentiation, particularly the *pks* genes. The expression level of these genes (estimated by quantitative real-time PCR) in sea urchin cells cultivated in different culture media is presented in relative units in Figure 9 (see the Experimental Section). The gene expression in sea urchin cells cultivated in the coelomic fluid of injured sea urchins increased more than 1.2-fold in comparison to the cells cultivated in seawater, whereas the *pks* expression in the cells cultivated in the coelomic fluid from intact sea urchins decreased by almost 1.4-fold; however, these differences were insignificant. The addition of a high concentration of ShA, a precursor of naphthoquinone pigments, (2 mM) to the sea urchin cells resulted in a marked intensification of the gene expression in all media tested (2.3–4.2-fold) compared to the control cells (Figure 9), but the effect was especially significant in seawater. In contrast, adding ShA of a lower concentration (0.5 mM) increased *pks* expression only in the cells cultivated in injured sea urchin coelomic fluid (2.2-fold). No apparent effect on *pks* expression was detected in the cells cultivated with 0.5 mM ShA in seawater or intact sea urchin coelomic fluid. These data can be explained by the presence of some specific components in the sea urchin coelomic fluids. The marked intensification of the *pks* expression (2.2-fold) in the presence of 0.5 mM ShA only in the cells cultivated in injured sea urchin coelomic fluid is additional evidence of a cellular reaction in response to injury signals. Our findings showing a significant difference between the number of sea urchin pigment cells cultivated in the coelomic fluids of injured and intact sea urchins support this suggestion.

Figure 9. Effect of Shikimic Acid (ShA) on the *pks* gene expression in a blastula-derived cell culture of the sea urchin *Strongylocentrotus intermedius.* ShA concentrations tested: 0.5 mM and 2.0 mM. *Y*-axis—relative units (r.u.). The cells were cultivated with ShA for 4 days in: seawater (SW), coelomic fluid of intact sea urchins (CFn), and coelomic fluid of injured sea urchins (CFreg). After this period, total RNA was isolated from cultivated cells for the following real-time PCR. Control cells are the cells cultivated without ShA. * $p < 0.05$; ** $p < 0.01$ *versus* values of *pks* expression in the appropriate control conditions.

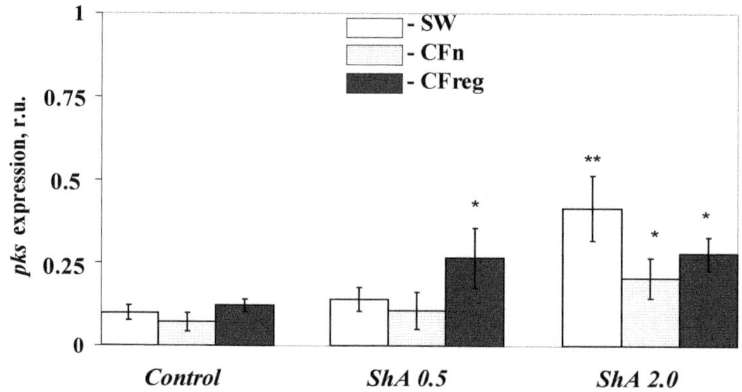

2.6. Discussion

Marine biotechnology progress is associated with the search, characterization, and development of optimal methods for obtaining new substances from marine organisms with pharmaceutically interesting biological activity, and many are of enormous scientific interest. On the one hand, marine organisms are a source of unique marine secondary metabolites that are the basis for developing new and improved natural products for commercial purposes [2]. On the other hand, they include a variety of phylogenetic groups that are important for understanding the evolutionary history of life. Some, such as the echinoderms, are members of a phylum of deuterostomate metazoans and occupy a key position in the early steps in chordate evolution [20,21]. As shown by Evans-Illidge *et al.* [1], the ancestral early-metazoans (Porifera), early eumetazoans (Cnidaria) and some deuterostomes (Echinodermata) have a very high concentration of bioactive compounds.

Here, we have shown that the culture medium composition affects the rate of appearance of pigment cells. In seawater, the number of dividing cells was higher than when the cells were cultured in the coelomic fluids, but the pigment cell differentiation occurred slowly and less extensively. An increase in cell pigmentation in the coelomic fluids probably leads to a decrease in the number of dividing cells in comparison to the cells incubated in seawater. In the early stages of cultivation, we detected dividing cells and a few pigment cells, but later during the pigment cell differentiation, we observed a decrease in the number of dividing cells and a corresponding increase in the number of pigment cells. This is not surprising: either proliferation or differentiation usually dominates in cell cultures. A significant decrease in the number of dividing cells can be explained by a general reduction of mitotic activity during cultivation that has been previously reported for many invertebrate cells after 7–10 days of cultivation [22,23].

We proposed that a significant difference in the proportion of pigment cells in the cultures grown in the coelomic fluids of intact and injured sea urchins in comparison to those cultivated in seawater was connected to the specific components of coelomic fluids that increase the adhesion of cultivated cells. Recently, approximately 30 proteins that may be involved in adhesion were uncovered by a large-scale proteomic analysis of sea urchin coelomic fluid [24]. Thus, far, a proteomic analysis has only been performed on normal sea urchin coelomic fluid.

In this study, we analyzed the coelomic fluid of both normal and injured sea urchins, and detected qualitative differences in their composition. Our results from the overview of integral MALDI mass spectra of sea urchin coelomic fluids have shown that new peaks appear in the coelomic fluid obtained from injured sea urchins. These new peaks may correspond to stress proteins produced after injury. As recently reported, relatively abundant proteins involved in stress responses were identified in the coelomic fluid of the sea urchin *S. purpuratus* [24]. Homologues of these proteins have been found throughout the animal kingdom and are often used as markers for environmental stress [25]. We have suggested that the new coelomic fluid proteins from the injured sea urchins may be related to stress-response proteins. The MALDI MS data together with the results for *pks* gene expression suggest that the specific components of sea urchin coelomic fluids play a role in pigment differentiation through the regulation of the genes implicated in naphthoquinone synthesis. Analyzing cells cultivated in the presence of 0.5 mM ShA, we showed that *pks* gene expression was two-fold higher in cells cultivated in the injured sea urchin coelomic fluid relative to cells cultivated in that from intact sea urchins. A higher proportion of cells containing naphthoquinones in response to different conditions of stress could be a consequence of a protective reaction and be important as a defense mechanism [26,27].

We found that echinochrome was produced only in the sand dollar cells and its content in cells grown in coelomic fluid was higher than that in cells grown in seawater. This finding suggests the existence of a specific regulatory factor that may induce pigmentation activity and subsequent naphthoquinone production in these coelomic fluids. Our morphological observations of the cultured cells correlated with the semi-quantitative data obtained using ESI MS. Unfortunately, not all of the experiments could be performed on the *S. mirabilis* cell cultures due to the difficulty in obtaining the sand dollar cells free of bacterial contamination.

3. Experimental Section

3.1. Animals

Adult sea urchins (*S. intermedius* and *S. mirabilis*, Echinoidae (Agassiz, 1863)), were collected during the breeding season (Vostok Bay, the Sea of Japan), and they were kept in aquaria filled with running, aerated seawater at 16 °C. Before experiments were conducted, the animals were rinsed free of any debris with UV-sterilized filtered seawater. Experiments were performed at the Marine

Biological Station «Vostok» (A.V. Zhirmunsky Institute of Marine Biology, Vladivostok, Russia FEB RAS). Gametes and embryos were obtained by the KCl injection method as described before [15]. The embryonic material was placed in tanks with UV-sterilized seawater (18 °C) throughout development and harvested at the mesenchymal blastula (16 h post fertilization, hpf).

3.2. Cell Culture

Developing sea urchin embryos were collected on a fine 30-μm nylon mesh and dissociated into single cells with 0.25% collagenase at 17 °C (for 20–30 min) as described earlier [15]. The resulting cell suspension containing all cell types was washed several times in seawater with antibiotics, and then sterile seawater supplemented with 2% fetal calf serum, FCS (Sigma, St. Louis, MO, USA) was added. Cell viability was estimated at different time points by a trypan blue exclusion test. The cells were seeded at the density of 6×10^6–8×10^6 cells/mL in plastic Petri dishes (Lux Culture Dishes, MP Biomedicals, Santa Ana, CA, USA), and after two days of cultivation, a subset of cells (after several strokes of gentle pipetting) at the density of 3×10^6–4×10^6 cells/mL was transferred into new plastic Petri dishes (Nunc, Nunclon Surface, Roskilde, Denmark) on the coverslips for further culturing. We used three types of the cell culture media (98%) supplemented with 2% FCS: seawater and the coelomic fluid preparations of intact and injured sea urchins (*S. intermedius*) that were obtained as described in [15]. In few, a day after damaging sea urchins, the coelomic fluids from control and injured sea urchins were collected by puncture in the area of Aristotle's lantern. After removal of the clotting of coelomocytes by centrifugation, the supernatant was sterilized by filtration (0.22 μm, Millipore, CA, USA). The cell cultures were maintained by replacement of 50% of the medium at 3–5-day intervals for 3–42 days at 17 °C. To observe primary cell cultures during the overall period we used an inverted microscope Axiovert 200M (Carl Zeiss, Jena, Germany).

3.3. Immunohistochemical Analysis

After cultivation on fibronectin-coated or non-coated coverslips for 12–72 h, the cells were fixed in 4% paraformaldehyde (Sigma) in PBS, pH 7.8, for 7–10 min at 4 °C and rinsed three times in cold PBS. The material was stored in PBS with 0.03% NaN_3 at 4 °C until needed. To reduce non-specific binding, the samples were incubated overnight in a blocking solution containing 10% normal goat serum (Sigma), 0.25% BSA, 0.1% Triton X-100, and 0.03% NaN_3 in PBS. The cells were first incubated for 8 h at 10 °C in blocking solution with the following primary antibodies (Abs): mouse monoclonal anti-α acetylated tubulin (Sigma, diluted 1:1000) and rabbit polyclonal anti-phospho-H3 histone (Merck, Millipore, Germany, diluted 1:500). As secondary Abs, we used goat anti-mouse (GAM) or goat anti-rabbit (GAR) secondary Abs conjugated to Alexa Fluor 488, 546 or 633 (Molecular Probes, Eugene, OR, USA) at a dilution of 1:1000–1:2000 in PBS for 2 h at RT. The specimens were then washed and embedded in the Vectashield mounting medium (Vector Laboratories, Burlingame, CA, USA) containing 0.1 μg/mL DAPI to reveal the nuclei. For negative controls, primary Abs were omitted from the staining protocol. Specimens were analyzed with a confocal laser scanning microscope (Zeiss LSM 780) equipped with a high sensitivity GaAsP detector (with the 63× or 100× oil-immersion objectives). For 3D reconstructions, ImageJ software (NIH) was used. The number of dividing cells in each type of culture medium was estimated by calculating the percentage of phospho-H3 histone-positive cells among the total number of examined cells (determined by DAPI staining of nuclei). At least 300–500 DAPI-stained cells were counted in each experiment for each type of culture medium tested.

3.4. MALDI and ESI MS

Coelomic fluids obtained from intact and injured sea urchins, as described previously [15] were selected for MALDI-MS analysis. The mass spectra were registered using an Ultraflex-III TOF/TOF mass spectrometer (Bruker Daltonics, Bremen, Germany), equipped with a smart-beam laser (355 nm) having a pulse frequency of 100 Hz and a pulse width of 3 ns (PIBOC, FEB RAS). The mass spectrometer was used in the linear mode with delayed ion extraction and with the matrix alpha-cyano-4-hydroxy-cinnamic acid (Bruker Daltonics). The Protein Calibration Standard I (Bruker Daltonics) (1 μL) was added as the internal standard.

Metabolites were extracted from cultivated sea urchin cells. Quinone pigments were extracted with 1 mL of acidified ethanol (1 part 25% HCl and 3 parts of 96% ethanol) for 24 h at 4 °C, as described [28]. Diethyl ether was added to supernatants, and an organic layer containing quinones was washed with 5% NaCl until the acid was almost removed. The ether solution was dried over anhydrous sodium sulfate. The extract including the pigments was stored at −25 °C in the dark, and it was then analyzed by ESI-MS using a mass spectrometer Agilent 6510 Q-TOF LC/MS (Agilent Technologies, Santa Clara, CA, USA).

The alcoholic extracts were obtained from cultivated cells (5×10^6 cells/mL) after 3 and 6 days in culture. The sample solution was membrane-filtered (0.45 μm, Millipore), and 2 μL aliquots were used for analysis.

The mass spectra were recorded in the negative-ion detection mode within the *m/z* mass range of 100–800 and at mass-to-charge ratio (*m/z*) of 70–500 for MS/MS spectra (scan time 1 s). The quinonoid pigments were identified by exact mass measurement of deprotonated molecula $[M - H]^-$ in the MS spectra and by comparisons of the characteristic MS/MS spectra with the MS/MS spectra of pure naphthoquinone standards. The MS/MS spectra were recorded in the MS/MS mode using a collision energy ranging from 20 V to 35 V; the precursor ions were isolated with an isolation width of 1.3 *m/z*. The mass spectrometer was calibrated using the ESI-L Low Concentration Tuning Mix (Agilent Technologies, USA). A high-resolution MS analysis was performed by adding the Reference Mix (Agilent Technologies, USA) through a reference sprayer in a Dual-ESI source. The data were analyzed with FlexAnalysis Version 3.0 (Bruker Daltonics, Bremen, Germany).

3.5. Quantitative Real-Time PCR (Q-Real-Time PCR)

Sterile solutions of ShA (Sigma) in seawater at the desired concentrations (0.5 mM and 2.0 mM) were added in a blastula-derived cell culture of the sea urchin *S. intermedius* after seeding, and the cells were incubated during four days. After this period, total RNAs from cultivated sea urchin cells were extracted with Yellow Solve reagent (Clonogen, St. Petersburg, Russia) and treated with DNase I (Sileks, Moscow, Russia) to remove genomic DNA. The first strand of cDNA was synthesized using 1.5 μg of total RNA as a template with the Reverse Transcription System (Sileks) in a 50 μL reaction volume as described previously [5]. The 0.5–2 μL samples of reverse transcription products were then amplified by PCR on the sea urchin actin gene using the primers 5′CAA CGG ATC CGG TAT GGT GAA GGC and 5′TCC AGA CGG AGG ATG GCG TGG GGA. The PCR reactions were made using iCycler thermocycler (Bio-Rad Laboratories, Inc., Hercules, CA, USA) with the conditions: one cycle of 2 min at 95 °C followed by 40 cycles of 15 s at 95 °C, 15 s at 50 °C and 35 s at 72 °C with a final extension cycle of 10 min at 72 °C. In the following Q-real-time PCR analyses, we used only those reverse transcription reactions that resulted in 500 bp PCR products for the *actin* gene. We discarded those reverse transcription reactions that resulted in both 500 and 700 bp PCR products for the *actin* gene, which indicated DNA contamination.

Quantitative real-time PCR was performed using the established protocol [15]. For TaqMan Q-real-time RT-PCR, cDNAs were amplified in 20 μL of the reaction mixture containing 1× TaqMan Buffer, 2.5 mM $MgCl_2$, 250 μM of each deoxynucleotide, 1 U Taq DNA polymerase, 0.5–2 μL cDNA samples, and 0.25 μM of each primer and probe (Real-Time PCR Kit, Syntol, Moscow, Russia). The amplification conditions consisted of one cycle of 2 min at 95 °C followed by 50 cycles of 10 s at 95 °C and 25 s at 62 °C. The TaqMan PCR assays were performed in an iCycler thermocycler supplied with the iQ5 Multicolor Real-Time PCR detection system (Bio-Rad Laboratories, Inc., Hercules, CA, USA) and data were analyzed with the iQ5 Optical System Software v.2.0 as described previously [6] and presented in relative units. The *S. intermedius actin* gene (GenBank accession number DQ229162) and *ubiquitin* gene (LOC754856) were used as an endogenous control to normalize variance in the quality and the amount of cDNA used in each Q-real-time RT-PCR experiment. The data were summarized from five independent experiments. Primers and TaqMan probes for the *actin, ubiquitin, pks* genes used in Q-real-time PCR were described previously [6,15].

3.6. Statistical Analysis

The data are presented as mean ± standard error of the mean (SE). Data from the experiments were subjected to a one-way analysis of variance (ANOVA) with Tukey's pairwise comparison test at $p = 0.05$ with the use of the Microsoft Office Excel 2003 program.

4. Conclusions

This study provides evidence for the production of naphthoquinone pigments in cultivated sea urchin cells. Echinochrome A and spinochrome E were produced by cultivated cells from the sand dollar *S. mirabilis* in all tested media, but only spinochromes were recorded in the cultivated cells from the sea urchin *S. intermedius*. Our data support the hypothesis that specific components of sea urchin coelomic fluids might act as inductive signals in pigment differentiation. The differentiation of pigment cells is accompanied by the active expression of genes involved in naphthoquinone synthesis and appears to be important for defense processes. Thus, our findings and the technology developed for directed pigment cell differentiation in culture may be instrumental in solving some practical tasks in marine biotechnology, including the generation of cell cultures producing complex bioactive compounds with therapeutic potential.

Acknowledgments

This study was supported by the Russian Foundation for Basic Research (12-04-00363a), the Program «Far East» of the Russian Academy of Sciences (11.2.1, 11.2.2), and the Program at the Far Eastern Federal University (11 G34.31.0010). The work was partly performed at the "Far Eastern Cooperative Center of Electron Microscopy" (A. V. Zhirmunsky Institute of Marine Biology, FEB RAS, Vladivostok, Russia). The authors express their thanks to Maria Maiorova for her help in obtaining of confocal microscope images and to the anonymous reviewers for their valuable comments which have improved the manuscript.

Author Contributions

N.A. and N.O. designed and coordinated the research, took part in the experiments with cell cultures and contributed equally in writing the MS; K.K. and N.A. carried out the real-time-PCR experiments and analyzed the data; P.D. performed the MALDI and ESI MS analysis. All authors read and approved the final manuscript.

References

1. Evans-Illidge, E.A.; Logan, M.; Doyle, J.; Fromont, J.; Battershill, C.N.; Ericson, G.; Wolff, C.W.; Muirhead, A.; Kearns, P.; Abdo, D.; *et al.* Phylogeny drives large scale patterns in Australian marine bioactivity and provides a new chemical ecology rationale for future biodiscovery. *PLoS One* **2013**, *8*, e73800; doi:10.1371/journal.pone.0073800.
2. Volkman, J.K. Australiasian research on marine natural products: Chemistry, bioactivity and ecology. *Mar. Freshw. Res.* **1999**, *50*, 761–779.
3. Koltsova, E.A.; Boguslavskaya, L.V.; Maximov, O.B. On the functions of quinonoid pigments in sea-urchin embryos. *Invertebr. Reprod. Dev.* **1981**, *4*, 17–23.
4. Service, M.; Wardlaw, A. Echinochrome-A as a bactericidal substance in the coelomic fluid of *Echinus esculentus* (L.). *Comp. Biochem. Physiol. Part B* **1984**, *79*, 161–165.
5. Castoe, T.A.; Stephens, T.; Noonan, B.P.; Calestani, C. A novel group of type I polyketide synthases (PKS) in animals and the complex phylogenomics of PKSs. *Gene* **2007**, *392*, 47–58.
6. Kiselev, K.V.; Ageenko, N.V.; Kurilenko, V.V. Involvement of the cell-specific pigment genes pks and sult in the bacteria defense response of the sea urchin *Strongylocentrotus intermedius. Dis. Aquat. Organ.* **2013**, *103*, 121–132.
7. Thompson, R.H. *Naturally Occurring Quinones*, 2nd ed.; Academic Press: New York, NY, USA, 1971; p. 734.
8. Singh, H.; Moore, R.E.; Scheuer, P.J. The distribution of quinone pigments in echinoderms. *Experientia* **1967**, *23*, 624–626.
9. Smith, L.C.; Rast, J.P.; Brocton, V.; Terwilleger, D.P.; Nair, S.V.; Bucley, K.M.; Majestke, A.J. The sea urchin immune system. *Invertebr. Surviv. J.* **2006**, *3*, 25–29.
10. Gibson, A.W.; Burke, R.D. The origin of pigment cells in embryos of the sea urchin *Strongylocentrotus purpuratus*. *Dev. Biol.* **1985**, *107*, 414–419.

11. Mishchenko, N.P.; Fedoreev, S.A.; Bagirova, V.L. Histochrome: A new original domestic drug. *Pharm. Chem. J.* **2003**, *37*, 48–52.
12. Kominami, T.; Takata, H. Process of pigment cell specification in the sand dollar, *Scaphechinus mirabilis. Dev. Growth Differ.* **2002**, *44*, 113–125.
13. Calestani, C.; Rast, J.P.; Davidson, E.H. Isolation of pigment cell specific genes in the sea urchin embryo by differential macroarray screening. *Development* **2003**, *130*, 4587–4596.
14. Beeble, A.; Calestani, C. Expression pattern of polyketide synthase-2 during sea urchin development. *Gene Expr. Patt.* **2012**, *12*, 7–10.
15. Ageenko, N.V.; Kiselev, K.V.; Odintsova, N.A. Expression of pigment cell-specific genes in the ontogenesis of the sea urchin *Strongylocentrotus intermedius. Evid.-Based Compl. Altern. Med.* **2011**, *2011*, doi:10.1155/2011/730356.
16. Odintsova, N.A.; Kiselev, K.V.; Bulgakov, V.P.; Koltsova, E.A.; Yakovlev, K.V. Influence of the activator of transcription gal4 on growth and development of embryos andembryonic cells in primary cultures of sand dollar. *Russ. J. Dev. Biol.* **2003**, *34*, 217–222.
17. Hans, F.; Dimitrov, S. Histone H3 phosphorylation and cell division. *Oncogene* **2001**, *20*, 3021–3027.
18. Sharlaimova, N.; Shabelnikov, S.; Petukhova, O. Small coelomic epithelial cells of the starfish *Asterias rubens* L. that are able to proliferate *in vivo* and *in vitro*. *Cell Tissue Res.* **2014**, doi:10.1007/s00441-013-1766-8.
19. Wagenaar, E.B. The timing of synthesis of proteins required for mitosis in the cell cycle of the sea urchin embryo. *Exp. Cell Res.* **1983**, *144*, 393–403.
20. Bromham, L.D.; Degnan, B.M. Hemichordates and deuterostome evolution: Robust molecular phylogenetic support for a hemichordate plus echinoderm clade. *Evol. Dev.* **1999**, *1*, 166–171.
21. Wray, G.A.; Lowe, C.J. Developmental regulatory genes and echinoderm evolution. *Syst. Biol.* **2000**, *49*, 28–51.
22. Peter, R.; Gschwentner, R.; Schürmann, W.; Rieger, R.M.; Ladurner, P. The significance of stem cells in free-living flatworms: One common source for all cells in the adult. *J. Appl. Biomed.* **2004**, *2*, 21–35.
23. Odintsova, N.A. Stem cells of marine invertebrates: Regulation of proliferation and induction of differentiation *in vitro*. *Cell Tissue Biol.* **2009**, *3*, 403–408.
24. Dheilly, N.M.; Raftos, D.A.; Haynes, P.A.; Courtney Smith, L.; Nair, S.V. Shotgun proteomics of coelomic fluid from the purple sea urchin, *Strongylocentrotus purpuratus*. *Dev. Comp. Immunol.* **2013**, *40*, 35–50.
25. Fink, A.L. Chaperone-mediated protein folding. *Physiol. Rev.* **1999**, *79*, 425–449.
26. Branco, P.C.; Pressinotti, L.N.; Borges, J.C.S.; Iunes, R.S.; Kfoury, J.R., Jr.; Silva, M.O.; Gonzalez, M.; Santos, M.F.; Peck, L.S. Cellular biomarkers to elucidate global warming effects on Antarctic sea urchin *Sterechinus neumayeri*. *Polar Biol.* **2012**, *35*, 221–229.
27. Branco, P.C.; Borges, J.C.S.; Santos, M.F.; Jensch, B.E., Jr.; Silva, M.O. The impact of rising sea temperature on innate immune parameters in the tropical subtidal sea urchin *Lytechinus variegatus* and the intertidal sea urchin *Echinometra lucunter*. *Mar. Environ. Res.* **2013**, *92*, 95–101.
28. Kuwahara, R.; Hatate, H.; Chkami, A.; Murata, H.; Kijidani, Y. Quantitative separation of antioxidant pigments in purple sea urchin shells using a reversed-phase high performance liquid chromatography. *LWT Food Sci. Technol.* **2010**, *43*, 1185–1190.

11

Identification and Biochemical Characterization of Halisulfate 3 and Suvanine as Novel Inhibitors of Hepatitis C Virus NS3 Helicase from a Marine Sponge

Atsushi Furuta [1,2], Kazi Abdus Salam [3], Idam Hermawan [4], Nobuyoshi Akimitsu [3], Junichi Tanaka [4], Hidenori Tani [5], Atsuya Yamashita [6], Kohji Moriishi [6], Masamichi Nakakoshi [7], Masayoshi Tsubuki [7], Poh Wee Peng [8], Youichi Suzuki [8], Naoki Yamamoto [8], Yuji Sekiguchi [2], Satoshi Tsuneda [1,*] and Naohiro Noda [1,2,*]

1 Department of Life Science and Medical Bioscience, Waseda University, 2-2 Wakamatsu-cho, Shinjuku-ku, Tokyo 162-8480, Japan; E-Mail: atsushi.5961@ruri.waseda.jp

2 Biomedical Research Institute, National Institute of Advanced Industrial Science and Technology (AIST), 1-1-1 Higashi, Tsukuba, Ibaraki 305-8566, Japan; E-Mail: y.sekiguchi@aist.go.jp

3 Radioisotope Center, The University of Tokyo, 2-11-16 Yayoi, Bunkyo-ku, Tokyo 113-0032, Japan; E-Mails: salam_bio26@yahoo.com (K.A.S.); akimitsu@ric.u-tokyo.ac.jp (N.A.)

4 Department of Chemistry, Biology and Marine Science, University of the Ryukyus, Nishihara, Okinawa 903-0213, Japan; E-Mails: damz_98@yahoo.com (I.H.); jtanaka@sci.u-ryukyu.ac.jp (J.T.)

5 Research Institute for Environmental Management Technology, National Institute of Advanced Industrial Science and Technology (AIST), 16-1 Onogawa, Tsukuba, Ibaraki 305-8569, Japan; E-Mail: h.tani@aist.go.jp

6 Department of Microbiology, Division of Medicine, Graduate School of Medicine and Engineering, University of Yamanashi, 1110 Shimokato, Chuo-shi, Yamanashi 409-3898, Japan; E-Mails: atsuyay@yamanashi.ac.jp (A.Y.); kmoriishi@yamanashi.ac.jp (K.M.)

7 Institute of Medical Chemistry, Hoshi University, 2-4-41 Ebara, Shinagawa-ku, Tokyo 142-8501, Japan; E-Mails: mnakako@hoshi.ac.jp (M.N.); tsubuki@hoshi.ac.jp (M.T.)

8 Department of Microbiology, Yong Loo Lin School of Medicine, National University of Singapore, Center for Translational Medicine, 14 Medical Drive, #15-02, Level 15, Singapore 117599, Singapore; E-Mails: micpwp@nus.edu.sg (P.W.P.); micys@nus.edu.sg (Y.S.); naoki_yamamoto@nuhs.edu.sg (N.Y.)

* Authors to whom correspondence should be addressed; E-Mails: stsuneda@waseda.jp (S.T.); noda-naohiro@aist.go.jp (N.N.)

Abstract: Hepatitis C virus (HCV) is an important etiological agent that is responsible for the development of chronic hepatitis, liver cirrhosis, and hepatocellular carcinoma. HCV nonstructural protein 3 (NS3) helicase is a possible target for novel drug development due to its essential role in viral replication. In this study, we identified halisulfate 3 (hal3) and suvanine as novel NS3 helicase inhibitors, with IC_{50} values of 4 and 3 μM, respectively, from a marine sponge by screening extracts of marine organisms. Both hal3 and suvanine inhibited the ATPase, RNA binding, and serine protease activities of NS3 helicase with IC_{50} values of 8, 8, and 14 μM, and 7, 3, and 34 μM, respectively. However, the dengue virus (DENV) NS3 helicase, which shares a catalytic core (consisting mainly of ATPase and RNA binding sites) with HCV NS3 helicase, was not inhibited by hal3 and suvanine, even at concentrations of 100 μM. Therefore, we conclude that hal3 and suvanine specifically inhibit HCV NS3 helicase via an interaction with an allosteric site in NS3 rather than binding to the catalytic core. This led to the inhibition of all NS3 activities, presumably by inducing conformational changes.

Keywords: marine organism; halisulfate 3; suvanine; hepatitis C virus; NS3 helicase; dengue virus

---

1. Introduction

An estimated 150 million people worldwide are chronically infected with the hepatitis C virus (HCV), a major etiological agent responsible for the development of chronic hepatitis, liver cirrhosis, and hepatocellular carcinoma (World Health Organization, 2013). The current standard therapy is based mainly on a triple combination of pegylated interferon-alfa, ribavirin, and a recently approved NS3 serine protease inhibitor (such as telaprevir), which increases the viral clearance rate to >70% [1,2]. However, because of severe side effects, the emergence of drug-resistant HCV mutations, and drug-drug interactions [3,4], the development of novel direct-acting antivirals that target the viral or host proteins involved in HCV replication are needed urgently. HCV nonstructural protein 3 (NS3) helicase has been considered as a novel antiviral target owing to its essential role in viral replication [5,6].

HCV is a member of the *Flaviviridae* family of positive-stranded RNA viruses. The viral genome contains a single open reading frame encoding a polyprotein that is processed by virus-encoded and host cellular proteases into structural and nonstructural proteins. The structural proteins (core protein [C], and the envelope glycoproteins E1 and E2) build up the virus particle, whereas the nonstructural proteins p7 and NS2 support particle assembly without being incorporated into the viral particles [7,8]. The remaining nonstructural proteins (NS3, NS4A, NS4B, NS5A, and NS5B) form a complex with viral RNA to support viral replication [9]. NS3 is a multifunctional enzyme with serine protease and NTPase/helicase domains at the *N*- and *C*-termini, respectively [10]. The NS3 helicase can unwind double-stranded RNA (dsRNA), double-stranded DNA, and RNA/DNA heteroduplexes in a 3′–5′ direction by using a nucleoside triphosphate as the energy source [11–14]. Although the exact role of NS3 helicase in the viral life cycle remains unclear, a fully functional NS3 helicase is required for replication of the HCV replicon [5] and for HCV replication in chimpanzees [15], suggesting that NS3 helicase inhibitors could be potential therapeutic agents. However, no HCV NS3 helicase inhibitors have yet been entered into clinical trials, at least in part due to similarities between NS3 and cellular RNA helicases [8].

HCV NS3 helicase is part of the family of viral DExH proteins; the NS3/NPH-II family that encompasses helicases from positive-stranded RNA viruses [16–18]. These closely related helicases share a catalytic core that consists mainly of NTPase and nucleic acid binding sites, as well as many other structural and functional features. Indeed, dengue virus (DENV) NS3 helicase, another viral DExH protein, and HCV NS3 helicase share highly conserved amino acid sequences, and consequently have similar conformational structures [19]. Thus, if a compound inhibits HCV NS3 helicase, it may also inhibit DENV NS3 helicase [20–22]. Assessing the inhibitory specificity can provide useful information to understand whether inhibitors target the NTPase, nucleic acid binding, or other allosteric sites of NS3 helicase.

HCV NS3 helicase inhibitors function by inhibiting NTP binding, nucleic acid binding, NTP hydrolysis or NDP release, the coupling of NTP hydrolysis to the translocation and unwinding of nucleic acids, or unwinding by sterically blocking helicase translocation [6]. In addition, owing to an interdependent linkage between NS3 helicase and serine protease activities [23–25], the inhibition of NS3 serine protease may also lead to the inhibition of NS3 helicase. Compounds that intercalate into the strands of double-stranded nucleic acids could also inhibit NS3 helicase [26].

Naturally occurring products are an important source of structurally diverse and biologically active secondary metabolites. The diversity of organisms in the marine environment has provided new drugs in almost all therapeutic areas [27–29]. To date, seven therapeutic agents derived from the marine environment are used as anticancer, antiviral, pain control, and hypertriglyceridemia agents [27]. The chemical structure has been isolated for two of these compounds, whereas the remaining five are synthetic agents based on marine products. An additional 13 agents are in phase 1, 2, or 3 clinical trials. Therefore, natural marine products include a number of highly significant lead compounds that are driving new drug development.

In this study, we screened extracts from marine organisms for NS3 helicase inhibitors using a fluorescence helicase assay based on photoinduced electron transfer (PET), as described in our previous study [30]. During purification, halisulfate 3 (hal3) and suvanine, which were isolated from marine sponge extracts, were identified as novel NS3 helicase inhibitors with IC_{50} values in the low micromolar range. The inhibitory effects of hal3 and suvanine against the other helicase-related activities of NS3 (ATPase, RNA binding, and serine protease activities) were also assessed. Finally, the inhibitory activities of hal3 and suvanine against DENV NS3 helicase were determined to characterize the binding sites of hal3 and suvanine.

2. Results and Discussion

To obtain novel NS3 helicase inhibitors, extracts from marine organisms were screened using a fluorescence helicase assay based on PET. Forty-three extracts prepared from marine organisms were screened, and 11 were identified that inhibited the helicase activity >50% (samples 4, 10, 13, 14, 17, 19, 21, 22, 25, 26, and 37) (Table 1), suggesting that these extracts contained NS3 helicase inhibitors. Of these extracts, sample 10 exhibited the strongest inhibition of NS3 helicase, and abolished its activity completely. Therefore, this extract was purified to isolate and concentrate the inhibitory components. After several purification steps, the inhibitory components were identified as hal3 and suvanine (Figure 1) by comparing their NMR spectra with those reported previously [31,32] for each compound (Supplementary Figures S1–S4). Hal3 and suvanine inhibited NS3 helicase activity in a dose-dependent manner, with IC_{50} values of 4 and 3 μM, respectively (Figure 2A,B).

Table 1. Inhibitory effects of extracts from marine organisms on hepatitis C virus (HCV) nonstructural protein 3(NS3) helicase activity.

No.	NS3 Helicase Activity (% of Control) *	Marine Organism	Species
1	92	Sponge	*Unidentified*
2	74	Soft coral	*Briareum*
3	57	Tunicate	*Unidentified*
4	36	Sponge	*Liosina*
5	54	Sponge	*Unidentified*
6	71	Sponge	*Xestospongia*
7	77	Sponge	*Epipolasis*
8	110	Sponge	*Unidentified*
9	86	Sponge	*Strongylophora*
10	**0**	**Sponge**	***Unidentified***
11	83	Sponge	*Stylotella aurantium*
12	78	Sponge	*Epipolasis*
13	25	Sponge	*Unidentified*
14	43	Sponge	*Hippospongia*
15	75	Sponge	*Unidentified*
16	85	Sponge	*Unidentified*
17	49	Sponge	*Xestospongia testudinaria*
18	69	Sponge	*Unidentified*
19	40	Sponge	*Theonella*
20	64	Sponge	*Unidentified*
21	44	Sponge	*Unidentified*
22	46	Sponge	*Petrosia*
23	72	Tunicate	*Unidentified*
24	61	Sponge	*Unidentified*
25	50	Tunicate	*Didemnum molle*
26	33	Sponge	*Unidentified*
27	67	Sponge	*Unidentified*
28	87	Soft coral	*Unidentified*
29	62	Sponge	*Unidentified*
30	60	Sponge	*Unidentified*
31	85	Sponge	*Cinachyra*
32	70	Sponge	*Liosina*
33	68	Sponge	*Unidentified*
34	58	Sponge	*Unidentified*
35	72	Sponge	*Stylotella*
36	57	Sponge	*Unidentified*
37	39	Sponge	*Unidentified*

Table 1. *Cont.*

38	72	Tunicate	*Didemnum*
39	62	Sponge	*Unidentified*
40	71	Jellyfish	*Unidentified*
41	74	Sponge	*Unidentified*
42	52	Tunicate	*Unidentified*
43	67	Annelid	*Unidentified*

* NS3 helicase activity in the presence of extract is expressed as a percentage of control in the absence of extract (100%); The sample with the strongest inhibition against NS3 helicase is in bold, underlined font; samples with relatively strong inhibition against NS3 helicase (<50%) are underlined.

Figure 1. Structures of halisulfate 3 (hal3) and suvanine.

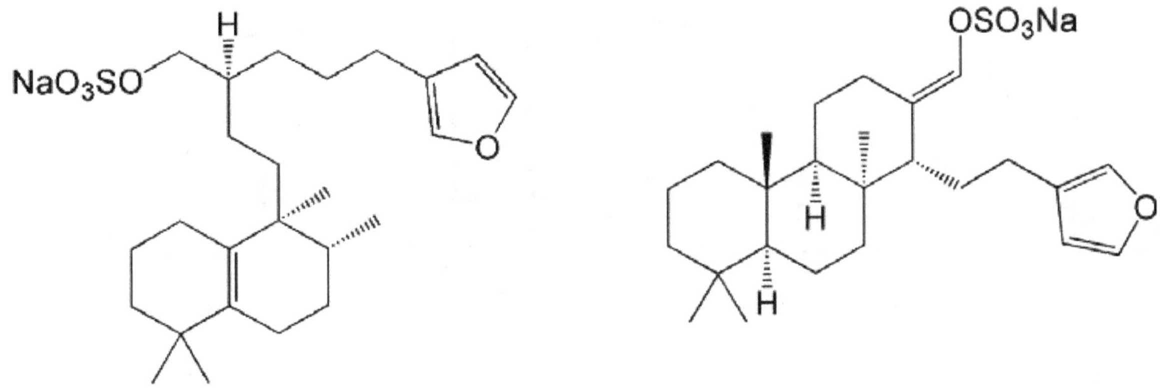

Halisulfa e

Suvanine

The inhibitory effects of hal3 and suvanine were confirmed using a gel-based helicase assay. The helicase activity was calculated as the ratio of the signal intensity derived from single-stranded (ssRNA) in the sample containing the inhibitor to the control sample (lacking the inhibitor but containing DMSO vehicle). Similar to the results of the fluorescence helicase assay, hal3 and suvanine inhibited helicase-catalyzed RNA unwinding in a dose-dependent manner (Figure 2C,D). Therefore, these data clearly indicate that hal3 and suvanine exert inhibitory effects. Hal3 and suvanine were identified in 1988 [33] and 1985 [34], respectively. They have similar distinguishing structural features of a sulfated side chain and a furan moiety at the terminus of the molecule (Figure 1). Although some bioactivities for hal3 and suvanine have been reported, this report is the first that identifies these compounds as helicase inhibitors. In addition, bioactive effects of hal3 alone have not been reported. A mixture of halisulfates 2–5 (hal3 and its analogues) showed antimicrobial activity against *S. aureus*, *C. albicans*, and *B. subtilis*. Moreover, a mixture of halisulfates 2–4 inhibited PMA-induced inflammation in a mouse ear edema assay and inhibited phospholipase A_2 [31]. Suvanine is a serine protease inhibitor [35] and an antagonist of the mammalian bile acid sensor farnesoid-X-receptor [36]. In addition, suvanine interferes with heat shock protein 60, a chaperone involved in the inflammatory response, giving evidence for its anti-inflammatory properties [37].

Figure 2. Inhibition of NS3 helicase-catalyzed RNA unwinding activity by hal3 and suvanine. (**A**,**B**) Inhibition curves of hal3 and suvanine generated using a fluorescence helicase assay. The NS3 helicase activities of samples containing inhibitor were calculated relative to control samples containing DMSO vehicle rather than inhibitor. The data are presented as mean ± standard deviation of three replicates; (**C**,**D**) Gel images representing the inhibitory effects of hal3 and suvanine in a gel-based helicase assay. Fluorescence-labeled ssRNA and dsRNA were applied to lanes *1* and *2*, respectively. The dsRNA was incubated with NS3 in the presence of increasing concentrations of inhibitor (lanes *3*–*7*, 0–100 μM). Lane *8* shows the control reaction in the absence of NS3.

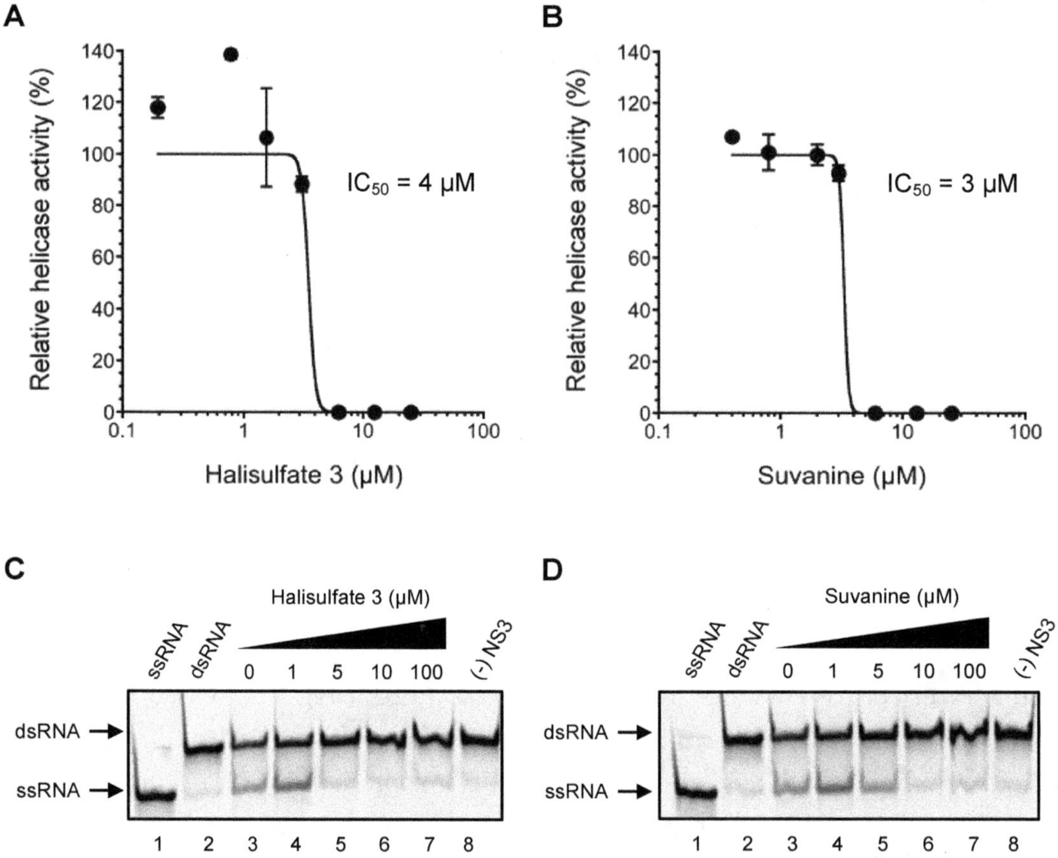

As the unwinding ability of NS3 helicase is dependent on ATP hydrolysis, the amount of inorganic phosphate (Pi) released from radioisotope-labeled ATP was measured to determine the effects of hal3 and suvanine on the ATPase activity of NS3 (Figure 3). The released Pi was separated by thin-layer chromatography and visualized using autoradiography. The density of the upper spots corresponding to Pi, which represents ATPase activity, decreased dose-dependently for both hal3 and suvanine. The ATPase activity was calculated as the ratio of the signal intensity derived from the released Pi in the sample containing inhibitor to that in the control sample (lacking the inhibitor but containing DMSO vehicle). The IC_{50} values of hal3 and suvanine were calculated to be 8 and 7 μM, respectively. As this concentration range is similar to that in which RNA unwinding was inhibited (Figure 2), it is likely that hal3 and suvanine inhibit NS3 helicase via the inhibition of ATPase activity.

Figure 3. Effects of hal3 and suvanine on NS3 ATPase activity demonstrated by autoradiography of an ATPase assay using [γ-^{32}P] ATP. Lane *1* contains the control reaction without NS3. Lanes *2–8* (**A**) and *2–9* (**B**) show the ATP hydrolysis reaction with poly(U) RNA at increasing concentrations (0–100 μM) of hal3 and suvanine, respectively.

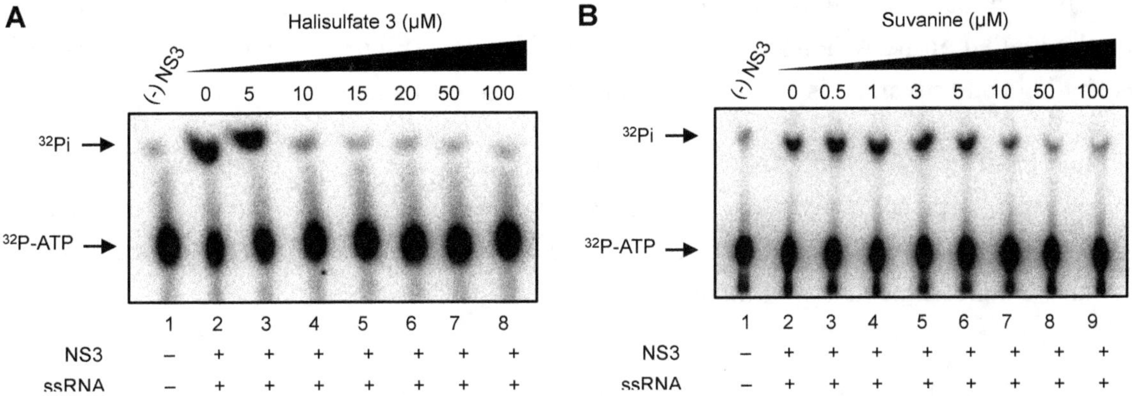

As RNA binding is required for NS3 helicase activity, the effects of hal3 and suvanine on NS3 RNA binding activity were examined by gel mobility shift assay (Figure 4). As a control, the non-specific binding of ssRNA to bovine serum albumin (BSA) was assessed (lane 2). The density of the upper bands corresponding to the NS3-ssRNA complex, which represents NS3 RNA binding activity, decreased dose-dependently in the presence of both hal3 and suvanine. RNA binding activity was calculated as the ratio of the signal intensity derived from the NS3-ssRNA complex in the sample containing the inhibitor to that in the control sample (lacking the inhibitor but containing DMSO vehicle). The IC_{50} values of hal3 and suvanine were calculated to be 8 and 3 μM, respectively. The data presented in Figures 2 and 4 reveal that the NS3 helicase and RNA binding activities decrease at similar inhibitor concentration ranges for hal3 and suvanine, suggesting that the inhibition of NS3 helicase by these compounds is associated with RNA binding activity.

Figure 4. Effects of hal3 and suvanine on NS3 RNA binding activity, assessed by autoradiography of a gel mobility shift assay using ^{32}P-labeled ssRNA. Lanes *1* and *2* contain control reactions consisting of heat-denatured ssRNA and 300 nM BSA instead of NS3, respectively. Lanes *3–9* (**A**) and *3–10* (**B**) show the RNA binding reaction with increasing concentrations (0–100 μM) of hal3 and suvanine, respectively.

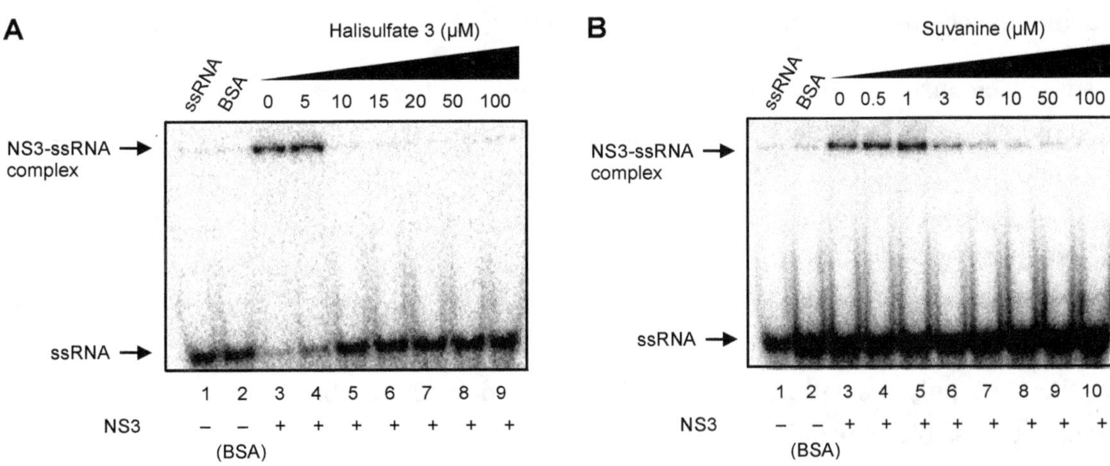

It was reported that the helicase activity of NS3 is interdependently linked to its serine protease activity [23–25]. Therefore, we examined the effects of hal3 and suvanine on NS3 serine protease activity using a fluorescence serine protease assay (Figure 5). Serine protease activity decreased in a dose-dependent manner in the presence of hal3 and suvanine, with IC_{50} values of 14 and 34 μM, respectively. Although the inhibition of the serine protease activity seems to be rather modest compared with that of the ATPase and RNA binding activities (Figures 3 and 4), the inhibition of NS3 helicase by hal3 and suvanine is likely to be also related to serine protease activity.

Figure 5. Effects of hal3 (**A**) and suvanine (**B**) on NS3 serine protease activity. The NS3 serine protease activity of samples containing inhibitor was calculated relative to control samples containing DMSO vehicle rather than inhibitor. The data are presented as means ± standard deviation of three replicates.

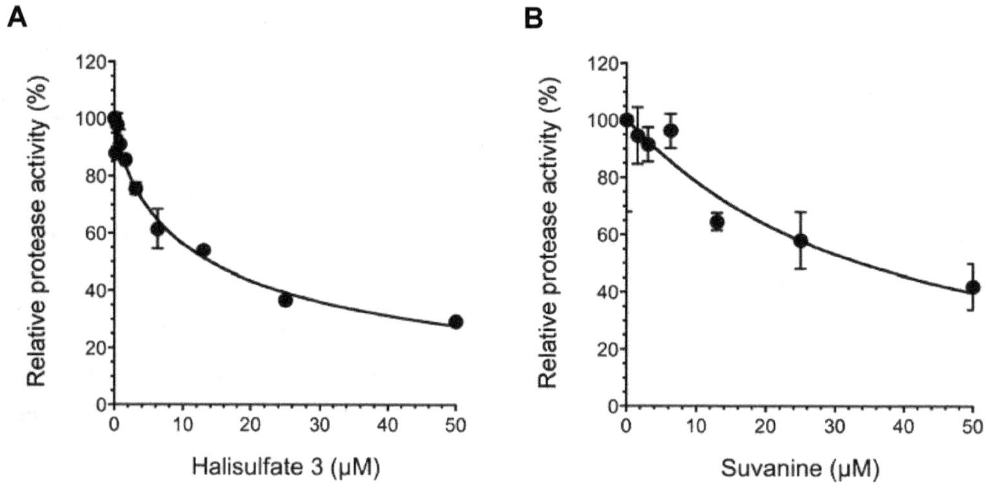

The catalytic cores of DENV and HCV NS3 helicases, which consist predominantly of ATPase and RNA binding sites, share almost identical folds and extensive structural similarity [38]. Because the substrate specificity of DENV and HCV NS3 helicases is similar [39], the dsRNA substrate and capture strand of the gel-based HCV NS3 helicase assay were also used for the gel-based DENV NS3 helicase assay (Figure 6), and helicase activity was calculated as described above. Hal3 and suvanine did not abolish DENV NS3 helicase activity, even in the presence of 100 μM of each inhibitor. This finding suggests that the inhibitory effects of hal3 and suvanine are specific to HCV NS3 helicase, and that these inhibitors bind less efficiently to any site in DENV NS3 helicase, including the catalytic core.

This study demonstrated that hal3 and suvanine inhibit the ATPase, RNA binding, and serine protease activities of NS3 (Figures 3–5). Taken together with observations that hal3 and suvanine did not inhibit DENV NS3 helicase (Figure 6), it is likely that these inhibitors do not bind to the catalytic core that contains the ATPase activity and RNA binding sites. Therefore, we conclude that hal3 and suvanine inhibit HCV NS3 helicase via interactions with allosteric sites of NS3. This likely induces conformational changes in NS3, inhibiting or abolishing its activities. Compounds with inhibitory activities against both helicase and serine protease activities have been reported previously [40]; however, there are only a small number of studies, and detailed inhibitory mechanisms are yet to be elucidated. The possible allosteric sites to which hal3 and suvanine bind could be an interface that forms between the helicase and protease domains of NS3. Indeed, a novel small-molecule binding site at the interface between these two domains was reported recently [41]. Furthermore, the inhibitory specificity of hal3 and suvanine against HCV NS3 helicase might be explained by structural differences between HCV and DENV NS3 helicases. A specific beta-strand tethers the *C* terminus of the helicase domain to the protease domain of HCV NS3, maintaining it in a compact conformation that differs from the extended conformation of DENV NS3 helicase [42]. As only HCV NS3 helicase forms an interface between the helicase and protease domains, the specificity of hal3 and suvanine for HCV NS3 helicase would be explained by the binding of hal3 and suvanine to the interface of HCV NS3.

Figure 6. Effects of hal3 (**A**) and suvanine (**B**) on DENV NS3 helicase activity, assessed using a gel-based helicase assay. Fluorescence-labeled ssRNA and dsRNA were applied to lanes *1* and *2*, respectively. The dsRNA was incubated with NS3 in the presence of increasing concentrations of inhibitor (lanes *3–7*, 0–100 μM). Lanes *8* and *9* contain the control reaction mixtures in the absence of NS3, and in the presence of 15 mM ATP-gamma-S as an inhibition control, respectively.

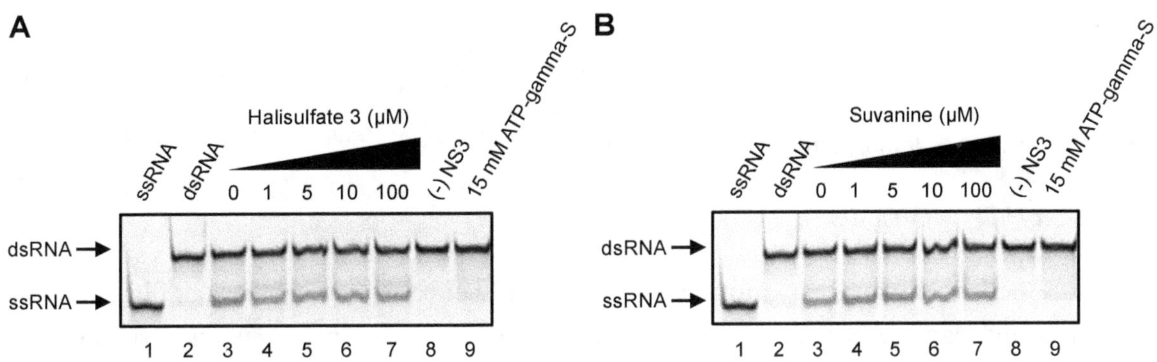

3. Experimental Section

3.1. Preparation of Extracts from Marine Organisms

Specimens of marine organisms were collected by scuba diving in Okinawa, Japan, and Sorong, Indonesia, and kept frozen until use. The specimens were chopped into small pieces, and soaked in acetone for 20 h followed by methanol for 6 h. The acetone and methanol solutions were then combined and concentrated, and residual materials were separated into ethyl acetate and aqueous layers; each layer was then dried to obtain residues.

3.2. Screening for HCV NS3 Helicase Inhibitors

The PET-based fluorescence helicase assay was performed as described previously [30]. The dsRNA substrate was prepared by annealing the 5′ BODIPY FL-labeled fluorescence strand (5′-CUAUUACCUCCACCCUCAUAACCUUUUUUUUUUUUUU-3′) to the quencher strand (5′-GGUUAUGAGGGUGGAGGUAAUAG-3′) at a 1:2 molar ratio. The dsRNA substrate contains the 3′-overhang that is necessary for the NS3 helicase to bind RNA prior to duplex unwinding. The capture strand (5′-CTATTACCTCCACCCTCATAACC-3′), which is complementary to the quencher strand, prevents the unwound duplexes from reannealing. None of the capture, quencher, or fluorescence strands are self-complementary. The fluorescence strand was purchased from J-Bio 21 Corporation (Tokyo, Japan), and was labeled with BODIPY FL at the 5′-end via an aminohexylphosphate linker with a six-carbon spacer. The quencher and capture strands were purchased from Japan Bio Services (Saitama, Japan). The reaction mixture contained 25 mM MOPS-NaOH (pH 6.5), 3 mM $MgCl_2$, 2 mM dithiothreitol, 4 U RNasin (Promega, Madison, WI, USA), 50 nM dsRNA substrate, 100 nM capture strand, 5 mM ATP, a marine organism extract, and 240 nM NS3 in a total reaction volume of 20 μL. Each marine organism extract diluted in DMSO was added to the reaction mixture at a final concentration of 17.5–32.5 μg/mL. The full-length HCV NS3 protein with serine protease and NTPase/helicase activities was expressed and purified as described previously [43].

The reaction was started by the addition of HCV NS3 helicase, and was performed at 37 °C for 30 min using a LightCycler 1.5 (Roche Diagnostics, Basel, Switzerland). The fluorescence intensity was recorded every 5 s from 0 to 5 min, and then every 30 s from 5 to 30 min. Helicase activity was calculated as the initial reaction velocity relative to control (in the absence of a marine extract, but presence of DMSO vehicle). The IC_{50} was calculated using KaleidaGraph (Synergy Software, Reading, PA, USA) by fitting plots of % activity *vs.* $[I]$ using Equation (1) unless otherwise noted [44]:

$$\% \text{ Activity} = \frac{100}{1+\left([I]/\text{IC}_{50}\right)^{h}} \tag{1}$$

where h is the Hill coefficient, and $[I]$ is the inhibitor concentration.

3.3. Gel-Based HCV NS3 Helicase Assay

A gel-based helicase assay was performed on HCV NS3 helicases using an Alexa Fluor 488-labeled dsRNA strand and capture strand with the same nucleic acid sequences described in Section 3.2. The dsRNA substrate was prepared by annealing the 5′ Alexa Fluor 488-labeled strand (5′-CUAUUACCUCCACCCUCAUAACCUUUUUUUUUUUUUU-3′) to the complementary strand (5′-GGUUAUGAGGGUGGAGGUAAUAG-3′) at a 1:2 molar ratio. The same capture strand described in Section 3.2 was used. All nucleic acid strands were purchased from Japan Bio Services (Saitama, Japan). The reaction mixture for HCV NS3 helicase contained the same components as described in Section 3.2, with increasing concentrations of hal3 or suvanine in a reaction volume of 20 μL. The reaction was started by the addition of HCV NS3 helicase, and performed at 37 °C for 60 min using a GeneAmp PCR System 2700 (Applied Biosystems, Foster City, CA, USA). The reaction was stopped by the addition of 5 μL of helicase termination buffer containing 10 mM Tris-HCl (pH 7.5), 50 mM EDTA, 30% glycerol, 0.06% bromophenol blue, and 0.12% Orange G. The inhibition of NS3 helicase was analyzed on a 20% native Tris/borate/EDTA (TBE) polyacrylamide gel, and labeled RNAs were visualized using Typhoon 9210 (GE Healthcare, Waukesha, WI, USA). The helicase activity was calculated as the ratio of the signal intensity derived from ssRNA in the sample containing inhibitor to that in the control sample containing DMSO vehicle instead of inhibitor.

3.4. ATPase Assay

NS3 ATPase activity was determined directly by monitoring [γ-^{32}P] ATP hydrolysis by thin-layer chromatography, as described previously [45,46]. The reaction mixture contained 25 mM MOPS-NaOH (pH 7.0), 1 mM dithiothreitol, 5 mM $MgCl_2$, 5 mM $CaCl_2$, 1 mM [γ-^{32}P] ATP (Muromachi Yakuhin, Tokyo, Japan), 300 nM NS3, 0.1 μg/μL poly (U) ssRNA (Sigma-Aldrich, St. Louis, MO, USA), and increasing concentrations of hal3 or suvanine in a volume of 10 μL. The reaction was conducted at 37 °C for 10 min, and stopped by the addition of 10 mM EDTA. Two microliters of each reaction mixture was then spotted onto a polyethyleneimine cellulose sheet (Merck, Darmstadt, Germany) and developed in 0.75 M LiCl/1 M formic acid solution for 20 min. The cellulose sheet was dried, and the released [γ-^{32}P] phosphoric acid was visualized using an Image Reader FLA-9000 and quantified using Multi Gauge software V 3.11 (Fujifilm, Tokyo, Japan). ATPase activity was calculated as the ratio of the signal intensity derived from the released Pi in the sample containing inhibitor to that in the control sample containing DMSO vehicle instead of inhibitor.

3.5. RNA Binding Assay

NS3 RNA binding activity was determined by gel mobility shift assay, as described previously [45,46]. The ssRNA (5′-UGAGGUAGUAGGUUGUAUAGU-3′) synthesized by Gene Design (Osaka, Japan) was labeled at the 5′-end with [γ-^{32}P] ATP (Muromachi Yakuhin, Tokyo, Japan) using T4 polynucleotide kinase (Toyobo, Osaka, Japan) at 37 °C for 60 min, and purified using the phenol-chloroform extraction method. The reaction mixture contained 30 mM Tris-HCl (pH 7.5), 100 mM NaCl, 2 mM $MgCl_2$, 1 mM dithiothreitol, 20 U RNasin Plus (Promega), 300 nM NS3, 0.1 nM ^{32}P-labeled ssRNA, and increasing concentrations of inhibitor in a volume of 20 μL. The reaction was performed at room temperature for 15 min. An equal volume of a dye solution containing 0.025% bromophenol blue and 10% glycerol in 0.5× TBE was then added to each reaction mixture, and samples were loaded onto a 6% native-polyacrylamide gel. The labeled RNA bands were visualized using an Image Reader FLA-9000 and quantified using Multi Gauge software V 3.11 (Fujifilm, Tokyo, Japan). RNA binding activity was calculated as the ratio of the signal intensity derived from the NS3-ssRNA complex in the sample containing hal3 or suvanine to that in the control sample containing DMSO vehicle rather than inhibitor.

3.6. Serine Protease Assay

A fluorescence NS3 serine protease assay, based on fluorescence resonance energy transfer, was conducted using reagents provided in a SensoLyte™ 520 HCV protease assay kit (AnaSpec, San Jose, CA, USA), as described previously [30]. Briefly, NS3 protein with a two-fold excess of the NS4A cofactor peptide Pep4AK was prepared in 1× assay buffer provided with the kit. HCV NS3/4A protease was mixed with increasing concentrations of inhibitor, and incubated at 37 °C for 15 min. The reaction was started by the addition of 5-FAM/QXL 520 substrate in a 20 μL total reaction volume containing 240 nM HCV NS3/4A protease and increasing concentrations of hal3 or suvanine. Reactions were then incubated at 37 °C for 120 min on a LightCycler 1.5 (Roche Diagnostics, Basel, Switzerland), and the fluorescence intensity was recorded every min for 120 min. NS3 serine protease activity was calculated as the initial reaction velocity in the sample containing inhibitor relative to the control sample containing DMSO vehicle rather than inhibitor.

3.7. Gel-Based DENV NS3 Helicase Assay

A gel-based helicase assay was performed using DENV NS3 helicases, and the Alexa Fluor 488-labeled dsRNA strand and capture strand with the same nucleic acid sequences described in the Section 3.3. DENV NS3 helicase requires a single stranded 3′ overhang to unwind dsRNA substrates in the 3′ to 5′ direction [39]; therefore, the substrate specificities of the DENV and HCV NS3 helicases are the same. The reaction mixture contained 50 mM Tris-HCl (pH 7.4), 1 mM DTT, 0.5% Tween 20, 0.25 μg/mL BSA, 2 mM $MgCl_2$, 4 U RNasin (Promega), 5 mM ATP, 50 nM dsRNA substrate, 300 nM capture strand, an inhibitor, and 240 nM DENV NS3 in a total volume of 20 μL. DENV NS3 helicase was prepared as described previously [47]. The reaction was started by the addition of DENV NS3 helicase, and was performed at 37 °C for 60 min using a GeneAmp PCR System 2700 (Applied Biosystems, Foster City, CA, USA). The reaction was then stopped by the addition of 5 μL helicase termination buffer that contained 10 mM Tris-HCl (pH 7.5), 50 mM EDTA, 30% glycerol, 0.06% bromophenol blue, and 0.12% Orange G. The inhibition of DENV NS3 helicase was analyzed on a 20% native TBE polyacrylamide gel, and the labeled RNAs were visualized using Typhoon 9210 (GE Healthcare, Waukesha, WI, USA). The helicase activity was calculated as the ratio of the signal intensity from ssRNA in the sample containing inhibitor to that in the control sample containing DMSO vehicle instead of inhibitor.

4. Conclusions

This study demonstrated that hal3 and suvanine isolated from a marine sponge inhibited NS3 helicase by suppressing the ATPase, RNA binding, and serine protease activities. Moreover, DENV NS3 helicase, which shares a catalytic core consisting mainly of ATPase and RNA binding activity sites with HCV NS3 helicase, was not inhibited by hal3 or suvanine. Therefore, it can be concluded that hal3 and suvanine inhibit HCV NS3 helicase specifically through interaction with an allosteric site of NS3 rather than the catalytic core, leading to the inhibition of all NS3 activities, presumably by inducing conformational changes. As such, it is possible that hal3 and suvanine are less likely to inhibit other cellular helicases that share a similar catalytic core to HCV NS3 helicase. This provides potentially useful information on advanced drug design strategies to identify novel NS3 helicase inhibitors that are expected to be more specific and less toxic. Experiments to address whether resistant HCV mutants emerge with the use of these compounds are underway in our laboratory.

Acknowledgments

The authors thank S. Nishikawa (AIST) for his kind gift of the expression plasmid pT7/His-NS3 containing the *N*-terminal His-tagged full-length HCV NS3. The Global COE Program "Center for Practical Chemical Wisdom" of the Ministry of Education, Culture, Sports, Science and Technology (MEXT) of Japan partially supported this study. This work was also partly supported by NUS SoM Start-up Grant (R-182-000-160-733, R-182-000-160-133) to NY.

References

1. Ghany, M.G.; Nelson, D.R.; Strader, D.B.; Thomas, D.L.; Seeff, L.B. An update on treatment of genotype 1 chronic hepatitis C virus infection: 2011 practice guideline by the American Association for the Study of Liver Diseases. *Hepatology* **2011**, *54*, 1433–1444.
2. Liang, T.J.; Ghany, M.G. Current and future therapies for hepatitis C virus infection. *N. Engl. J. Med.* **2013**, *368*, 1907–1917.
3. Sarrazin, C.; Hézode, C.; Zeuzem, S.; Pawlotsky, J.-M. Antiviral strategies in hepatitis C virus infection. *J. Hepatol.* **2012**, *56*, S88–S100.
4. Scheel, T.K.H.; Rice, C.M. Understanding the hepatitis C virus life cycle paves the way for highly effective therapies. *Nat. Med.* **2013**, *19*, 837–849.
5. Lam, A.M.I.; Frick, D.N. Hepatitis C virus subgenomic replicon requires an active NS3 RNA helicase. *J. Virol.* **2006**, *80*, 404–411.
6. Kwong, A.D.; Rao, B.G.; Jeang, K.-T. Viral and cellular RNA helicases as antiviral targets. *Nat. Rev. Drug Discov.* **2005**, *4*, 845–853.
7. Bartenschlager, R.; Penin, F.; Lohmann, V.; André, P. Assembly of infectious hepatitis C virus particles. *Trends Microbiol.* **2011**, *19*, 95–103.
8. Bartenschlager, R.; Lohmann, V.; Penin, F. The molecular and structural basis of advanced antiviral therapy for hepatitis C virus infection. *Nat. Rev. Microbiol.* **2013**, *11*, 482–496.
9. Lohmann, V.; Körner, F.; Koch, J.-O.; Herian, U.; Theilmann, L.; Bartenschlager, R. Replication of subgenomic hepatitis C virus RNAs in a hepatoma cell line. *Science* **1999**, *285*, 110–113.
10. Gallinari, P.; Brennan, D.; Nardi, C.; Brunetti, M.; Tomei, L.; Steinkühler, C.; de Francesco, R. Multiple enzymatic activities associated with recombinant NS3 protein of hepatitis C virus. *J. Virol.* **1998**, *72*, 6758–6769.
11. Kim, D.W.; Gwack, Y.; Han, J.H.; Choe, J. *C*-terminal domain of the hepatitis C virus NS3 protein contains an RNA helicase activity. *Biochem. Biophys. Res. Commun.* **1995**, *215*, 160–166.
12. Tai, C.-L.; Chi, W.-K.; Chen, D.-S.; Hwang, L.-H. The helicase activity associated with hepatitis C virus nonstructural protein 3 (NS3). *J. Virol.* **1996**, *70*, 8477–8484.
13. Gwack, Y.; Kim, D.W.; Han, J.H.; Choe, J. Characterization of RNA binding activity and RNA helicase activity of the hepatitis C virus NS3 protein. *Biochem. Biophys. Res. Commun.* **1996**, *225*, 654–659.
14. Gwack, Y.; Kim, D.W.; Han, J.H.; Choe, J. DNA helicase activity of the hepatitis C virus nonstructural protein 3. *Eur. J. Biochem.* **1997**, *250*, 47–54.
15. Kolykhalov, A.A.; Mihalik, K.; Feinstone, S.M.; Rice, C.M. Hepatitis C virus-encoded enzymatic activities and conserved RNA elements in the 3′ nontranslated region are essential for virus replication *in vivo*. *J. Virol.* **2000**, *74*, 2046–2051.
16. Hall, M.C.; Matson, S.W. Helicase motifs: The engine that powers DNA unwinding. *Mol. Microbiol.* **1999**, *34*, 867–877.
17. Gorbalenya, A.E.; Koonin, E.V. Helicases: Amino acid sequence comparisons and structure-function relationships. *Curr. Opin. Struct. Biol.* **1993**, *3*, 419–429.
18. Jankowsky, E. *RNA Helicases*; Royal Society of Chemistry: London, UK, 2010; pp. 168–188.

19. Frick, D.N.; Lam, A.M.I. Understanding helicases as a means of virus control. *Curr. Pharm. Des.* **2006**, *12*, 1315–1338.
20. Borowski, P.; Deinert, J.; Schalinski, S.; Bretner, M.; Ginalski, K.; Kulikowski, T.; Shugar, D. Halogenated benzimidazoles and benzotriazoles as inhibitors of the NTPase/helicase activities of hepatitis C and related viruses. *Eur. J. Biochem.* **2003**, *270*, 1645–1653.
21. Zhang, N.; Chen, H.-M.; Koch, V.; Schmitz, H.; Liao, C.-L.; Bretner, M.; Bhadti, V.S.; Fattom, A.I.; Naso, R.B.; Hosmane, R.S.; *et al.* Ring-expanded ("fat") nucleoside and nucleotide analogues exhibit potent *in vitro* activity against *Flaviviridae* NTPases/helicases, including those of the West Nile virus, hepatitis C virus, and Japanese encephalitis virus. *J. Med. Chem.* **2003**, *46*, 4149–4164.
22. Borowski, P.; Heising, M.V.; Miranda, I.B.; Liao, C.-L.; Choe, J.; Baier, A. Viral NS3 helicase activity is inhibited by peptides reproducing the Arg-rich conserved motif of the enzyme (motif VI). *Biochem. Pharmacol.* **2008**, *76*, 28–38.
23. Frick, D.N.; Ginzburg, O.; Lam, A.M.I. A method to simultaneously monitor hepatitis C virus NS3 helicase and protease activities. *Methods Mol. Biol.* **2010**, *587*, 223–233.
24. Dahl, G.; Sandström, A.; Akerblom, E.; Danielson, U.H. Effects on protease inhibition by modifying of helicase residues in hepatitis C virus nonstructural protein 3. *FEBS J.* **2007**, *274*, 5979–5986.
25. Frick, D.N. The hepatitis C virus NS3 protein: A model RNA helicase and potential drug target. *Curr. Issues Mol. Biol.* **2007**, *9*, 1–20.
26. Belon, C.A.; Frick, D.N. Helicase inhibitors as specifically targeted antiviral therapy for hepatitis C. *Future Virol.* **2009**, *4*, 277–293.
27. Mayer, A.M.S.; Glaser, K.B.; Cuevas, C.; Jacobs, R.S.; Kem, W.; Little, R.D.; McIntosh, J.M.; Newman, D.J.; Potts, B.C.; Shuster, D.E. The odyssey of marine pharmaceuticals: A current pipeline perspective. *Trends Pharmacol. Sci.* **2010**, *31*, 255–265.
28. Gerwick, W.H.; Moore, B.S. Lessons from the past and charting the future of marine natural products drug discovery and chemical biology. *Chem. Biol.* **2012**, *19*, 85–98.
29. Newman, D.J.; Cragg, G.M. Natural products as sources of new drugs over the 30 years from 1981 to 2010. *J. Nat. Prod.* **2012**, *75*, 311–335.
30. Furuta, A.; Salam, K.A.; Akimitsu, N.; Tanaka, J.; Tani, H.; Yamashita, A.; Moriishi, K.; Nakakoshi, M.; Tsubuki, M.; Sekiguchi, Y.; *et al.* Cholesterol sulfate as a potential inhibitor of hepatitis C virus NS3 helicase. *J. Enzym. Inhib. Med. Chem.* **2013**, doi:10.3109/14756366.2013.766607.
31. Müller, E.L.; Faulkner, D.J. Absolute configuration of halisulfate 3 from the sponge *Ircinia* sp. *Tetrahedron* **1997**, *53*, 5373–5378.
32. Manes, L.V.; Crews, P.; Kernan, M.R.; Faulkner, D.J.; Fronczek, F.R.; Gandour, R.D. Chemistry and revised structure of suvanine. *J. Org. Chem.* **1988**, *53*, 570–575.
33. Kernan, M.R.; Faulkner, D.J. Sesterterpene sulfates from a sponge of the family Halichondriidae. *J. Org. Chem.* **1988**, *53*, 4574–4578.
34. Manes, L.V.; Naylor, S.; Crews, P.; Bakus, G.J. Suvanine, a novel sesterterpene from an *Ircinia* marine sponge. *J. Org. Chem.* **1985**, *50*, 284–286.

35. Kimura, J.; Ishizuka, E.; Nakao, Y.; Yoshida, W.Y.; Scheuer, P.J.; Kelly-Borges, M. Isolation of 1-methylherbipoline salts of halisulfate-1 and of suvanine as serine protease inhibitors from a marine sponge, *Coscinoderma mathewsi*. *J. Nat. Prod.* **1998**, *61*, 248–250.
36. Di Leva, F.S.; Festa, C.; D'Amore, C.; de Marino, S.; Renga, B.; D'Auria, M.V.; Novellino, E.; Limongelli, V.; Zampella, A.; Fiorucci, S. Binding mechanism of the farnesoid X receptor marine antagonist suvanine reveals a strategy to forestall drug modulation on nuclear receptors. Design, synthesis, and biological evaluation of novel ligands. *J. Med. Chem.* **2013**, *56*, 4701–4717.
37. Cassiano, C.; Monti, M.C.; Festa, C.; Zampella, A.; Riccio, R.; Casapullo, A. Chemical proteomics reveals heat shock protein 60 to be the main cellular target of the marine bioactive sesterterpene suvanine. *ChemBioChem* **2012**, *13*, 1953–1958.
38. Singleton, M.R.; Dillingham, M.S.; Wigley, D.B. Structure and mechanism of helicases and nucleic acid translocases. *Annu. Rev. Biochem.* **2007**, *76*, 23–50.
39. Wang, C.-C.; Huang, Z.-S.; Chiang, P.-L.; Chen, C.-T.; Wu, H.-N. Analysis of the nucleoside triphosphatase, RNA triphosphatase, and unwinding activities of the helicase domain of dengue virus NS3 protein. *FEBS Lett.* **2009**, *583*, 691–696.
40. Ndjomou, J.; Kolli, R.; Mukherjee, S.; Shadrick, W.R.; Hanson, A.M.; Sweeney, N.L.; Bartczak, D.; Li, K.; Frankowski, K.J.; Schoenen, F.J.; *et al.* Fluorescent primuline derivatives inhibit hepatitis C virus NS3-catalyzed RNA unwinding, peptide hydrolysis and viral replicase formation. *Antivir. Res.* **2012**, *96*, 245–255.
41. Saalau-Bethell, S.M.; Woodhead, A.J.; Chessari, G.; Carr, M.G.; Coyle, J.; Graham, B.; Hiscock, S.D.; Murray, C.W.; Pathuri, P.; Rich, S.J.; *et al.* Discovery of an allosteric mechanism for the regulation of HCV NS3 protein function. *Nat. Chem. Biol.* **2012**, *8*, 920–925.
42. Ding, S.C.; Kohlway, A.S.; Pyle, A.M. Unmasking the active helicase conformation of nonstructural protein 3 from hepatitis C virus. *J. Virol.* **2011**, *85*, 4343–4353.
43. Tani, H.; Akimitsu, N.; Fujita, O.; Matsuda, Y.; Miyata, R.; Tsuneda, S.; Igarashi, M.; Sekiguchi, Y.; Noda, N. High-throughput screening assay of hepatitis C virus helicase inhibitors using fluorescence-quenching phenomenon. *Biochem. Biophys. Res. Commun.* **2009**, *379*, 1054–1059.
44. Copeland, R.A. *Evaluation of Enzyme Inhibitors in Drug Discovery*; John Wiley & Sons: New York, NY, USA, 2005; pp. 111–140.
45. Salam, K.A.; Furuta, A.; Noda, N.; Tsuneda, S.; Sekiguchi, Y.; Yamashita, A.; Moriishi, K.; Nakakoshi, M.; Tsubuki, M.; Tani, H.; *et al.* Inhibition of hepatitis C virus NS3 helicase by manoalide. *J. Nat. Prod.* **2012**, *75*, 650–654.
46. Salam, K.A.; Furuta, A.; Noda, N.; Tsuneda, S.; Sekiguchi, Y.; Yamashita, A.; Moriishi, K.; Nakakoshi, M.; Tsubuki, M.; Tani, H.; *et al.* Psammaplin A inhibits hepatitis C virus NS3 helicase. *J. Nat. Med.* **2013**, *67*, 765–772.
47. Takahashi, H.; Takahashi, C.; Moreland, N.J.; Chang, Y.-T.; Sawasaki, T.; Ryo, A.; Vasudevan, S.G.; Suzuki, Y.; Yamamoto, N. Establishment of a robust dengue virus NS3-NS5 binding assay for identification of protein-protein interaction inhibitors. *Antivir. Res.* **2012**, *96*, 305–314.

12

Effect of *Salicornia herbacea* on Osteoblastogenesis and Adipogenesis *in Vitro*

Fatih Karadeniz [1], Jung-Ae Kim [1,2], Byul-Nim Ahn [3], Myeong Sook Kwon [2] and Chang-Suk Kong [2,*]

[1] Marine Biotechnology Center for Pharmaceuticals and Foods, Silla University, Busan 617-736, Korea; E-Mails: f_karadeniz@yahoo.com (F.K.); jale8469@gmail.com (J.-A.K.)

[2] Department of Food and Nutrition, College of Medical and Life Science, Silla University, Busan 617-736, Korea; E-Mail: 1506audtnr@hanmail.net

[3] Department of Organic Material Science and Engineering, Pusan National University, Busan 609-735, Korea; E-Mail: icetwig@naver.com

* Author to whom correspondence should be addressed; E-Mail: cskong@silla.ac.kr

External Editor: Bernie Degnan

Abstract: Bone-related complications are among the highest concerning metabolic diseases in the modern world. Bone fragility and susceptibility to fracture increase with age and diseases like osteoporosis. Elevated adipogenesis in bone results in osteoporosis and loss of bone mass when coupled with lack of osteoblastogenesis. In this study the potential effect of *Salicornia herbacea* extract against osteoporotic conditions was evaluated. Adipogenesis inhibitory effect of *S. herbacea* has been evidenced by decreased lipid accumulation of differentiating cells and expression levels of crucial adipogenesis markers in 3T3-L1 pre-adipocytes. *S. herbacea* treatment reduced the lipid accumulation by 25% of the control. In addition, mRNA expression of peroxisome proliferator-activated receptor (PPAR)γ, CCAAT/enhancer-binding protein (C/EBP)α and sterol regulatory element binding protein (SREBP)1c were inhibited by the presence of *S. herbacea*. Bone formation enhancement effect of *S. herbacea* was also confirmed in MC3T3-E1 pre-osteoblasts. The presence of *S. herbacea* significantly elevated the alkaline phosphatase (ALP) activity by 120% at a concentration of 100 μg/mL in differentiating osteoblasts. *S. herbacea* also significantly increased the expression of osteoblastogenesis indicators, ALP, bone morphogenetic protein (BMP)-2, osteocalcin and collagen type I (collagen-I). In conclusion,

S. herbacea possess potential to be utilized as a source of anti-osteoporotic agent that can inhibit adipogenesis while promoting osteoblastogenesis.

Keywords: *Salicornia herbacea*; osteoporosis; adipogenesis; osteoblastogenesis

1. Introduction

Natural sources that can be used for the treatment or prevention of diseases have been of much interest during the past decades. In this context, extracts and compounds from plants are among the main targets of studies [1,2]. Several studies in the past reported various biologically active plants with marine plants allocating a large part [3]. Due to the harsh conditions of marine environments and the need for strong protection, marine organisms produce unique substances. Plant extracts and compounds have been intensely studied in order to fully understand their action mechanism against common complications such as oxidative stress and inflammation [4–6], as well as diseases with high morbidity and mortality rates including diabetes, cancer, obesity and AIDS [7–9]. The halophyte *Salicornia herbacea* is endemic to the western coast of the Korean peninsula and a part of folk medicine due to its effect on constipation, diabetes, obesity, *etc.*, [10]. As expected, a number of experiments have credited *S. herbacea* for various bioactivities including antioxidative, antiinflamatory, antihyperglycemic and antihyperlidemic [11–13]. It was proposed that *S. herbacea* contained active flavonoids [14–16]. However, very few investigations have been carried out in order to evaluate the effect of *S. herbacea* on osteoporosis.

Bone-related diseases, namely osteoporosis and age-related osteopenia are reported to be associated with bone mass loss due to lack of osteoblastogenesis [17]. Studies showed that glucocorticoid treatment also caused an increase in bone adipocytes, which finally resulted in fractures and osteoporosis [18]. Several mechanisms have been suggested for possible cause of bone mass imbalance. Peroxisome proliferator activated receptor (PPAR)γ is confirmed to play crucial roles in the outcome of bone marrow mesenchymal cell differentiation [19]. Both *in vitro* and *in vivo* mechanism studies demonstrated that the activation of PPARγ promotes adipogenesis. Likewise, suppression of PPARγ pathway was shown to inhibit adipogenesis and stimulate osteoblast differentiation depending on the binding ligand [20]. Several diabetic drugs as ligands of PPARγ activate adipogenesis and lower blood glucose. However, this activation also causes problems in bone mass by favoring adipogenesis of bone mesenchymal cells and deteriorating the bone mass balance. Diabetic drugs, obesity-related factors and long chain fatty acids were confirmed to be activating ligands for PPARγ [21]. Overall, in this study *S. herbacea* was tested for its potential effect on adipogenesis of pre-adipocytes and pre-osteoblast differentiation with a possible intervention in PPARγ pathway.

2. Results

2.1. Effect of *S. herbacea* on Differentiation of 3T3-L1 Pre-Adipocytes

Prior to assays, the cytotoxicity of samples was tested *in vitro* using the 3T3-L1 and MC3T3-E1 cells which were used in further experiments. The cytotoxic effect of *S. herbacea* extracts in 3T3-L1

(Figure 1A) and MC3T3-E1 (Figure 1B) cells are presented. At a concentration of 10 μg/mL, there was no cytotoxicity for both cell lines while the two highest concentrations (50 and 100 μg/mL) caused slight decrease in cell viability of 3T3-L1 cells, indicating a small effect on cell viability at higher concentrations. However, no cytotoxic effect was observed on MC3T3-E1 cells at concentrations of 50 and 100 μg/mL.

Figure 1. Cytotoxicity levels of *S. herbacea* on the viability of 3T3-L1 (**A**) and MC3T3-E1 (**B**) cells. Cell viability was observed by MTT assay and presented as percentage value of untreated undifferentiated blank cells. Values are means ± SD (n = 3). [a,b] Means with different letters are significantly different ($p < 0.05$) by Duncan's multiple range test. (Con: Differentiated untreated cells).

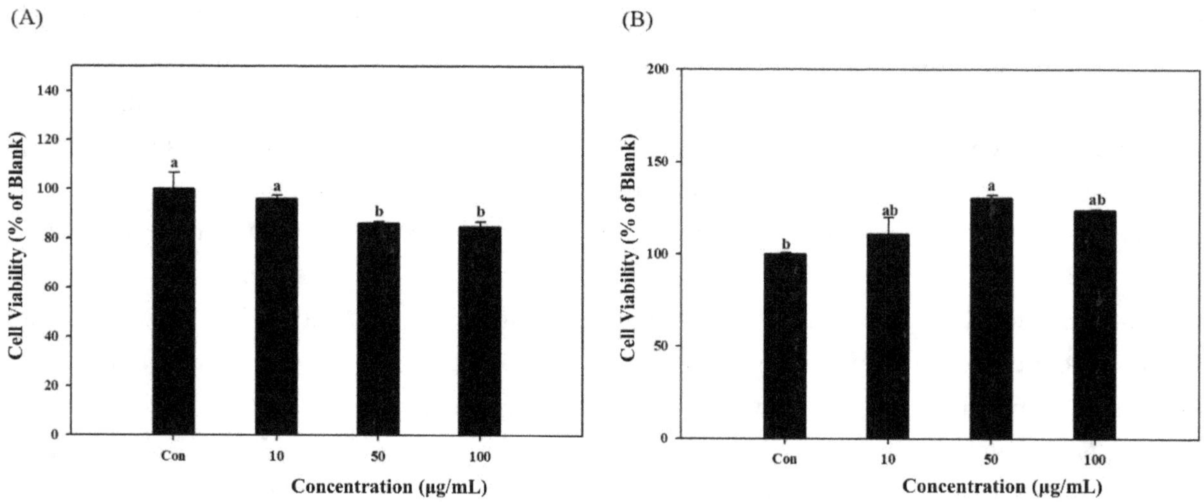

In order to evaluate the effect of *S. herbacea* on adipogenesis, 3T3-L1 mouse pre-adipocytes were differentiated into mature adipocytes in the presence and absence of *S. herbacea* extract. Lipid accumulation which is a key indicator of successful maturation into adipocytes was used to analyze the anti-adipogenesis effect. Lipid accumulation is observed by staining of intracellular triglycerides with Oil-Red O. Cell images after staining indicated that the presence of *S. herbacea* decreased the intracellular lipid amount (as red colored areas in cell images) in a dose-dependent manner (Figure 2A). Observable lipid accumulation stained by Oil-Red O was at a minimum compared to all other cases, in the case of the highest concentration treated (100 μg/mL). Inhibited lipid accumulation was quantified by the elution of accumulated Oil Red O stain. Absorbance values of intracellular accumulated stain are shown in Figure 2B. Quantification of lipid accumulation was in accordance with the cell images. The presence of *S. herbacea* lowered the intracellular stain amount as an indicator of suppressed lipid accumulation. In order to assess the mechanism behind the lowering of lipid amounts, glycerol release into the culture medium was quantified. Excessive glycerol secretion of adipocytes is a key marker for triglyceride lipolysis. In the presence of *S. herbacea*, glycerol secretion of differentiated cells increased from 3.09 ± 0.51 to 4.64 ± 0.62, 5.26 ± 0.40 and 5.79 ± 0.32 at concentrations of 10, 50 and 100 μg/mL, respectively (Figure 2C).

Next, in order to evaluate whether *S. herbacea* affects the expression of key transcription factors for adipogenesis, RT-PCR experiments were carried out. Possible changes in the mRNA expressions of

PPARγ, differentiation-dependent factor 1/sterol regulatory element-binding protein (SREBP1c) and CCAAT/enhancer-binding proteins (C/EBPα) were observed. Different solvent fractions were tested for their ability to suppress PPARγ mRNA expression. Among four tested fractions (*n*Hexane, 85% MeOH, *n*BuOH and H_2O), the *n*BuOH fraction decreased the PPARγ expression most compared to control (Figure 3A). In addition, treatment significantly suppressed the expression of PPARγ, SREBP1c and C/EBPa, compared to fully differentiated control adipocytes (Figure 3B). Hence, *n*BuOH was considered for further activity based-isolation of compounds. Other fractions also showed an observable decrease in mRNA expression, the H_2O extract being the least effective.

Figure 2. Effect of *S. herbacea* on adipogenic differentiation as stained cell images following Oil Red O staining of intracellular lipids (**A**), quantification of lipid accumulation by the absorbance of eluted intracellular stain (**B**) and glycerol secretion into culture medium during final differentiation (**C**) of mature 3T3-L1 adipocytes. Values are means ± SD (n = 6). [a–c] Means with different letters are significantly different ($p < 0.05$) by Duncan's multiple range test.

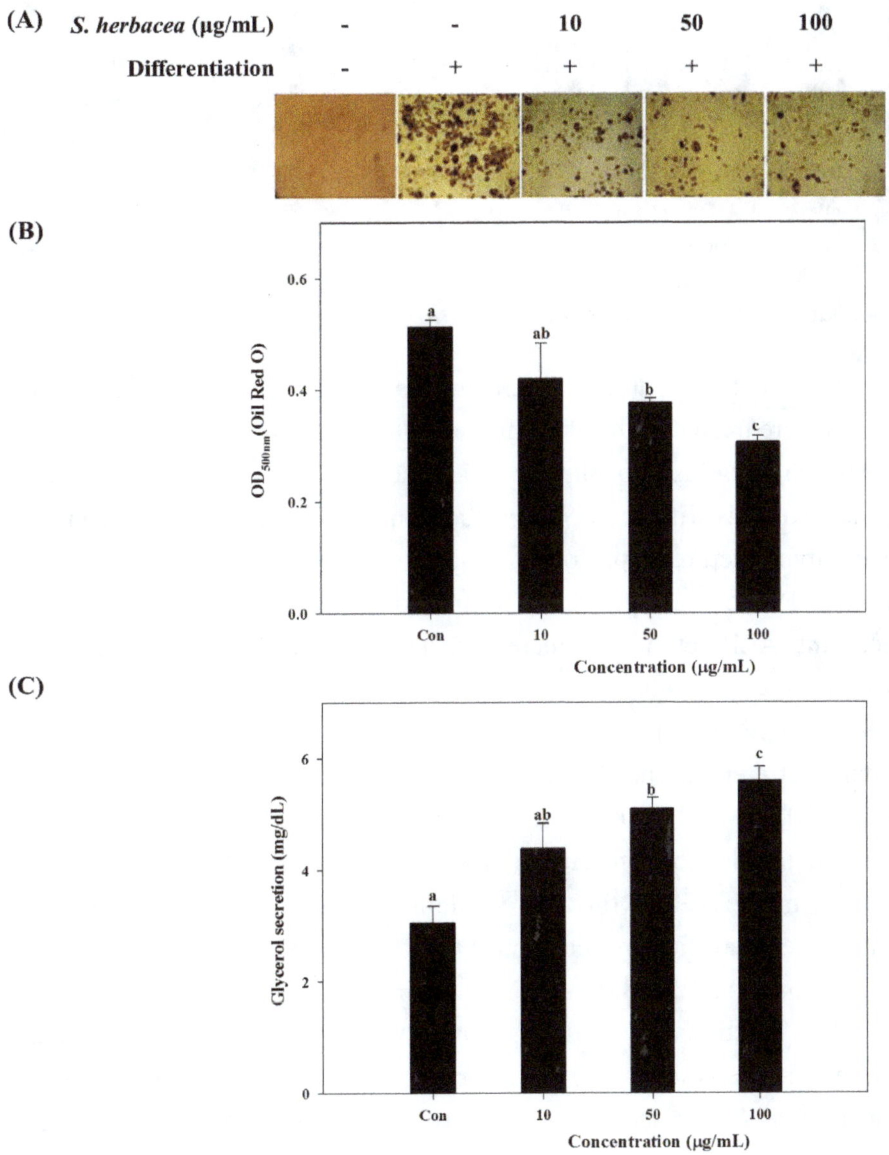

Figure 3. Effect of *S. herbacea* extract fractions (**A**) and the crude extract (**B**) on the mRNA expression levels of adipogenic target genes in 3T3-L1 adipocytes assessed by RT-PCR and band quantification. Values are means ± SD (n = 3). [a–e] Means with different letters are significantly different ($p < 0.05$) by Duncan's multiple range test.

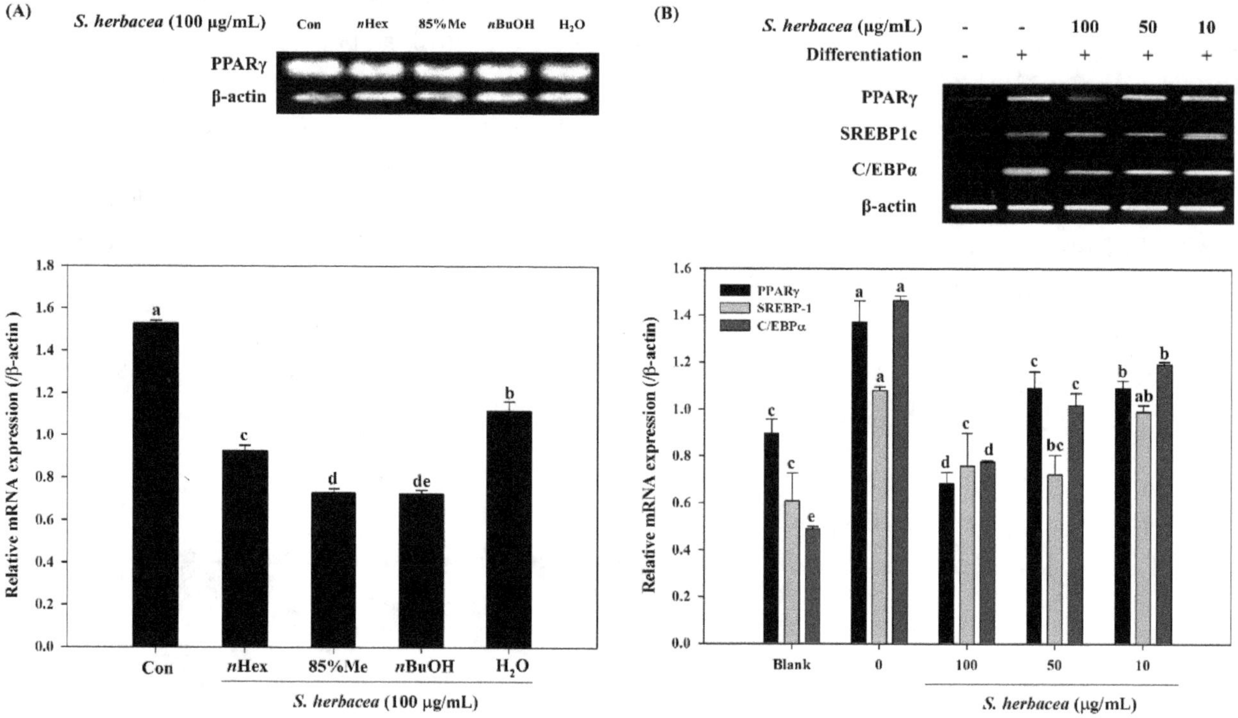

2.2. *Effect of* S. herbacea *on Osteogenic Differentiation of MC3T3-E1 Pre-Osteoblasts*

S. herbacea was tested for its ability to enhance osteoblast differentiation in mouse MC3T3-E1 pre-osteoblasts. Cells were treated with different concentrations of *S. herbacea* during the differentiation process. Measurement of alkaline phosphatase (ALP) activity of MC3T3-E1 osteoblasts was used as a marker for osteoblastogenesis. Effect of *S. herbacea* on ALP activity was measured by a commercial kit according to the manufacturer's protocol. Results suggested an enhancing ability of *S. herbacea* towards ALP-indicated osteoblastogenesis. Compared to differentiated control cells, *S. herbacea* treatment enhanced the ALP activity by increasing the absorbance values of reaction mixtures from 0.681 ± 0.092 to 0.814 ± 0.081, 0.896 ± 0.094 and 1.049 ± 0.102 at concentrations of 10, 50 and 100 μg/mL, respectively (Figure 4).

During osteogenic differentiation, MC3T3-E1 pre-osteoblasts were treated with *S. herbacea* extract at concentrations of 10, 50 and 100 μg/mL. In order to detect and evaluate the differentiation of pre-osteoblasts, following incubation during 14 days of differentiation, gene expression of key osteogenic markers, namely ALP, BMP-2, osteocalcin, collagen-I, and intracellular levels of ALP and collagen-I proteins were quantified by real-time PCR and Western blotting, respectively. *S. herbacea* enhanced the expression of ALP, BMP-2, osteocalcin and collagen-I mRNA in a dose-dependent manner (Figure 5A). Fractions of extract were shown to enhance the mRNA expression of osteogenesis markers in a similar way to suppressing adipogenesis indicators (Figure 5B). As expected, treatment with *S. herbacea* notably increased intracellular levels of ALP and collagen-I proteins in a similar fashion to mRNA levels (Figure 6). According to the results, the presence of *S. herbacea* enhanced the expression of osteogenic marker proteins in MC3T3-E1 cells.

Figure 4. Effect of *S. herbacea* on alkaline phosphatase (ALP) activity in MC3T3-E1 cells after 14 days of differentiation. Values are means ± SD (n = 6). [a–c] Means with different letters are significantly different ($p < 0.05$) by Duncan's multiple range test.

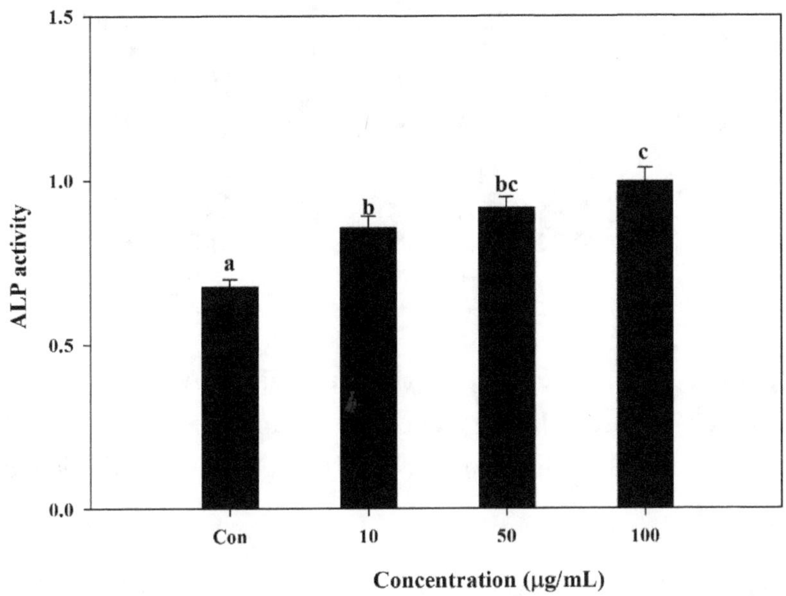

Figure 5. Effect of *S. herbacea* extract (**A**) and fractions (**B**) on mRNA expression of key osteblastogenesis markers in MC3T3-E1 osteoblasts according to real time PCR assay. Values are means ± SD (n = 3). [a–e] Means with different letters are significantly different ($p < 0.05$) by Duncan's multiple range test.

(A)

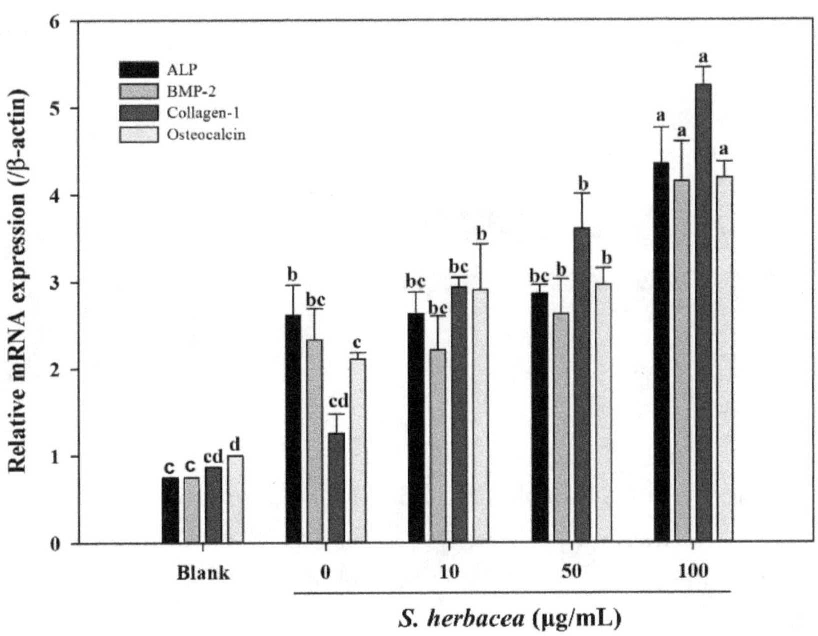

Figure 5. *Cont.*

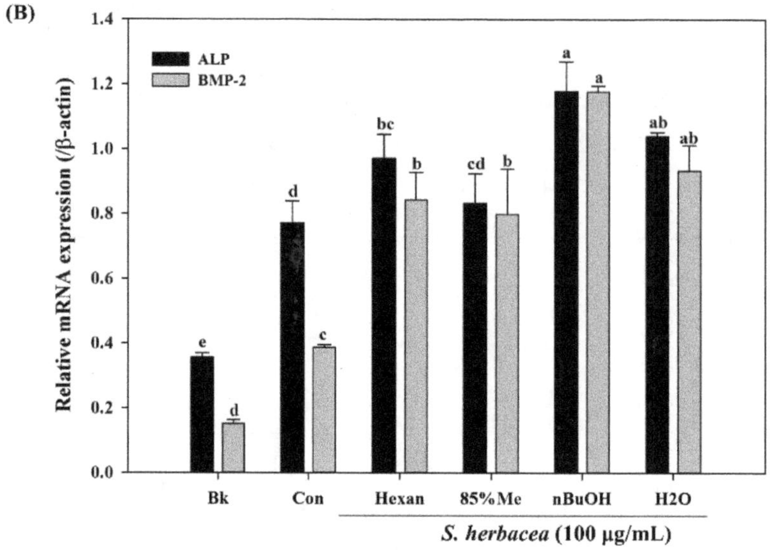

Figure 6. Effect of *S. herbacea* extract on protein levels of ALP (**A**) and collagen (**B**) in differentiated MC3T3-E1 osteoblasts. Values are means ± SD (n = 3). [a–d] Means with different letters are significantly different ($p < 0.05$) by Duncan's multiple range test.

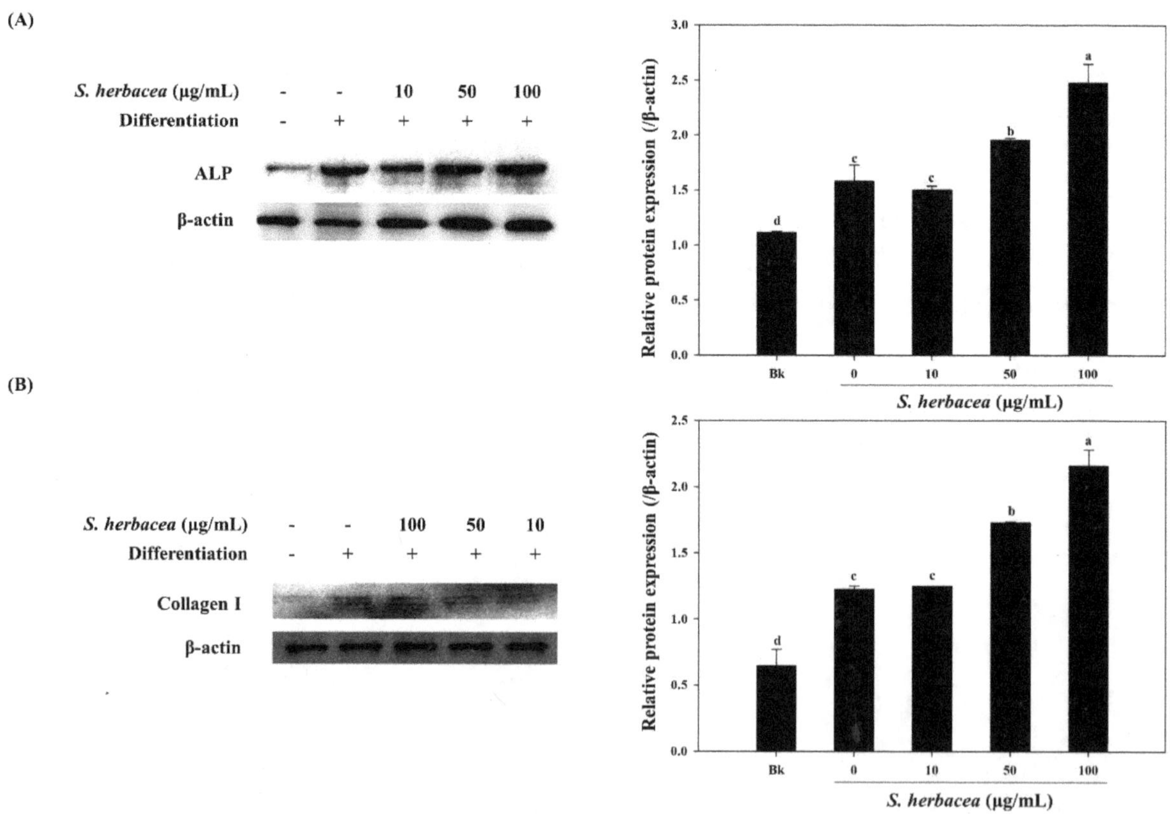

Following a bioactivity-directed isolation method, two formerly known flavonoid glycosides were isolated from the *n*BuOH fraction of *S. herbacea* as a part of ongoing research (Figure 7). In total accordance with our earlier report, the isolated flavonoid glycosides were confirmed by comparison of basic structural data (data not shown) [15].

Figure 7. Isolated flavonoid glycosides as anti-osteoporotic ingredients of *S. herbacea*.

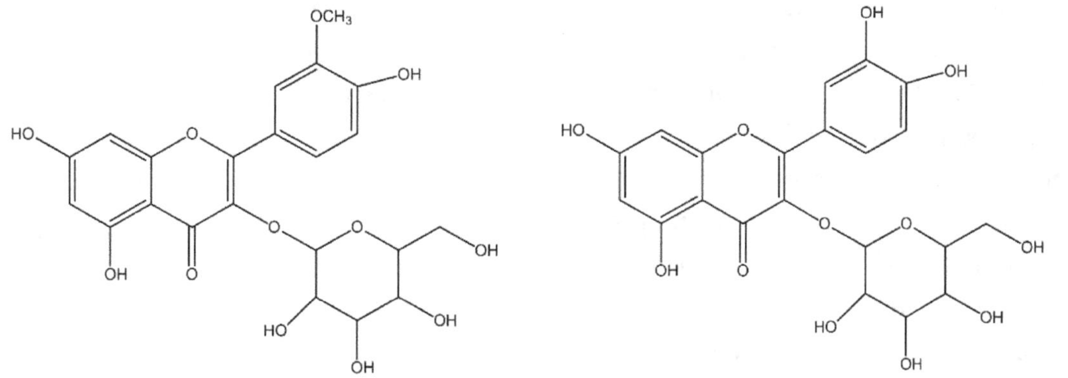

$C_{22}H_{22}O_{12}$
Exact Mass : 478.1111
Mol. Wt. : 646.8080
C, 55.23; H, 4.64; O, 40.13

$C_{21}H_{20}O_{12}$
Exact Mass : 464.0955
Mol. Wt. : 464.3763
C, 54.31; H, 4.34; O, 41.34

3. Discussion

The main age-related metabolic diseases are the major causes of diminished life quality for the elderly. Additionally, recent studies indicate an increasing trend for the younger generation to be diagnosed with obesity, type-2 diabetes and osteoporosis [16,22,23]. Scientists are turning their attention to natural products to use to intervene with the progression of the aforementioned diseases due to bioavailability, biodegradation and fewer side effects. In this context, marine-based organisms are of high interest to pharmaceutical researchers as most marine plants and animals live in extreme conditions, which results in the need for unique compounds for survival. By means of this promising situation, past decades have been fruitful in ways of natural bioactive substance development. Numerous marine plants have been studied intensively in order to discover bioactive chemicals that can act against cancer, tumor growth, oxidative stress, diabetes, obesity and osteoporosis [5,24,25]. Halophytes are also plants that survive under extreme conditions such as high concentrations of salt and harsh climate changes. In this context, the halophyte *S. herbacea* was examined here for possible compounds with health beneficiary effects. Reports had already stated that *S. herbacea* contains flavonoid glycosides and polysaccharides, which are natural antioxidants [15].

In this study, we examined the ability of *S. herbacea* to relieve osteoporosis conditions with a regulatory mechanism towards adipogenesis/osteoblastogenesis imbalance. Uncontrolled inducement of adipogenesis was shown to create a bone mass imbalance which resulted in elevated fragility and susceptibility. Differentiation of both pre-adipocyte and mesenchymal cells is partly regulated by PPARγ signaling [26]. Obesity conditions were recently linked to increased osteoporosis progression in proportional to elevated adipokines [27]. Adipokines and triglycerides trigger in a similar way an increase in adipogenesis, which is required to be in balance with osteoblastogenesis for healthy bones. Regulation of this imbalance is crucial to prevent and treat osteoporosis, especially when developed under obesity.

Our results showed a strong inhibition on lipid accumulation of differentiated 3T3-L1 adipocytes by *S. herbacea* treatment. Glycerol assay results showed that cells treated with *S. herbacea* released more glycerol to the culture medium. Furthermore, a possible mechanism behind the anti-adipogenic effect of *S. herbacea* was evaluated by assessing expression levels of adipogenesis regulator factors, PPARγ, SREBP1c and C/EBPα. In addition to lowering lipid accumulation, *S. herbacea* also lowered the expression of the aforementioned adipogenesis factors. Results indicated that *S. herbacea* inhibited not only lipid accumulation, but also adipogenesis. In order to be used for further bioactivity-directed isolation, comparison of *S. herbacea* solvent fractions was carried out. Among all fractions, *n*BuOH was found to be the most active to inhibit PPARγ mRNA expression. In addition, Kong *et al.* [14] also reported that two glycosides isolated from *S. herbacea* were shown to possess chemoprotective effect against cancer through matrix metalloproteinase inhibition.

Increase in bone adipocytes is accompanied by severe fragility that defines osteoporosis. In order to relieve the deteriorated stem cell differentiation, inducing differentiating cells towards osteoblastogenesis is considered to be a crucial treatment step. Our results showed that *S. herbacea* inhibited adipogenesis by possible interaction with the PPAR-γ pathway and lipolysis. Results also indicated that *S. herbacea* might attenuate the imbalance of bone cell differentiation towards osteoblastogenesis.

In osteoporosis, mesenchymal cell perseverance towards adipogenesis is also accompanied by diminished osteoblastogenesis [28,29]. If an increase in adipogenesis is coupled with a decrease in osteoblastogenesis, bone loses its sturdiness and tends to be more fragile.

We examined the ability of *S. herbacea* extract to enhance osteoblast differentiation in MC3T3-E1 pre-osteoblasts. Differentiating MC3T3-E1 pre-osteoblasts were introduced to *S. herbacea* in different concentrations. Following a full differentiation, alkaline phosphatase (ALP) activity of cell lysates was evaluated. The alkaline phosphatase level is known to be an indicator of successful osteoblast differentiation, as it plays a crucial role in the mineralization of bone [30]. In this regard, results indicated that the presence of *S. herbacea* elevated the ALP activity suggesting a role in the mineralization of osteoblasts at all treated concentrations, indicating an enhancement towards bone formation.

Bone formation is carried out by a distinctive and well-studied pathway of factors and proteins. In this cascade of signaling, bone morphogenetic protein (BMP) 2, 4 and 7 and osteocalcin are some of the key factors elevated at gene expression levels [31]. During osteoblastogenesis, osteocalcin has also been reported to be a cell marker for the final differentiation state. On the other hand, BMPs are known to enhance expression of alkaline phosphatase (ALP), type I collagen (collagen-I) and other non-collagenous bone proteins as indicators for successful osteoblastogenesis [30]. Therefore, mRNA expression of ALP, BMP-2, osteocalcin and collagen-I were assessed in the absence and presence of *S. herbacea*. After full maturation into osteoblasts, RT-PCR experiments suggested an increasing trend towards expression of the aforementioned markers in cells treated with *S. herbacea*, a result confirmed for ALP and collagen-I by western blot. In the light of these results, it was suggested that *S. herbacea* enhanced osteoblast differentiation while inhibiting adipogenesis in pre-adipocytes. Ha *et al.* [32] also reported that *S. herbacea* might be a potential source of antioxidant agents because of its effect on ovariectomy-induced oxidative stress, which is considered to cause age-related diseases including osteoporosis. Taken together, our results on *S. herbacea* extracts provide evidence for an inhibitory effect on adipogenesis while enhancing osteoblast differentiation. Therefore, *S. herbacea* is proposed as a promising source of bioactive agents for the effective prevention and treatment of osteoporosis.

In this regard, following an earlier reported isolation method adapted for anti-osteoporosis bioactivity-directed flow, two formerly known flavonoid glycosides were isolated [14]. These glycosides, namely isorhamnetin 3-*O*-β-D-glucoside and quercetin 3-*O*-β-D-glucoside are suggested to be possible bioactive reagents of *S. herbacea*, responsible for relieving the effect of osteoporosis through regulation of adipogenesis/osteoblastogenesis imbalance by inhibiting PPARγ pathway and enhancing bone formation. Isolation of glycosides from the *n*BuOH fraction was expected as this fraction was the most active for inhibiting adipogenesis and enhancing osteoblastogenesis. Several other flavonoid glycosides were isolated and evaluated for their anti-adipogenic activities. Studies also showed the possible absorption and bioactivity mechanisms of dietary flavonoids through the small intestine. Our previous results [14,15] were also in accordance with current assays and suggested that isorhamnetin 3-*O*-β-D-glucoside and quercetin 3-*O*-β-D-glucoside are strong bioactive substituents of *S. herbacea*. On the other hand, isorhamnetin 3-*O*-β-D-glucoside and quercetin 3-*O*-β-D-glucoside were also shown to act on inhibition of differentiation of 3T3-L1 cells through AMPK/MAPK pathways. Notoya *et al.* [33] suggested an inhibitory effect of the flavonoid, quercetin, on proliferation, differentiation and mineralization of osteoblasts. However, Kim *et al.* [34] also stated that quercetin was able to inhibit proliferation while elevating the osteogenic differentiation of adipose stromal cells. In such a case, quercetin derivatives might also show the same distinct bioactivity on adipogenesis and osteoblastogenesis as our preliminary results indicated. The isolated derivative of quercetin, quercetin 3-*O*-β-D-glucoside has the quercetin backbone with a glucoside side-chain containing -OH branches similar to phloroglucinol derivatives. Considering that phloroglucinol derivatives have been reported to have anti-adipogenesis and pro-osteoblastogenesis activities, it could be suggested that these flavonoids might be responsible for the reported bioactivity of S. herbacea. Nonetheless, the *n*BuOH fraction of *S. herbacea* showed strong effects on adipogenesis and osteoblastogenesis and two flavonoid glycosides were isolated from the fraction. However, further studies to reveal the real action mechanisms of isolated compounds and their efficiency on adipogenic and osteogenic differentiation are needed. According to the results, the *n*BuOH fraction of *S. herbacea* might contain other bioactive constituents that are responsible for anti-adipogenic and pro-osteoblastogenic effects. In the future, coupled with *in vivo* assays, results of this study will help to evaluate the true potential of *S. herbacea*, as a potential source of compounds against obesity-related osteoporosis.

4. Experimental Section

4.1. Plant Materials

The whole plant of *S. herbacea* was briefly dried under shade and kept at −25 °C until use. The air-dried sample of *S. herbacea* was chopped into small pieces and extracted for 24 h with CH_2Cl_2 (3 L × 2) at room temperature. After removal of the solvent, the residue was re-extracted for 24 h with MeOH (3 L × 2) at room temperature.

The combined crude extracts (50 g) were suspended between CH_2Cl_2 and water. The organic layer was further partitioned between 85% aqueous MeOH and *n*-hexane and then the aqueous layer was fractioned with *n*-BuOH and H_2O, respectively, to afford the *n*-hexane (3.2 g), 85% aq. MeOH (12.1 g), *n*-BuOH (16.3 g) and water (15.5 g) fractions.

4.2. Cell Culture and Adipocyte/Osteoblast Differentiations

Murine 3T3-L1 pre-adipocytes were seeded in 6-well plates at a density of 2×10^5 cells/well prior to experiments and grown to confluence in Dulbecco's modified Eagle's medium (DMEM) with 10% fetal bovine serum (FBS) at 37 °C in a humidified atmosphere of 5% CO_2. At 1 day postconfluence (designated "day 0"), a mixture of 3-isobutyl-1-methylxanthine (0.5 mM), dexamethasone (0.25 M) and insulin (5 μg/mL) in DMEM containing 10% FBS was introduced into cells in order to induce cell differentiation. After 48 h (day 2), DMEM containing 10% FBS supplemented with insulin (5 μg/mL) was introduced to cells following removal of induction medium. While replacing this medium with a fresh one every two days, *S. herbacea* extract was administered to the culture medium from day 0 to day 6 for Oil Red O and RT-PCR experiments and day 6 to day 8 for the glycerol secretion assay.

Murine osteoblast-like MC3T3-E1 cells were seeded in 6-well plates at a density of 1×10^5 cells/well and grown to confluence in α-Modified minimal essential medium (αMEM) supplemented with 10% heat-inactivated fetal bovine serum (FBS), 1 mM sodium pyruvate, 100 units/L penicillin and 100 mg/L streptomycin at 37 °C in a humidified atmosphere of 5% CO_2. The cells were induced into osteoblastogenesis by adding of ascorbic acid and β-glycerophosphate into medium for five days under the conditions of the earlier report [35]. Following confluence, the cell differentiation was initiated with culture medium containing 50 μg/mL ascorbic acid and 10 mM β-glycerophosphate for three days. Then, the induction medium was removed and the cell monolayer was washed twice with phosphate buffered saline (PBS). *S. herbacea* extract was administered to the culture medium prior to further incubation of 48 h.

4.3. Cytotoxicity Determination Using MTT Assay

Cytotoxic levels of the *S. herbacea* on cultured cells were measured using MTT (3-(4,5-dimethylthiazol-2-yl)-2,5-diphenyltetrazolium bromide) assay, which is based on the conversion of MTT to MTT-formazan by mitochondrial enzyme. The cells were grown in 96-well plates at a density of 5×10^3 cells/well. After 24 h, the cells were washed with fresh medium and were treated with control medium or the medium supplemented with *S. herbacea*. After incubation for seven days while changing the medium and re-treating the samples every two days, cells were rewashed and 100 μL of MTT solution (1 mg/mL) was added and incubated for 4 h. Finally, 100 μL of DMSO was added to solubilize the formed formazan crystals and the amount of formazan crystal was determined by measuring the absorbance at 540 nm using a GENios® microplate reader (Tecan Austria GmbH, Grödig, Austria). Relative cell viability was determined by the amount of MTT converted into formazan crystal. Viability of cells was quantified as a percentage compared to the control and dose response curves were developed.

4.4. Oil-Red O Staining and Glycerol Release Assay

Fully differentiated cells were fixed with 10% fresh formaldehyde in PBS for 1 h at room temperature and stained with filtered Oil-Red O solution (60% isopropanol and 40% water) for at least 1 h. After incubation, the wells were emptied of Oil-Red O staining solution, washed with distilled water and air dried. Images of lipid droplets in 3T3-L1 adipocytes were collected by an Olympus microscope (Tokyo, Japan). Finally, dye retained in the cells was eluted with isopropanol and quantified by measuring the optical absorbance at 500 nm using a microplate reader (Tecan Qustria GmbH, Grödig, Austria).

The glycerol levels were determined using the enzymatic reagent, free glycerol reagent (Sigma, St. Louis, MO, USA), directed by the protocol of GPO-TRINDER (Sigma, St. Louis, MO, USA).

4.5. Cellular ALP Activity

Cellular ALP activity of *S. herbacea*-treated and control cells was measured following incubation of 14 days. The cell monolayer was gently washed twice with PBS and lysed using 0.1% Triton X-100 and 25 mM carbonate buffer. The lysates were centrifuged at 4 °C 12,000× *g* for 15 min. Enzyme assay buffer (15 mM ρ-nitrophenyl phosphate, 1.5 mM $MgCl_2$ and 200 mM carbonate buffer) was used to measure the ALP activity of the supernatants. The absorbance of reactive solution was measured at 405 nm.

4.6. RNA Extraction and Reverse Transcription-Polymerase Chain Reaction Analysis

Total RNA was isolated from 3T3-L1 and D1 adipocytes and MC3T3-E1 osteoblasts in the presence/absence of *S. herbacea* using Trizol reagent (Invitrogen Co., Carlsbad, CA, USA). For synthesis of cDNA, RNA (2 µg) was added to RNase-free water and oligo (dT), denaturated at 70 °C for 5 min and cooled immediately. RNA was reverse transcribed in a master mix containing 1× RT buffer, 1 mM dNTPs, 500 ng oligo (dT), 140 U M-MLV reserve transcriptase and 40 U RNase inhibitor at 42 °C for 60 min and at 72 °C for 5 min using an automatic T100 Thermo Cycler (Bio-Rad, Hertfordshire, UK). The target cDNA was amplified using the following sense and antisense primers: forward 5′-TTT-TCA-AGG-GTG-CCA-GTT-TC-3′ and reverse 5′-AAT-CCT-TGG-CCC-TCT-GAG-AT-3′ for PPARγ; forward 5′-TGT-TGG-CAT-CCT-GCT-ATC-TG-3′ and reverse 5′-AGG-GAA-AGC-TTT-GGG-GTC-TA-3′ for SREBP1c; forward 5′-TTA-CAA-CAG-GCC-AGG-TTT-CC-3′ and reverse 5′-GGC-TGG-CGA-CAT-ACA-GTA-CA-3′ for C/EBPα; forward 5′-CCA-CAG-CTG-AGA-GGG-AAA-TC-3′ and reverse 5′-AAG-GAA-GGC-TGG-AAA-AGA-GC-3′ for β-actin. The amplification cycles were carried out at 95 °C for 45 s, 60 °C for 1 min and 72 °C for 45 s. Final PCR products were separated by electrophoresis on 1.5% agarose gel for 30 min at 100 V after 30 cycles. Gels were then stained with 1 mg/mL ethidium bromide visualized by UV light using Davinch-Chemi imager™ (CAS-400SM, Wako Co., Osaka, Japan).

4.7. Real-Time RT-PCR Analysis of mRNA Expression

Gene expression was measured by real time RT-PCR in a Thermal Cycler Dice® Real Time System TP800 (Takara Bio Inc., Ohtsu, Japan) following the manufacturer's protocol. Briefly 1.0 µL of DNA sample and 12.5 µL of Maxima® SYBR Green qPCR Master Mix (Fermentas, Waltham, MA, USA) containing Taq DNA polymerase, dNTP and reaction buffer were mixed. The target cDNA was amplified using the following sense and antisense primers: forward 5′-CCA-GCA-GGT-TTC-TCT-CTT-GG-3′ and reverse 5′-CTG-GGA-GTC-TCA-TCC-TGA-GC-3′ for ALP; forward 5′-GGA-CCC-GCT-GTC-TTC-TAG-TG-3′ and reverse 5′-GCC-TGC-GGT-ACA-GAT-CTA-GC-3′ for BMP-2;

forward 5′-GCT-GTG-TTG-GAA-ACG-GAG-TT-3′ and reverse 5′-CAT-GTG-GGT-TCT-GAC-TGG-TG-3′ for Osteocalcin; forward 5′-GAG-CGG-AGA-GTA-CTG-GAT-CG-3′ and reverse 5′-TAC-TCG-AAC-GGG-AAT-CCA-TC-3′ for Collagen I; forward 5′-CCA-CAG-CTG-AGA-GGG-AAA-TC-3′ and reverse 5′-AAG-GAA-GGC-TGG-AAA-AGA-GC-3′ for β-actin. The PCR amplification was carried out for an initial denaturation at 95 °C for 10 min, followed by 40 PCR cycles. Each cycle proceeded at 95 °C for 15 s, 60 °C for 60 s. Relative quantification was calculated using the 2-(ΔΔCT) method. β-Actin was used as an internal control.

4.8. Western Blot Analysis

Western blotting was performed according to standard procedures. Briefly, cells were lysed in RIPA lysis buffer (Sigma-Aldrich Corp., St. Louis, MO, USA) at 4 °C for 30 min. Cell lysates (35 μg) were separated by 12% SDS-polyacrylamide gel electrophoresis, transferred onto a polyvinylidene fluoride membrane (Amersham Pharmacia Biotech., Amersham, England, UK), blocked with 5% skimmed milk and hybridized with primary antibodies (diluted 1:1000) against ALP and collagen I. After incubation with horseradish-peroxidase-conjugated secondary antibody at room temperature, immunoreactive proteins were detected using a chemiluminescece ECL assay kit (Amersham Pharmacia Biosciences, England, UK) according to the manufacturer's instructions. Western blot bands were visualized using a Davinch-Chemi imager™ (CAS-400SM, Wako Co., Osaka, Japan).

4.9. Statistical Analysis

The data were presented as mean ± SD. Differences between the means of the individual groups were analyzed using the analysis of variance (ANOVA) procedure of Statistical Analysis System, SAS v9.1 (SAS Institute, Cary, NC, USA) with Duncan's multiple range tests. The significance of differences was defined at the $p < 0.05$ level.

5. Conclusions

In this study, *Salicornia herbacea* was extracted and experimented with, in order to evaluate its ability as a potential bioactive agent against osteoporosis. Results clearly indicated that, *S. herbacea* was able to inhibit adipogenesis of both pre-adipocytes and bone marrow mesenchymal cells. Inhibiting adipogenesis is important in relieving the fragile bone mass of osteoporosis which is caused by elevated adipogenesis of mesenchymal cells. In addition, results also indicated that *S. herbacea* can enhance osteoblastogenesis which is important to affect the change of fragile bone towards strong bone formation. Finally, following a bioactivity-directed isolation, two flavonoid glycosides were suggested to be responsible for the anti-osteoporosis effect of *S. herbacea* but have yet to be evaluated for their exact mechanism of action. Nonetheless, *S. herbacea* is a promising lead source for bioactive substances that can be utilized against osteoporosis.

Acknowledgments

This research was supported by Basic Science Research Program through the National Research Foundation of Korea (NRF) funded by the Ministry of Education, Science and Technology

(NRF-2012R1A1A3014642). Also, this research was financially supported by the Ministry of Education (MOE) and National Research Foundation of Korea (NRF) through the Human Resource Training Project for Regional Innovation (No. NRF-2013H1B8A2032201).

Author Contributions

All authors contributed to this report extensively. J.-A.K. and C.-S.K. conceived the idea and designed the experiments and protocols. F.K. interpreted the data and wrote the manuscript. J.-A.K and F.K. carried out the adipogenesis experiments. Immunoblotting, mineralization and cytotoxicity experiments were performed by B.-N.A. and M.S.K. Finally, C.-S.K. supervised the work.

References

1. Qin, G.-W.; Xu, R.-S. Recent Advances on Bioactive Natural Products from Chinese Medicinal Plants. *Med. Res. Rev.* **1998**, *18*, 375–382.
2. Gurib-Fakim, A. Medicinal Plants: Traditions of Yesterday and Drugs of Tomorrow. *Mol. Aspects Med.* **2006**, *27*, 1–93.
3. Rinehart, K.L.; Holt, T.G.; Fregeau, N.L.; Keifer, P.A.; Wilson, G.R.; Perun, T.J.; Sakai, R.; Thompson, A.G.; Stroh, J.G.; Shield, L.S.; *et al.* Bioactive Compounds from Aquatic and Terrestrial Sources. *J. Nat. Prod.* **1990**, *53*, 771–792.
4. Ngo, D.-H.; Wijesekara, I.; Vo, T.-S.; van Ta, Q.; Kim, S.-K. Marine Food-Derived Functional Ingredients as Potential Antioxidants in the Food Industry: An Overview. *Food Res. Int.* **2011**, *44*, 523–529.
5. Mayer, A.M.S.; Rodríguez, A.D.; Berlinck, R.G.S.; Fusetani, N. Marine Pharmacology in 2007–2008: Marine Compounds with Antibacterial, Anticoagulant, Antifungal, Anti-Inflammatory, Antimalarial, Antiprotozoal, Antituberculosis, and Antiviral Activities; Affecting the Immune and Nervous System, and Other Miscellaneous Mechanisms of Action. *Comp. Biochem. Physiol. Part C Toxicol. Pharmacol.* **2011**, *153*, 191–222.
6. Abad, M.J.; Bedoya, L.M.; Bermejo, P. Natural Marine Anti-Inflammatory Products. *Mini Rev. Med. Chem.* **2008**, *8*, 740–754.
7. Tarling, C.A.; Woods, K.; Zhang, R.; Brastianos, H.C.; Brayer, G.D.; Andersen, R.J.; Withers, S.G. The Search for Novel Human Pancreatic a-Amylase Inhibitors: High-Throughput Screening of Terrestrial and Marine Natural Product Extracts. *ChemBioChem* **2008**, *9*, 433–438.
8. Cragg, G.M.; Grothaus, P.G.; Newman, D.J. Impact of Natural Products on Developing New Anti-Cancer Agents. *Chem. Rev.* **2009**, *109*, 3012–3043.
9. Vo, T.-S.; Kim, S.-K. Potential Anti-HIV Agents from Marine Resources: An Overview. *Mar. Drugs* **2010**, *8*, 2871–2892.
10. Kang, S.; Kim, D.; Lee, B.; Kim, M.-R.; Chiang, M.; Hong, J. Antioxidant Properties and Cytotoxic Effects of Fractions from Glasswort (*Salicornia herbacea*) Seed Extracts on Human Intestinal Cells. *Food Sci. Biotechnol.* **2011**, *20*, 115–122.
11. Im, S.-A.; Kim, G.-W.; Lee, C.-K. Immunomodulatory Activity of *Salicornia herbacea* L. Components. *Nat. Prod. Sci.* **2003**, *9*, 273–277.

12. Lee, K.Y.; Lee, M.H.; Chang, I.Y.; Yoon, S.P.; Lim, D.Y.; Jeon, Y.J. Macrophage Activation by Polysaccharide Fraction Isolated from *Salicornia herbacea*. *J. Ethnopharmacol.* **2006**, *103*, 372–378.
13. Hwang, J.Y.; Lee, S.K.; Jo, J.R.; Kim, M.E.; So, H.A.; Cho, C.W.; Seo, Y.W.; Kim, J.I. Hypolipidemic Effect of *Salicornia herbacea* in Animal Model of Type 2 Diabetes Mellitus. *Nutr. Res. Pract.* **2007**, *1*, 371–375.
14. Kong, C.-S.; Kim, Y.A.; Kim, M.-M.; Park, J.-S.; Kim, J.-A.; Kim, S.-K.; Lee, B.-J.; Nam, T.J.; Seo, Y. Flavonoid Glycosides Isolated from *Salicornia herbacea* Inhibit Matrix Metalloproteinase in HT1080 Cells. *Toxicol. In Vitro* **2008**, *22*, 1742–1748.
15. Kong, C.-S.; Kim, J.-A.; Qian, Z.-J.; Kim, Y.A.; Lee, J.I.; Kim, S.-K.; Nam, T.J.; Seo, Y. Protective Effect of Isorhamnetin 3-*O*-β-D-Glucopyranoside from *Salicornia herbacea* against Oxidation-Induced Cell Damage. *Food Chem. Toxicol.* **2009**, *47*, 1914–1920.
16. Lee, Y.S.; Lee, S.; Lee, H.S.; Kim, B.-K.; Ohuchi, K.; Shin, K.H. Inhibitory Effects of Isorhamnetin-3-*O*-β-D-Glucoside from *Salicornia herbacea* on Rat Lens Aldose Reductase and Sorbitol Accumulation in Streptozotocin-Induced Diabetic Rat Tissues. *Biol. Pharm. Bull.* **2005**, *28*, 916–918.
17. Muruganandan, S.; Roman, A.A.; Sinal, C.J. Adipocyte Differentiation of Bone Marrow-Derived Mesenchymal Stem Cells: Cross Talk with the Osteoblastogenic Program. *Cell. Mol. Life Sci.* **2009**, *66*, 236–253.
18. Fraser, L.A.; Adachi, J.D. Glucocorticoid-Induced Osteoporosis: Treatment Update and Review. *Ther. Adv. Musculoskelet. Dis.* **2009**, *1*, 71–85.
19. Harsløf, T.; Tofteng, C.L.; Husted, L.B.; Nyegaard, M.; Børglum, A.; Carstens, M.; Stenkjær, L.; Brixen, K.; Eiken, P.; Jensen, J.E.; *et al.* Polymorphisms of the Peroxisome Proliferator-Activated Receptor γ (PPARγ) Gene Are Associated with Osteoporosis. *Osteoporosis Int.* **2011**, *22*, 2655–2666.
20. Nuttall, M.E.; Gimble, J.M. Controlling the Balance between Osteoblastogenesis and Adipogenesis and the Consequent Therapeutic Implications. *Curr. Opin. Pharm.* **2004**, *4*, 290–294.
21. Spiegelman, B.M. PPAR-Γ: Adipogenic Regulator and Thiazolidinedione Receptor. *Diabetes* **1998**, *47*, 507–514.
22. Dietz, W.H. Health Consequences of Obesity in Youth: Childhood Predictors of Adult Disease. *Pediatrics* **1998**, *101*, 518–525.
23. Cha, E.; Umpierrez, G.; Kim, K.H.; Bello, M.K.; Dunbar, S.B. Characteristics of American Young Adults with Increased Risk for Type 2 Diabetes: A Pilot Study. *Diabetes Educ.* **2013**, *39*, 454–469.
24. Harvey, A.L. Plant Natural Products in Anti-Diabetic Drug Discovery. *Curr. Org. Chem.* **2010**, *14*, 1670–1677.
25. Newman, D.J.; Cragg, G.M. Natural Products as Sources of New Drugs over the 30 Years from 1981 to 2010. *J. Nat. Prod.* **2012**, *75*, 311–335.
26. De Ugarte, D.A.; Morizono, K.; Elbarbary, A.; Alfonso, Z.; Zuk, P.A.; Zhu, M.; Dragoo, J.L.; Ashjian, P.; Thomas, B.; Benhaim, P.; *et al.* Comparison of Multi-Lineage Cells from Human Adipose Tissue and Bone Marrow. *Cells Tissues Organs* **2003**, *174*, 101–109.

27. Zhao, L.J.; Jiang, H.; Papasian, C.J.; Maulik, D.; Drees, B.; Hamilton, J.; Deng, H.W. Correlation of Obesity and Osteoporosis: Effect of Fat Mass on the Determination of Osteoporosis. *J. Bone Miner. Res.* **2008**, *23*, 17–29.
28. Rodriguez, J.P.; Astudillo, P.; Rios, S.; Pino, A.M. Involvement of Adipogenic Potential of Human Bone Marrow Mesenchymal Stem Cells (MSCs) in Osteoporosis. *Curr. Stem Cell Res. Ther.* **2008**, *3*, 208–218.
29. Rosen, C.J.; Bouxsein, M.L. Mechanisms of Disease: Is Osteoporosis the Obesity of Bone? *Nat. Clin. Pract. Rheumatol.* **2006**, *2*, 35–43.
30. Rawadi, G.; Vayssière, B.; Dunn, F.; Baron, R.; Roman-Roman, S. BMP-2 Controls Alkaline Phosphatase Expression and Osteoblast Mineralization by a Wnt Autocrine Loop. *J. Bone Miner. Res.* **2003**, *18*, 1842–1853.
31. Ryoo, H.-M.; Lee, M.-H.; Kim, Y.-J. Critical Molecular Switches Involved in BMP-2-Induced Osteogenic Differentiation of Mesenchymal Cells. *Gene* **2006**, *366*, 51–57.
32. Ha, B.J.; Lee, S.H.; Kim, H.J.; Lee, J.Y. The Role of *Salicornia herbacea* in Ovariectomy-Induced Oxidative Stress. *Biol. Pharm. Bull.* **2006**, *29*, 1305–1309.
33. Notoya, M.N.; Tsukamoto, Y.; Nishimura, H.; Woo, J.T.; Nagai, K.; Lee, I.-S.; Hagiwara, H. Quercetin, a Flavonoid, Inhibits the Proliferation, Differentiation, and Mineralization of Osteoblasts *in vitro*. *Eur. J. Pharmacol.* **2004**, *485*, 89–96.
34. Kim, Y.J.; Bae, Y.C.; Suh, K.T.; Jung, J.S. Quercetin, a Flavonoid, Inhibits Proliferation And Increases Osteogenic Differentiation in Human Adipose Stromal Cells. *Biochem. Pharmacol.* **2006**, *72*, 1268–1278.
35. Fatokun, A.A.; Stone, T.W.; Smith, R.A. Hydrogen Peroxide-induced Oxidative Stress in MC3T3-E1 Cells: The Effects of Glutamate and Protection by Purines. *Bone* **2006**, *39*, 542–551.

13

Identification of the Major ACE-Inhibitory Peptides Produced by Enzymatic Hydrolysis of a Protein Concentrate from Cuttlefish Wastewater

Isabel Rodríguez Amado [1], José Antonio Vázquez [1,*], Pilar González [1], Diego Esteban-Fernández [2], Mónica Carrera [3] and Carmen Piñeiro [4]

[1] Group of Recycling and Valorisation of Waste Materials (REVAL), Marine Research Institute (IIM-CSIC), r/Eduardo Cabello, 6. Vigo, Galicia 36208, Spain; E-Mails: sabelara@iim.csic.es (I.R.A.); pgonzalez@iim.csic.es (P.G.)

[2] Department of Chemistry, Humboldt-Universitaet zu Berlin, Brook-Taylor Strasse 2, Berlin 12489, Germany; E-Mail: esteband@chemie.hu-berlin.de

[3] Institute of Molecular Systems Biology (IMSB), ETH Zürich, Zürich 8093, Switzerland; E-Mail: mcarrera@iim.csic.es

[4] Scientific Instrumentation Service (SICIM), Marine Research Institute (IIM-CSIC), r/ Eduardo Cabello, 6. Vigo, Galicia 36208, Spain; E-Mail: cpineiro@iim.csic.es

* Author to whom correspondence should be addressed; E-Mail: jvazquez@iim.csic.es

---

Abstract: The aim of this work was the purification and identification of the major angiotensin converting enzyme (ACE) inhibitory peptides produced by enzymatic hydrolysis of a protein concentrate recovered from a cuttlefish industrial manufacturing effluent. This process consisted on the ultrafiltration of cuttlefish softening wastewater, with a 10 kDa cut-off membrane, followed by the hydrolysis with alcalase of the retained fraction. Alcalase produced ACE inhibitors reaching the highest activity (IC_{50} = 76.8 ± 15.2 μg mL^{-1}) after 8 h of proteolysis. Sequential ultrafiltration of the 8 h hydrolysate with molecular weight cut-off (MWCO) membranes of 10 and 1 kDa resulted in the increased activity of each permeate, with a final IC_{50} value of 58.4 ± 4.6 μg mL^{-1}. Permeate containing peptides lower than 1 kDa was separated by reversed-phase high performance liquid chromatography (RP-HPLC). Four fractions (A–D) with potent ACE inhibitory activity were isolated and their main peptides identified using high performance liquid chromatography coupled to an electrospray ion trap Fourier transform ion cyclotron resonance-mass spectrometer (HPLC-ESI-IT-FTICR) followed by comparison with databases and *de novo* sequencing. The amino acid sequences of the identified peptides contained at least one hydrophobic and/or a proline together with positively charged residues in at least one of the three *C*-terminal positions. The IC_{50} values of the fractions ranged from 1.92 to 8.83 μg mL^{-1}, however this study fails to identify which of these peptides are ultimately responsible for the potent antihypertensive activity of these fractions.

Keywords: ultrafiltration; proteolysis; ACE inhibitory peptides; cuttlefish byproducts; peptide identification; HPLC-ESI-MS

---

1. Introduction

Cardiovascular diseases (CVD) are the main cause of death globally, accounting for approximately 17 million deaths a year, nearly one third of the total [1]. According to the World Health Organization, most cardiovascular diseases can be prevented by addressing risk factors, such as tobacco use, unhealthy diet and obesity, physical inactivity, high blood pressure, diabetes, and raised lipids [2]. Among all CVD, complications of hypertension account for 9.4 million deaths worldwide every year [3] and are the most common risk factors of heart diseases. Thus, hypertension is now considered a major health problem worldwide.

Blood pressure is regulated by the renin-angiotensin system, wherein, the angiotensin converting enzyme (ACE) hydrolyses biologically inactive angiotensin I to the potent vasoconstrictor angiotensin II, responsible for the increase in the arterial pressure. At present, the renin-angiotensin system has become a key target for drugs combating hypertension [4] and, more specifically, various synthetic ACE inhibitors are widely used to treat cardiovascular disorders. However, these synthetic drugs have been described to have some side effects [5] and, thus, natural sources of ACE inhibitors are being investigated as a milder but effective alternative for the control of high blood pressure. Among the origins of these molecules are food-derived peptides from different sources, such as cheese whey [6], gelatin [7], and different meat and fish proteins [8].

Marine biodiversity is a valuable source of molecules with diverse biological activities. Hence, interest has focused on the search of bioactive compounds, including ACE inhibitors, from marine sources. Peptides with ACE-inhibitory activity have been isolated from different marine species including shrimp [9], sole [10], cuttlefish [11,12], and salmon [13], among others.

However, a more sustainable alternative providing high-value bioactive compounds is the use of fish byproducts. Among waste materials, fish processing wastewaters have been recently studied as starting materials to obtain ACE inhibitory peptides. Recently a functional concentrate from industrial shrimp (*Penaeus* spp.) cooking juice with antioxidative and antihypertensive (ACE-inhibitory) capacity has been obtained by Pérez-Santín *et al.* [14]. In addition, a screening study of different cuttlefish industrial processing wastewaters demonstrated the great potential of ultrafiltration (UF)-fractionation followed by proteolysis of cuttlefish protein concentrates as a source of peptide mixtures with antihypertensive and antioxidant activity [15].

The purpose of the present study was to purify and identify the major ACE inhibitory peptides produced by enzymatic hydrolysis of proteins recovered by UF from cuttlefish softening wastewater. For that purpose, the ACE inhibitory peptides from a cuttlefish byproduct hydrolysate were fractionated by sequential ultrafiltration using 10 and 1 molecular weight cut-off (MWCO) membranes. The most active fractions in the 1 kDa permeate were isolated by reversed-phase high performance liquid chromatography (RP-HPLC), their ACE-inhibitory activity quantified and the main peptides were identified by mass spectrometry.

2. Results and Discussion

2.1. Protein Recovery by Ultrafiltration-Diafiltration (UF-DF)

The wastewater used in this work was the result of successive cooking operations using the same water in order to obtain a high initial protein concentration. For this reason, the protein content was significantly higher than others previously reported for cuttlefish processing effluents [16,17]. Protein concentration in cuttlefish softening wastewater was 11.5 ± 0.34 g L^{-1} and reached 95.42 ± 1.22 g L^{-1} of protein content after ultrafiltration-diafiltration (UF-DF) at 10 kDa. As can be seen in Figure 1, Equation (1) was able to simulate the DF process and parameter determinations were always significant (Student's t test, $\alpha = 0.05$). Results of UF-DF using this cut-off membrane showed protein retentions higher than 95% of the initial protein content in the retentate and an overall concentration factor (fc) above 4.0.

Figure 1. Protein recovery from cuttlefish wastewater by UF-DF at 10 kDa molecular weight cut-off (MWCO). Left: concentration of retained protein in linear relation with the volumetric concentration factor (fc) showing experimental data (points) and theoretical profiles (discontinuous line). Right: Progress of protein (○) retention with the increase of diavolume from DF process (D). For clarity, confidence intervals (in all cases less than 5% of the experimental mean value; $\alpha = 0.05$; $n = 2$) were omitted. Equation (1) was used to fit the experimental data.

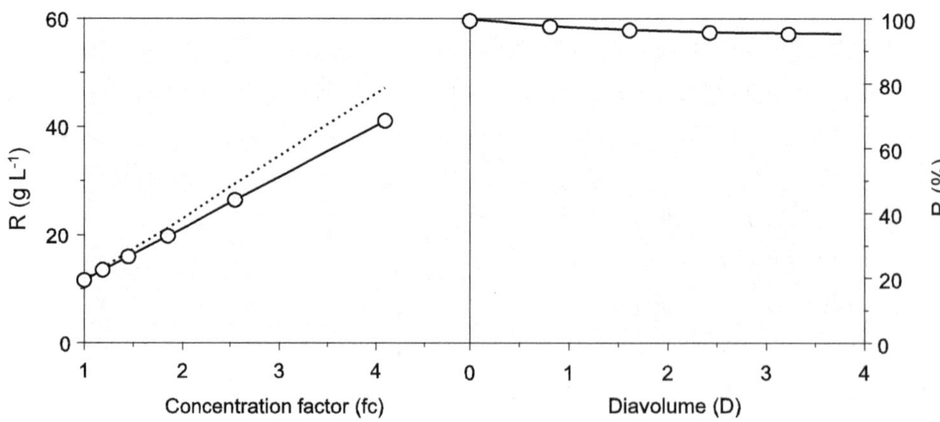

This result slightly improves previously published data in the treatment of cuttlefish effluents using membrane technology. An 85% reduction in the chemical oxygen demand (COD) of cuttlefish processing wastewater was observed after nanofiltration using a 550 Da cellulose acetate membrane [16]. On the other hand, a similar protein retention rate (78%) was reported in the microfiltration associated to ultrafiltration (15 kDa MWCO) of cuttlefish effluents using ceramic membranes [17]. All these studies consistently indicate a great efficiency of membrane technology for protein concentration. The use of these byproducts as a source of protein contributes to improving depuration of industrial wastewaters, reducing the treatment costs, and decreasing their contaminating effects.

2.2. ACE Inhibitory Activity of Cuttlefish Hydrolysate and Their UF Fractions

To obtain the ACE inhibitors, the cuttlefish protein concentrate was hydrolysed using alcalase according to optimal conditions previously reported for this substrate [12]. The extent of protein degradation by the proteolytic enzyme was estimated by assessing the degree of hydrolysis (DH) (Equation (2)). According to this calculation a high DH was observed from the beginning of the digestion, reaching a final value of 37.8% ± 0.4% after 8 h of proteolysis (Table 1).

Table 1. Degrees of hydrolysis (DH) after 0.5, 2, and 8 h digestion with alcalase of a cuttlefish wastewater protein concentrate. Angiotensin converting enzyme (ACE) inhibitory activity (I_{ACE} (%)) and IC_{50} values (μg mL^{-1}) of these hydrolysates, as well as retentates obtained by 10 and 1 kDa membrane and permeate of 1 kDa from the 8 h hydrolysate. I_{ACE} (%) was calculated by Equation (3). Mean values and standard deviations from triplicate samples are shown.

Values of DH, I_{ACE} (%) and IC_{50} for Hydrolysis and UF Processes				
		DH (%)	**I_{ACE} (%)**	**IC_{50} (μg mL^{-1})**
		–	**37.8 ± 0.4**	**292.5 ± 20.3**
	0.5 h	17.5 ± 0.2	76.2 ± 3.2	214.4 ± 27.3
Hydrolysis	2 h	23.8 ± 1.8	80.4 ± 2.3	122.5 ± 13.1
	8 h	36.4 ± 0.9	92.1 ± 0.6	76.8 ± 15.2
	10 kDa retentate	–	83.3 ± 0.6	273.9 ± 30.1
Ultrafiltration	1 kDa retentate	–	94.6 ± 2.9	235.6 ± 20.2
	1 kDa permeate	–	68.0 ± 1.4	58.4 ± 4.6

Indeed, the highest levels of hydrolysis were greater than those obtained from cuttlefish muscle [18] and other seafood species [13] using this enzyme. As reported before [12], this difference might be attributed to the thermal treatment of proteins recovered by UF, which tends to increase the susceptibility of proteins to enzymatic hydrolysis due to their partial denaturation.

ACE inhibitory activity improved with increasing time of hydrolysis, reaching maximal values (92.1% ± 0.6%) after 8 h of proteolysis (Table 1). In addition, a direct relationship between ACE inhibitory activity and DH was observed in a study evaluating the ACE inhibitory activities of cuttlefish muscle protein hydrolysates prepared by treatment with various bacterial proteases [18]. In fact, the authors obtained a lower inhibitory activity (51.5% ± 1.5%) from a cuttlefish protein hydrolysate prepared using alcalase with a DH of 12.5%, *i.e.*, three-fold lower than the DH reported in the present study. However, it must be noted that hydrolysis conditions and calculation methods for the DH and ACE inhibitory varied between both studies, making difficult the direct comparison of the results.

To calculate IC_{50} values, dose-response curves (Figure 2) were plotted and accurately fitted by Equation (4), with coefficients of determination greater than 0.98.

Figure 2. Left: Experimental data (symbols) for ACE-inhibitory activity of cuttlefish softening wastewater protein concentrate before (●) and after hydrolysis with alcalase for 0.5 (○), 2 (△), and 8 h (□). Right: Experimental data (symbols) for ACE-inhibitory activity of 10 kDa (▽) and 1 kDa (▼) retentates and 1 kDa permeate (▲) from the 8 h hydrolysate. The dose-response curves (lines) were obtained by modeling experimental data to the Equation (4).

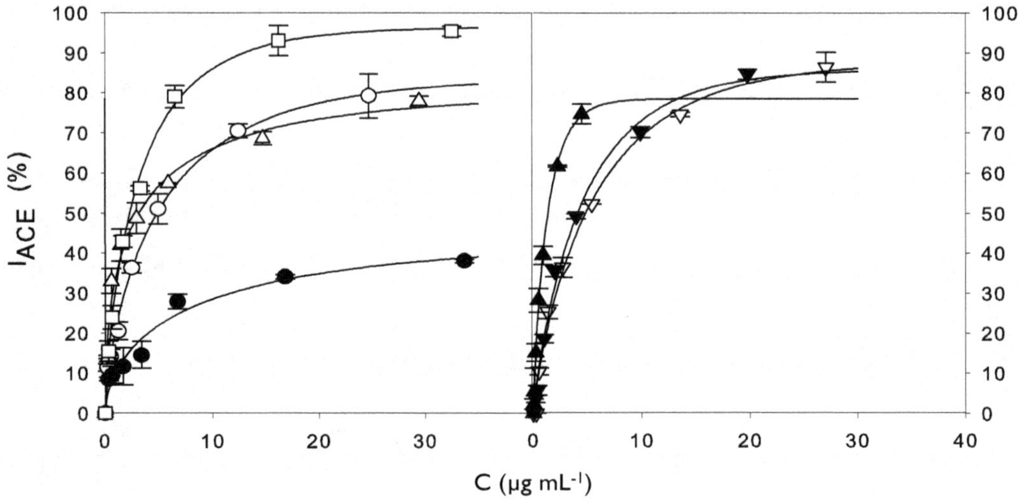

Equation (4) allowed the direct determination of IC_{50} values using mathematical nonlinear fitting. The estimation of IC_{50} values using this model is more precise than direct interpolation or linear regression and allows the objective comparison between samples. This equation was consistent (Fisher's F test; $p < 0.05$) and parameter estimations were always significant (Student's t test, $\alpha = 0.05$) (data not shown). According to this approach, IC_{50} decreased with increasing time of hydrolysis reaching the lowest value (maximal ACE inhibitory activity) after 8 h of hydrolysis. These differences revealed that the use of alcalase was efficient in releasing peptides with ACE inhibitory activity from cuttlefish protein concentrate. In fact, this endoprotease has been widely employed for the production of potent marine derived ACE inhibitors from sources as diverse as the algae *Chlorella ellipsoidea* [19], salmon byproducts [20], seaweed pipefish muscle protein [21], the sea cucumber *Acaudina molpadioidea* [22], and lizard fish [23]. This enzyme is also less expensive than others commercially available, thus, being a good candidate for the industrial production of bioactive hydrolysates from marine byproducts. Values of IC_{50} for ACE inhibitory activity of whole cuttlefish hydrolysates have being previously reported for various digestive proteases [11], including alcalase [18]. These values ranged from 1.19 to 2.31 mg mL^{-1}, however the direct comparison with IC_{50} values in the present study (Table 1) is not possible due to differences in the hydrolysis conditions but, also, in the methodology employed for ACE inhibitory activity determination and calculation.

The 8 h hydrolysate, which led to the highest ACE inhibitory activity (lowest IC_{50}), was fractionated in an attempt to isolate the active peptides. The use of UF has proven to be a valuable resource for hydrolysate fractionation, significantly decreasing the IC_{50} value of peptide mixtures obtained from different sources [6,24–26]. For this reason, the 8 h hydrolysate was sequentially ultrafiltered through membranes at 10 and 1 kDa MWCO (Figure 3). Then, permeate and retentates ACE inhibitory activity was determined. As summarized in Table 1, the lowest IC_{50} value was associated to 1 kDa permeate (58.4 ± 4.6 µg mL^{-1}), indicating that the maximum contribution to ACE-inhibitory activity is due to the peptides with MW below 1 kDa. Similarly, the ACE inhibitory activity was significantly higher in the lowermost MW fraction (<1 kDa) from collagenous residues from squid skins hydrolyzed with esperase [25].

Figure 3. Process of ACE inhibitory peptide production from a protein enriched fraction recovered from cuttlefish wastewater by UF-DF at 10 kDa (Step 1) followed by hydrolysis with alcalase (0.4% v/v) for 8 h at 55 °C (Step 2). Then, the hydrolysate was fractionated using a sequentially UF through membranes of 10 (Step 3) and 1 kDa (Step 4). UFP: Permeates of UF; UFR: retentates of UF.

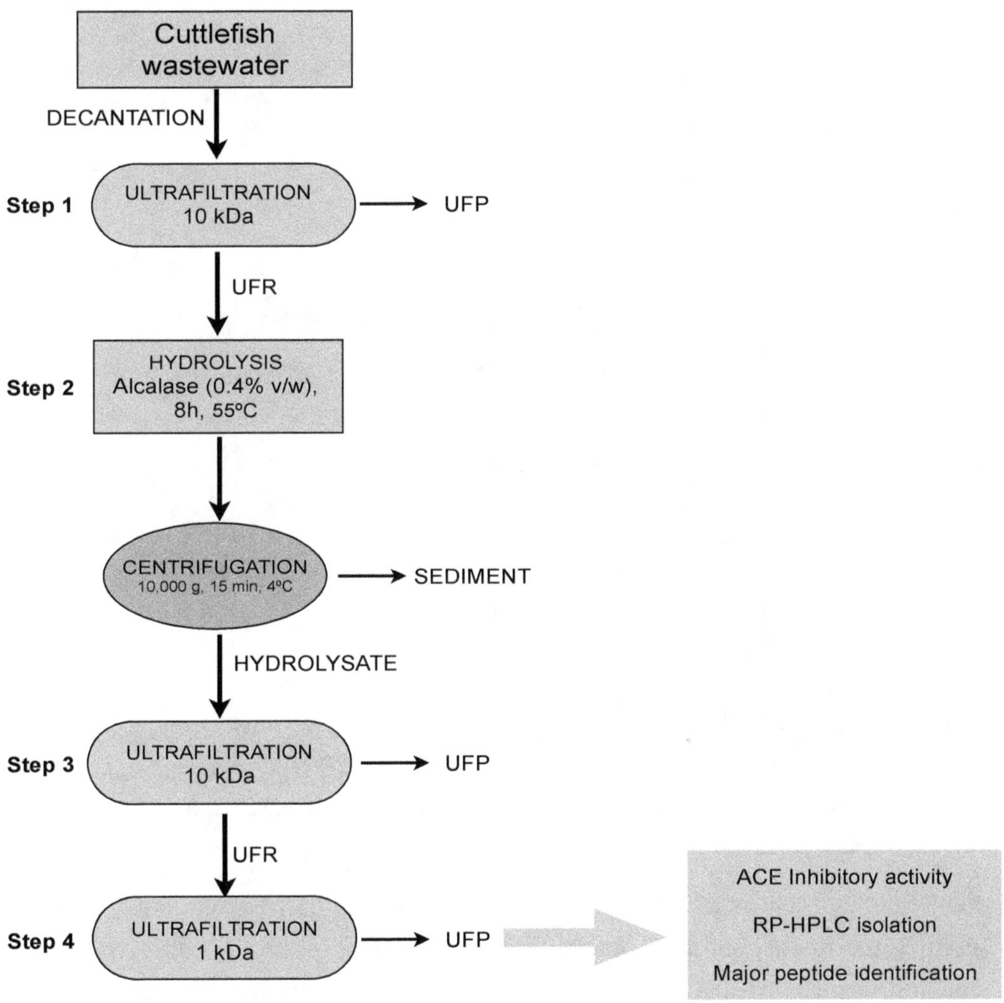

This observation is in good agreement with previous studies reporting greater *in vitro* activity of low than high molecular weight peptides from different protein origin [13,18].

2.3. Identification of Major ACE Inhibitory Peptides

The ACE inhibitory peptides in the 1 kDa permeate from cuttlefish hydrolysate were separated by RP-HPLC on a C18 column. The chromatogram (Figure 4) was divided into sixty-five fractions (2 mL) that were freeze-dried and their ACE inhibitory activity determined after resuspension in milliQ water and pH adjustment to 7.0. Four fractions showing the highest activity were collected separately under optimized separation conditions that allowed a better resolution of the peaks, *i.e.*, lower flow rate and a narrower acetonitrile gradient. The separation procedure was successively repeated to obtain a sufficient protein concentration for further analysis. The most active fractions (Figure 4A–D) were freeze-dried and their ACE inhibitory activity and IC_{50} values determined (Table 2).

Figure 4. Peptide profile of the 8 h hydrolysate that was permeated through the 1 kDa ultrafiltration membrane after RP-HPLC separation. Lines indicate the peaks corresponding to the four active fractions (A–D).

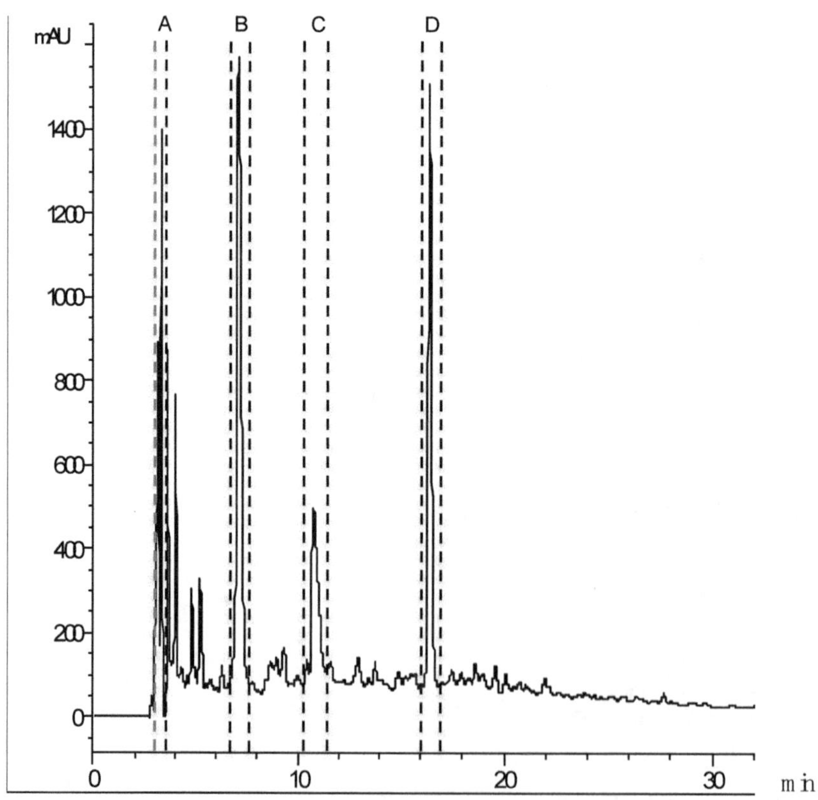

The chemical characterization of each RP-HPLC fraction was carried out using mass spectrometry, in order to identify the main peptides contributing to the ACE inhibitory activity measured in these samples. According to these results (Table 2), among the four fractions only two (C and D) matched the protein databases and the other two fractions (A and B) had to be sequenced *de novo*.

Table 2. Major identified peptides corresponding to the four more active fractions after semi-preparative fractionation by reversed-phase high performance liquid chromatography (RP-HPLC) of the 1 kDa ultrafiltration permeate from the 8 h hydrolysate. The IC_{50} ($\mu g\ mL^{-1}$) value and the protein source of these fractions are shown.

Fraction	Peptides	Protein Source	Mass (Da)	IC_{50} ($\mu g\ mL^{-1}$)
A	GNALVFLLP		943.57	1.92 ± 0.40
	FALASSLVN		921.53	
	SPSIAPAL		755.40	
	GLFAAHRK		899.54	
	FTPESLAI		877.50	
	DAAIKTLTK		960.60	
	MMQLRKA		877.50	
	RALKIPAM		899.54	
	FTVKPVSR		933.59	
	SSSKAKKKP		960.60	
	KAVPIAIL		824.58	
	TRAASCP		705.30	
	INKAVTGLK		943.57	
	PEGSIRP		755.40	
B	VEDAEVGKK		974.52	4.00 ± 0.19
	FAGDDAPRA		919.43	
	AGSVNKSK		790.44	
	KAGSELGL		774.44	
	QSLEVSK		790.43	
	KEAEVSK		790.43	
	GAEVTVSK		790.43	
	QEVSLSK		790.43	
	AGEVSLSK		790.43	
	NFGCSVK		754.35	
	SVCGGFGK		754.35	
	AGDDPAR		701.32	
C	ADQSEGALQK	Myosin HC	1045.5	8.63 ± 0.86
	ISEQEASQR	Tropmyosin	1046.5	
	LGEGGRSTHE	Myosin HC	1041.5	
	RLKEAENR	Actin	981.5	
	TDQLGEGGRS	Myosin HC	1018.5	
D	DEDATGVIR	Tn C	974.47	7.18 ± 2.44
	GTDPEDALRN	Myosin RLC	1086.49	
	GVNLEDAKRS	Myosin HC	1087.56	
	LTEAPLNPK	Actin	981.55	
	REDIDGNIK	Myosin LCK	1058.54	

The peptides identified in the fractions C and D are derived from muscular protein fragments of the cuttlefish mantle. Softening of fish muscle consists on the heating and addition of salts to solubilize myofibrillar proteins, improving the water-fat-retention and leading to acceptable rigidity/elasticity of the muscle gels [27]. Thus, as expected and since the cuttlefish hydrolysate was prepared from a protein concentrate recovered from softening wastewater, myofibrillar proteins, such as troponin, actin, and myosin were identified as protein sources from the identified peptides (Table 2). Antihypertensive peptides isolated from other squid proteins hydrolysates, such as collagen [27] and gelatin [7,28], have also been reported. These studies identified peptides sequences of the major ACE inhibitory fractions with a high content in proline, glycine, and leucine residues and suggesting the presence of this last amino acid could play an important role in the ACE-inhibitory activity [28].

As is known, peptide sequence greatly contributes to its biological activity. The structure–function relationship for ACE inhibitory peptides is not fully characterized, although recently published papers deal with the structure-activity relationships of di and tripeptides [29,30]. Some of the information available about this topic indicates that the three amino acids in the *C*-terminal peptide sequence are particularly important for their binding to ACE, as they might interact with the active site of the enzyme [31]. For instance, this enzyme prefers hydrophobic amino acids, including aromatic residues (phenylalanine, tryptophan, and tyrosine) and branched chain amino acids (valine, leucine, and isoleucine) in one of the three positions of the *C*-terminal [32]. In this study, all of the peptides identified in the four potent ACE inhibitory fractions contained at least one hydrophobic residue, being the most abundant branched chain amino acids (Table 2).

On the other hand, ACE has low affinity for peptides with dicarboxylic amino acids in this position or those containing the amino acid proline in the penultimate position [33], although proline in final or penultimate position of the *C*-terminal peptide sequence promotes enzyme binding [34]. In fact, several identified ACE inhibitory peptides had a proline residue in one of these positions of *C*-terminal, although this is neither sufficient nor essential to confer activity [18]. Furthermore, the positively charged amino acids, such as arginine and lysine at the *C*-terminal sequences were also reported to contribute towards ACE inhibitory or antihypertensive effect [35]. As can be seen in Table 2, among the identified peptides many of them contained one of these residues in the *C*-terminal. This was especially evident in fraction B, where ten of the twelve peptides identified contained lysine or arginine in last position. In agreement with other published papers, these findings confirm that peptide sequences are greatly important to define their ACE inhibitory activity (IC_{50} values). However, the differences in the composition of the peptides identified in this study fail to explain the differences found in the ACE inhibitory activity and, thus, further studies are needed to define which of these peptides are ultimately responsible for the potent antihypertensive activity of these fractions.

3. Experimental Section

3.1. Cuttlefish Processing Wastewater

Wastewater from the softening treatment during industrial manufacturing of cuttlefish was kindly provided by Frinova S.A., (Porriño, Galicia, Spain). The average content of suspended solids of this water was ranged among 0.3–0.7 g L^{-1} and the solid residue was around 10–15 g L^{-1}. The water was decanted to discard the particulate matter, sampled for analytical determinations, and stored at −20 °C until further use.

3.2. Total Protein Analysis

Protein concentration was determined by the bicinchoninic acid assay (BCA, Pierce, Rockford, IL, USA) using bovine serum albumin as standard. Briefly, 200 μL of the BCA working reagent were mixed with 25 μL of sample. The mixture was incubated for 30 min at 37 °C and the absorbance taken at 562 nm using a Multiskan Spectrum Microplate Spectrophotometer (Thermo Scientific, Waltham, MA, USA).

3.3. Ultrafiltration-Diafiltration Process

Cleaned cuttlefish wastewater was subjected to ultrafiltration-diafiltration using a Prep/Scale-TFF polyethersulfone cartridge (Millipore Corporation, Bedford, MA, USA) of 10 kDa molecular weight cut-off (MWCO). The operation mode was the following: an initial phase of ultrafiltration (UF) at 40 °C with total recirculation of retentate was performed, immediately followed by a diafiltration (DF) process. During UF, the inlet pressure remained constant to determine the flow rate drops due to the increased concentration of protein in the retentate and to possible membrane fouling. The final retentate (after DF) was frozen at −20 °C for further hydrolysis and permeate from the DF phase was discarded.

For the modeling of the membrane process, we assumed that in the DF with constant volume (filtration flow = water intake flow), the concentration (or the total amount) of a permeable solute in the retentate followed first order kinetics [15]:

$$R = R_f + R_0 \exp\left[-(1-s)D\right] \quad (1)$$

where, R is the concentration of permeable protein in the retentate (% from the level at initial DF), R_0 the initial concentration (%), R_f is the final and asymptotic concentration (%), D is the relative diavolume (volume of added water/constant retentate volume) and s is the specific retention of protein with variation between 0 (the solute is filtered as the solvent) and 1 (the solute is totally retained). Thus, using normalized values (%): $R_0 + R_f = 100$, being $R_0 = 0$ if all protein is permeable.

3.4. Preparation of Enzymatic Hydrolysates

The cuttlefish wastewater concentrate (retentate) was pre-incubated at 55 °C for 10 min prior to enzymatic hydrolysis using Alcalase 2.4 L from Novo Co. (Novozyme Nordisk, Bagsvaerd, Denmark). The enzyme to substrate ratio (E/S) was 0.4% (v/w) in relation to retentate weight and the reaction was performed in a temperature-controlled water bath under constant magnetic stirring [12]. Samples were taken after 0.5, 2, and 8 h of hydrolysis, boiled for 15 min to inactivate the protease and then cooled in an ice-water bath for at least 30 min. Hydrolysates were centrifuged at 10,000× g for 15 min (4 °C) in an Avanti J-26 XP centrifuge (Beckman Coulter, Inc., Miami, FL, USA) and supernatants were stored at −20 °C until further analysis of their degree of hydrolysis and antihypertensive activity.

3.5. Degree of Hydrolysis Determination

The degree of hydrolysis (DH) was defined as the ratio of the released tyrosine to the initial tyrosine in the retentate [12], according to the following expression:

$$DH = \frac{C_t - C_0}{C_0} \times 100 \tag{2}$$

where, C_t is the tyrosine concentration (g L^{-1}) of the hydrolysate at time t and C_0 is the tyrosine concentration (g L^{-1}) of the sample at time zero. Tyrosine was measured by the method described in Barker and Worgan [36].

3.6. Angiotensin I-Converting Enzyme (ACE) Inhibition Assay

The antihypertensive activity was determined using a modified version of the method described by Shalaby *et al.* [37] with *N*-[3-(2-Furyl) acryloyl]-L-phenylalanyl-glycyl-glycine (FAPGG) as substrate. Briefly, 10 μL of sample were mixed with 10 μL of ACE solution (0.5 U mL^{-1}) in each well of a 96-well microtiter plate. The reaction was started by adding 150 μL of substrate (0.88 mM FAPGG in 50 mM Tris-HCl, pH 7.5, 0.3 M NaCl) pre-heated at 37 °C. The absorbance at 340 nm (A340) was acquired at time intervals of 30 s for 30 min in a Multiskan Spectrum Microplate Spectrophotometer (Thermo Scientific, Waltham, MA, USA). The control consisted of samples containing 10 μL of buffer (50 mM Tris-HCl, pH 7.5, 0.3 M NaCl) instead of protein hydrolysate. Samples and controls were performed in triplicate.

ACE-inhibitory capacity (I_{ACE}) was calculated as a function of the average slope of decrease in absorbance with time and expressed as percent of enzyme inhibition according to the following expression:

$$I_{ACE}(\%) = \left[1 - \frac{\rho A_h}{\rho A_c}\right] \times 100 \tag{3}$$

where, I_{ACE} is the ACE-inhibitory capacity (%), ρA_h is the slope of decrease in A340 in the presence of inhibitor (hydrolysate) and ρA_c is the slope of decrease in A340 in the absence of inhibitor (control).

For the calculation of the protein concentration causing a 50% ACE inhibition (IC_{50}), dose-response curves were obtained assaying different concentrations of hydrolysates. IC_{50} values were calculated by fitting the dose-response curves of I_{ACE} *vs.* protein concentration to a mechanistic model developed by Estévez *et al.* [6]:

$$I_{ACE} = 100K\left(1 - \frac{(K_m + S)}{\left(S + K_m\left(1 + \frac{C(K_m + S)}{IC_{50}K_m(2K - 1)}\right)\right)}\right) \tag{4}$$

where, K is the maximum I_{ACE} (%), C is the protein concentration (μg mL^{-1}) and IC_{50} is the concentration for semi-maximum response (μg mL^{-1}). K_m is the Michaelis–Menten constant of FAPGG and S is the concentration of this substrate in the assay, respectively. Both K_m and S are constants under these experimental conditions, with values of 182 μg mL^{-1} and 310 μg mL^{-1}, respectively [6].

3.7. Isolation of Major ACE-Inhibitory Peptides

The 8-h-hydrolysate was further ultrafiltered to remove peptides with high molecular weight. UF were carried out in a stirred UF cell module (Millipore Corporation, Bedford, MA, USA) under a nitrogen flow pressure of 40 psi at room temperature using separately 10 and 1 kDa MWCO membranes (Millipore Corporation, Bedford, MA, USA). Firstly, the hydrolysate was ultrafiltrate using a 10 kDa MWCO membrane followed by further fractionation of the 10 kDa permeate through a 1 kDa MWCO membrane. Samples from 10 and 1 kDa retentates and 1 kDa permeate were collected and protein concentration and ACE-inhibitory activity (IC_{50}) were determined.

The 1 kDa permeate was separated by RP-HPLC on an Agilent 1200 series system (Agilent, Waldbronn, Germany) using an ACE 5 C18 column (10 × 250 mm, 5 μm). Peptides were eluted and monitored at 220 nm using water containing 0.1% trifluoric acetic acid (TFA) (mobile phase A) and acetonitrile with 0.1% TFA (mobile phase B), with a linear gradient of acetonitrile (0%–50% v/v) during 50 min. The ACE inhibitory activity was determined in 2 mL fractions collected at a flow rate of 2.5 mL min^{-1}. Those fractions showing the highest ACE inhibitory activity were collected using an Agilent Zorbax Eclipse C18 column (4.6 × 250 mm, 5 μm) with a linear gradient of acetonitrile (5%–30% v/v) at a flow rate of 0.75 mL min^{-1} during 30 min. According to the chromatographic peptide profile obtained, four fractions from ten samples were defined and collected manually, frozen at −80 °C, and lyophilized for further protein quantitation, antihypertensive activity determination, and mass spectrometry identification analysis.

3.8. Mass Spectrometry Analysis of ACE-Inhibitory Peptides

The freeze-dried fractions were diluted in 1 mL of 1% ACN, 0.1% formic acid (FA) and 2 μL aliquots were analyzed using a C18 column coupled to an electrospray ion trap Fourier transform ion cyclotron resonance-mass spectrometer (ESI-IT-FTICR, Thermo Fisher Scientific, Bremen, Germany). A data dependent fragmentation method consisting of cycles of high resolution full scan spectra followed by ion trap fragmentation and detection of the five most intense masses was used for identification purposes. The chromatographic system (Agilent 1200 series, Waldbronn, Germany) was equipped with an Agilent SB C-18 column (0.5 × 150 mm, 5 μm). A very slow elution program, namely, 0/5, 6/5, 7/10, 32/40, 34/90 (min/%B, being mobile phase A 5% ACN, 0.1% FA and mobile phase B pure ACN, 0.1% FA), at a flow of 20 μL min^{-1}, was optimized to enhance the separation of the low hydrophobic species fractionated in the previous semi-preparative chromatographic dimension.

3.9. Mass Spectrometry Data Processing

MS/MS spectra were searched using SEQUEST-SORCERER™ 2 package (Sage-N Research Inc., Milpitas, CA, USA), against the UniProt/SwissProt database (release 2012_10; 455.545 entries), which also included their respective decoy sequences. The following constraints were used for the searches: semi-tryptic cleavage with up to two missed cleavage sites and tolerances 1.0 Da for precursor ions and 0.5 Da for MS/MS fragments ions. The variable modifications allowed were methionine oxidation (m), carbamidomethylation of Cys and acetylation of the *N*-terminus of the protein (*N*-Acyl). The database search results were subjected to statistical analysis with the PeptideProphet algorithm (v.4.4) [38]. The FDR was kept below 1%.

De novo sequencing was performed by manual interpretation of the ion series of the spectra with aid of the software packages: DeNovoX (Thermo Scientific, Waltham, MA, USA) and PEAKS Studio 6.0 (Bioinformatics Solutions Inc., Waterloo, Ontario, Canada). The parameters used for both programs were as follows: selection of the peptide charge (1+, 2+, 3+), tolerances of 0.3–0.5 Da for precursor and fragments ions, use or not of trypsin or Glu-C as proteases and three variable modifications: Methionine oxidation (m), carbamidomethylation of Cys and acetylation of the *N*-terminus of the protein (*N*-Acyl).

4. Conclusions

In this study, ACE inhibitory peptides were produced from cuttlefish byproduct protein hydrolysate using a combination of membrane technology (UF-DF) and enzyme proteolysis using alcalase. The sequential ultrafiltration with two membranes cut-offs (10 and 1 kDa) led to recover a remarkable antihypertensive activity. The most bioactive peptides were subsequently separated by means of RP-HPLC and characterized by HPLC-ESI-IT-FTICR. The procedure developed in the present work demonstrated to be a feasible biotechnological approach to produce antihypertensive peptides from marine wastes and to help in the depuration and valorization of wastewaters generated by marine food processing. Nevertheless, further studies (peptide synthesis, re-chromatography of the HPLC fractions, *etc.*) are needed in order to identify which among these peptide mixtures are responsible for the observed activity.

Acknowledgements

We wish to thank to Margarita Nogueira and Ana Durán for their excellent technical assistance. We also thank to Almudena Pena for the kindly supply of the cuttlefish wastewater. The company Frinova S.A. funded this study (Contract N° 20090732 co-financed by CDTI). The authors also thank the Unit of Information Resources for Research (URICI-CSIC) for the co-funding of this publication in Open Access format.

Author Contributions

I.R.A. performed the experiments and wrote the text; J.A.V. supervised the experiments and corrected the manuscript; P.G. supervised the project; D.E.F. characterized the peptides by HPLC-ESI-IT-FTICR; M.C. processed the mass spectrometry data and C.P. separated the peptides by RP-HPLC and corrected the manuscript.

References

1. World Health Organization. *Causes of Death 2008: Data Sources and Methods*; World Health Organization: Geneva, Switzerland, 2011. Available online: http://www.who.int/healthinfo/global_burden_disease/cod_2008_sources_methods.pdf (accessed on 20 October 2013).

2. World Health Organization. *A Global Brief on Hypertension. Silent Killer, Global Public Health Crisis*; World Health Organization: Geneva, Switzerland, 2013. Available online: http://apps.who.int/iris/bitstream/10665/79059/1/WHO_DCO_WHD_2013.2_eng.pdf (accessed on 14 December 2013).
3. Lim, S.S.; Vos, T.; Flaxman, A.D.; Danaei, G.; Shibuya, K.; Adair-Rohani, H.; Al Mazroa, M.A.; Amann, M.; Anderson, H.R.; Andrews, K.G.; *et al.* A comparative risk assessment of burden of disease and injury attributable to 67 risk factors and risk factor clusters in 21 regions, 1990–2010: A systematic analysis for the Global Burden of Disease Study 2010. *Lancet* **2012**, *380*, 2224–2260.
4. Fang, H.; Luo, M.; Sheng, Y.; Li, Z.; Wu, Y.; Liu, C. The antihypertensive effect of peptides: A novel alternative to drugs? *Peptides* **2008**, *29*, 1062–1071.
5. Israili, Z.H.; Hall, W.D. Cough and angioneurotic edema associated with angiotensin-converting enzyme inhibitor therapy. A review of the literature and pathophysiology. *Ann. Int. Med.* **1992**, *117*, 234–242.
6. Estévez, N.; Fuciños, P.; Sobrosa, A.C.; Pastrana, L.; Pérez, N.; Rúa, M.L. Modeling the angiotensin-converting enzyme inhibitory activity of peptide mixtures obtained from cheese whey hydrolysates using concentration-response curves. *Biotechnol. Prog.* **2012**, *28*, 1197–1206.
7. Alemán, A.; Pérez-Santín, E.; Bordenave-Juchereau, S.; Arnaudin, I.; Gómez-Guillén, M.C.; Montero, P. Squid gelatin hydrolysates with antihypertensive, anticancer and antioxidant activity. *Food Res. Int.* **2011**, *44*, 1044–1051.
8. Ryan, J.T.; Ross, R.P.; Bolton, D.; Fitzgerald, G.F.; Stanton, C. Bioactive peptides from muscle sources: Meat and fish. *Nutrients* **2011**, *3*, 765–791.
9. He, H.; Chen, X.; Sun, C.; Zhang, Y.; Gao, P. Preparation and functional evaluation of oligopeptide-enriched hydrolysate from shrimp (*Acetes chinensis*) treated with crude protease from *Bacillus* sp. SM98011. *Bioresour. Technol.* **2006**, *97*, 385–390.
10. Jung, W.-K.; Mendis, E.; Je, J.-Y.; Park, P.-J.; Son, B.W.; Kim, H.C.; Choi, Y.K.; Kim, S.-K. Angiotensin I-converting enzyme inhibitory peptide from yellowfin sole (*Limanda aspera*) frame protein and its antihypertensive effect in spontaneously hypertensive rats. *Food Chem.* **2006**, *94*, 26–32.
11. Balti, R.; Nedjar-Arroume, N.; Bougatef, A.; Guillochon, D.; Nasri, M. Three novel angiotensin I-converting enzyme (ACE) inhibitory peptides from cuttlefish (*Sepia officinalis*) using digestive proteases. *Food Res. Int.* **2010**, *43,* 1136–1143.
12. Amado, I.R.; Vázquez, J.A.; González, M.P.; Murado,M.A. Production of antihypertensive and antioxidant activities by enzymatic hydrolysis of protein concentrates recovered by ultrafiltration from cuttlefish processing wastewaters. *Biochem. Eng. J.* **2013**, *76*, 43–54.
13. Kristinsson, H.G.; Rasco, B.A. Biochemical and functional properties of Atlantic salmon (*Salmo salar*) muscle proteins hydrolyzed with various alkaline proteases. *J. Agric. Food Chem.* **2000**, *48*, 657–666.
14. Pérez-Santín, E.; Calvo, M.M.; López-Caballero, M.E.; Montero, P.; Gómez-Guillén, M.C. Compositional properties and bioactive potential of waste material from shrimp cooking juice. *LWT-Food Sci. Technol.* **2013**, *54*, 87–94.

15. Murado, M.A.; Fraguas, J.; Montemayor, M.I.; Vázquez, J.A.; González, P. Preparation of highly purified chondroitin sulphate from skate (*Raja clavata*) cartilage by-products. Process optimization including a new procedure of alkaline hydroalcoholic hydrolysis. *Biochem. Eng. J.* **2010**, *49*, 126–132.
16. Ferjani, E.; Ellouze, E.; Ben Amar, R. Treatment of seafood processing wastewaters by ultrafiltration-nanofiltration cellulose acetate membranes. *Desalination* **2005**, *177*, 43–49.
17. Ellouze, E.; Ben Amar, R.; Ben Salah, A.H. Cross-flow microfiltration using ceramic membranes applied to the cuttlefish effluents treatment: Effect of operating parameters and the addition of pre or post-treatment. *Desalination* **2005**, *177*, 229–240.
18. Balti, R.; Nedjar-Arroume, N.; Adjé, E.Y.; Guillochon, D.; Nasri, M. Analysis of novel angiotensin I-converting enzyme inhibitory peptides from enzymatic hydrolysates of cuttlefish (*Sepia officinalis*) muscle proteins. *J. Agric. Food Chem.* **2010**, *58*, 3840–3846.
19. Ko, S.-C.; Kang, N.; Kim, E.-A.; Kang, M.C.; Lee, S.-H.; Kang, S.-M.; Lee, J.-B.; Jeon, B.-T.; Kim, S.-K.; Park, S.-J.; *et al.* A novel angiotensin I-converting enzyme (ACE) inhibitory peptide from a marine *Chlorella ellipsoidea* and its antihypertensive effect in spontaneously hypertensive rats. *Process Biochem.* **2012**, *47*, 2005–2011.
20. Ahn, C.-B.; Jeon, Y.-J.; Kim, Y.-T.; Je, J.-Y. Angiotensin I converting enzyme (ACE) inhibitory peptides from salmon byproduct protein hydrolysate by alcalase hydrolysis. *Process Biochem.* **2012**, *47*, 2240–2245.
21. Wijesekara, I.; Qian, Z.-J.; Ryu, B.; Ngo, D.-H.; Kim, S.-K. Purification and identification of antihypertensive peptides from seaweed pipefish (*Syngnathus schlegeli*) muscle protein hydrolysate. *Food Res. Int.* **2011**, *44*, 703–707.
22. Zhao, Y.; Li, B.; Dong, S.; Liu, Z.; Zhao, X.; Wang, J.; Zeng, M. A novel ACE inhibitory peptide isolated from *Acaudina molpadioidea* hydrolysate. *Peptides* **2009**, *30*, 1028–1033.
23. Wu, S.; Sun, J.; Tong, Z; Lan, X.; Zhao, Z.; Liao, D. Optimization of hydrolysis conditions for the production of angiotensin-I converting enzyme-inhibitory peptides and isolation of a novel peptide from lizard fish (*Saurida elongata*) muscle protein hydrolysate. *Mar. Drugs* **2012**, *10*, 1066–1080.
24. Hyun, C.-K.; Shin, H.-K. Utilization of bovine blood plasma proteins for the production of angiotensin I converting enzyme inhibitory peptides. *Process Biochem.* **2000**, *36*, 65–71.
25. Je, J.-Y.; Park, P.-J.; Kwon, J.Y.; Kim, S.-K. A novel Angiotensin I converting enzyme inhibitory peptide from Alaska Pollack (*Theragra chalcogramma*) frame protein hydrolysate. *J. Agric. Food Chem.* **2004**, *52*, 7842–7845.
26. Alemán, A.; Gómez-Guillén, M.C.; Montero, P. Identification of ACE-inhibitory peptides from squid skin collagen after *in vitro* gastrointestinal digestion. *Food Res. Int.* **2013**, *54*, 790–795.
27. Gordon, A.; Barbut, S. Effect of chloride salts on protein extraction and interfacial protein film formation in meat batters. *J. Sci. Food Agric.* **1992**, *58*, 227–238.
28. Alemán, A.; Giménez, B.; Pérez-Santin, E.; Gómez-Guillén, M.C.; Montero, P. Contribution of Leu and Hyp residues to antioxidant and ACE-inhibitory activities of peptide sequences isolated from squid gelatin hydrolysate. *Food Chem.* **2011**, *125*, 334–341.

29. He, R.; Ma, H.; Zhao, W.; Qu, W.; Zhao, J.; Luo, L.; Zhu, W. Modeling the QSAR of ACE-inhibitory peptides with ANN and its applied illustration. *Int. J. Pept.* **2012**, *2012*, 1–9.
30. Wu, J.; Aluko, R. E.; Nakai, S. Structural requirements of angiotensin I converting enzyme inhibitory peptides: Quantitative structure-activity relationship study of Di- and Tripeptides. *J. Agric. Food Chem.* **2006**, *54*, 732–738.
31. Ondetti, M.A.; Cushman, D.W. Enzymes of the renin-angiotensin system and their inhibitors. *Ann. Rev. Biochem.* **2012**, *51*, 283–308.
32. Mullally, M.M.; Meisel, H.; FitzGerald, R.J. Identification of a novel angiotensin-I-converting enzyme inhibitory peptide corresponding to a tryptic fragment of bovine β-lactoglobulin. *FEBS Lett.* **1997**, *402*, 99–101.
33. FitzGerald, R.J.; Meisel, H. Milk protein-derived peptide inhibitors of angiotensin-I-converting enzyme. *Br. J. Nutr.* **2000**, *84*, S33–S37.
34. Haque, E.; Chand, R. Antihypertensive and antimicrobial bioactive peptides from milk proteins. *Eur. Food Res. Technol.* **2008**, *227*, 7–15.
35. Cheung, H.-S.; Wang, F.-L.; Ondetti, M.; Sabo, E.; Cushman, D. Binding of peptide substrates and inhibitors of angiotensin-converting enzyme: Importance of the COOH-terminal dipeptide sequences. *J. Biol. Chem.* **1980**, *255*, 401–407.
36. Barker, T.W.; Worgan, J.T. The utilisation of palm oil processing effluents as substrates for microbial protein production by the fungus *Aspergillus oryzae*. *Appl. Microbiol. Biotechnol.* **1981**, *11*, 234–240.
37. Shalaby, S.M.; Zakora, M.; Otte, J. Performance of two commonly used angiotensin-converting enzyme inhibition assays using FA-PGG and HHL as substrates. *J. Dairy Res.* **2006**, *73*, 178–186.
38. Keller, A.; Nesvizhskii, A.I.; Kolker, E.; Aebersold, R. Empirical statistical model to estimate the accuracy of peptide identifications made by MS/MS and database search. *Anal. Chem.* **2002**, *74*, 5383–5392.

PERMISSIONS

All chapters in this book were first published in MDPI; hereby published with permission under the Creative Commons Attribution License or equivalent. Every chapter published in this book has been scrutinized by our experts. Their significance has been extensively debated. The topics covered herein carry significant findings which will fuel the growth of the discipline. They may even be implemented as practical applications or may be referred to as a beginning point for another development.

The contributors of this book come from diverse backgrounds, making this book a truly international effort. This book will bring forth new frontiers with its revolutionizing research information and detailed analysis of the nascent developments around the world.

We would like to thank all the contributing authors for lending their expertise to make the book truly unique. They have played a crucial role in the development of this book. Without their invaluable contributions this book wouldn't have been possible. They have made vital efforts to compile up to date information on the varied aspects of this subject to make this book a valuable addition to the collection of many professionals and students.

This book was conceptualized with the vision of imparting up-to-date information and advanced data in this field. To ensure the same, a matchless editorial board was set up. Every individual on the board went through rigorous rounds of assessment to prove their worth. After which they invested a large part of their time researching and compiling the most relevant data for our readers.

The editorial board has been involved in producing this book since its inception. They have spent rigorous hours researching and exploring the diverse topics which have resulted in the successful publishing of this book. They have passed on their knowledge of decades through this book. To expedite this challenging task, the publisher supported the team at every step. A small team of assistant editors was also appointed to further simplify the editing procedure and attain best results for the readers.

Apart from the editorial board, the designing team has also invested a significant amount of their time in understanding the subject and creating the most relevant covers. They scrutinized every image to scout for the most suitable representation of the subject and create an appropriate cover for the book.

The publishing team has been an ardent support to the editorial, designing and production team. Their endless efforts to recruit the best for this project, has resulted in the accomplishment of this book. They are a veteran in the field of academics and their pool of knowledge is as vast as their experience in printing. Their expertise and guidance has proved useful at every step. Their uncompromising quality standards have made this book an exceptional effort. Their encouragement from time to time has been an inspiration for everyone.

The publisher and the editorial board hope that this book will prove to be a valuable piece of knowledge for researchers, students, practitioners and scholars across the globe.

LIST OF CONTRIBUTORS

Jie Lei, Christopher Franco and Wei Zhang
Centre for Marine Bioproducts Development, Flinders University, Adelaide, SA 5042, Australia
Department of Medical Biotechnology, School of Medicine, Flinders University, Adelaide, SA 5042, Australia

Mohammad Ferdous Mehbub
Centre for Marine Bioproducts Development, Flinders University, Adelaide, SA 5042, Australia
Department of Medical Biotechnology, School of Medicine, Flinders University, Adelaide, SA 5042, Australia
Department of Fisheries Technology, Faculty of Fisheries, Hajee Mohammad Danesh Science and Technology University, Dinajpur 5200, Bangladesh

Jabin R. Watson, Bernard M. Degnan and Sandie M. Degnan
School of Biological Science, University of Queensland, Brisbane, Queensland 4072, Australia

Timothy C. R. Brennan
Systems and Synthetic Biology Group, Australian Institute for Bioengineering and Nanotechnology, University of Queensland, Brisbane, Queensland 4072, Australia

Jens O. Krömer
Centre for Microbial Electrosynthesis (CEMES), Advanced Water Management Centre, University of Queensland, Brisbane, Queensland 4072, Australia

Stefano Varrella, Giovanna Romano, Adrianna Ianora, Nadia Ruocco and Maria Costantini
Stazione Zoologica Anton Dohrn, Villa Comunale, Naples 80121, Italy

Matt G. Bentley
Dove Marine Laboratory, School of Marine Science and Technology, Newcastle University, Newcastle upon Tyne, UK

Swarna Oli and Tanja Schirmeister
Institute of Pharmacy and Biochemistry, Johannes Gutenberg-University of Mainz, Staudinger Weg 5, Mainz 55128, Germany

Ute Hentschel
Department of Botany II, Julius-von-Sachs-Institute for Biological Sciences, University of Würzburg, Julius-von-Sachs Platz 3, Würzburg 97082, Germany

Usama Ramadan Abdelmohsen
Department of Botany II, Julius-von-Sachs-Institute for Biological Sciences, University of Würzburg, Julius-von-Sachs Platz 3, Würzburg 97082, Germany
Department of Pharmacognosy, Faculty of Pharmacy, Minia University, Minia 61519, Egypt

Eric H. Andrianasolo, Liti Haramaty, Richard Lutz and Paul Falkowski
Center for Marine Biotechnology, Institute of Marine and Coastal Sciences, Rutgers, The State University of New Jersey, New Brunswick, NJ 08901, USA

Eileen White
The Cancer Institute of New Jersey, 195 Little Albany Street, New Brunswick, NJ 08903, USA

Ning Wu
Institute of Oceanology, Chinese Academy of Sciences, 7 Nanhai Road, Qingdao 266071, China

Ge Liu
Institute of Oceanology, Chinese Academy of Sciences, 7 Nanhai Road, Qingdao 266071, China
The University of Chinese Academy of Sciences, Beijing 100049, China

Ming Liu
Key Laboratory of Marine Drugs, Ministry of Education, School of Medicine and Pharmacy, Ocean University of China, Qingdao 266003, China

Jianteng Wei
Lanzhou Institute of Chemical Physics, Chinese Academy of Sciences, Lanzhou 730000, China

Haijuan Huang and Yuyan Zhang
Qingdao University of Science and Technology, Qingdao 266042, China

Jin Zhao
Department of Biotechnology, Zhengzhou University, Zhengzhou 450001, China

Lin Xiao
Qingdao Agricultural University, Qingdao 266109, China

Lanhong Zheng
Yellow Sea Fisheries Research Institute, Chinese Academy of Fishery Sciences, Qingdao 266071, China

Xiukun Lin
Department of Pharmacology, Capital Medical University, Beijing 100069, China

Yuichi Kamo and Wahyudiono
Department of Chemical Engineering, Nagoya University, Furo-cho, Chikusa-ku, Nagoya 464-8603, Japan

Hideki Kanda
Department of Chemical Engineering, Nagoya University, Furo-cho, Chikusa-ku, Nagoya 464-8603, Japan
Japan Science and Technology Agency, Chiyoda, Tokyo 102-0076, Japan

Motonobu Goto
Japan Science and Technology Agency, Chiyoda, Tokyo 102-0076, Japan

Siti Machmudah
Department of Chemical Engineering, Nagoya University, Furo-cho, Chikusa-ku, Nagoya 464-8603, Japan
Department of Chemical Engineering, Sepuluh Nopember Institute of Technology, Kampus ITS Sukolilo, Surabaya 60111, Indonesia

Massimo Monteleone
STAR-Agroenergy Research Group, University of Foggia, via Gramsci 89/91, Foggia 71121, Italy

Massimo Franchi
Institute of Marine Science, National Research Council, via Pola 4, Lesina 71010, Italy

Matteo Francavilla
STAR-Agroenergy Research Group, University of Foggia, via Gramsci 89/91, Foggia 71121, Italy
Institute of Marine Science, National Research Council, via Pola 4, Lesina 71010, Italy

Carmela Caroppo
Institute for Coastal Marine Environment, National Research Council, via Roma 3, Taranto 74121, Italy

Si-Mei Long, Feng-Yin Liang, Qi Wu, Xi-Lin Lu, Xiao-Li Yao and Zhong Pei
Department of Neurology, National Key Clinical Department and Key Discipline of Neurology, Guangdong Key Laboratory for Diagnosis and Treatment of Major Neurological Diseases, The First Affiliated Hospital, Sun Yat-sen University, Guangzhou 510080, China

Shi-Chang Li, Jing Li and Ji-Yan Pang
School of Chemistry & Chemical Engineering, Sun Yat-sen University, Guangzhou 510275, China

Huanxing Su
State Key Laboratory of Quality Research in Chinese Medicine, Institute of Chinese Medical Sciences, University of Macau, Macao 999078, China

Natalya V. Ageenko and Nelly A. Odintsova
Cytotechnology Laboratory, A.V. Zhirmunsky Institute of Marine Biology, FEB RAS, Vladivostok 690041, Russia
Far Eastern Federal University, Sukhanova Str. 8, Vladivostok 690950, Russia

Konstantin V. Kiselev
Laboratory of Biotechnology, Institute of Biology and Soil Sciences, FEB RAS, Vladivostok 690022, Russia

Pavel S. Dmitrenok
Laboratory of Instrumental and Radioisotope Methods of Analysis, G.B. Elyakov Pacific Institute of Bioorganic Chemistry, FEB RAS, Vladivostok 690022, Russia

Satoshi Tsuneda
Department of Life Science and Medical Bioscience, Waseda University, 2-2 Wakamatsu-cho, Shinjuku-ku, Tokyo 162-8480, Japan

Yuji Sekiguchi
Biomedical Research Institute, National Institute of Advanced Industrial Science and Technology (AIST), 1-1-1 Higashi, Tsukuba, Ibaraki 305-8566, Japan

Atsushi Furuta and Naohiro Noda
Department of Life Science and Medical Bioscience, Waseda University, 2-2 Wakamatsu-cho, Shinjuku-ku, Tokyo 162-8480, Japan
Biomedical Research Institute, National Institute of Advanced Industrial Science and Technology (AIST), 1-1-1 Higashi, Tsukuba, Ibaraki 305-8566, Japan

Kazi Abdus Salam and Nobuyoshi Akimitsu
Radioisotope Center, The University of Tokyo, 2-11-16 Yayoi, Bunkyo-ku, Tokyo 113-0032, Japan

Idam Hermawan and Junichi Tanaka
Department of Chemistry, Biology and Marine Science, University of the Ryukyus, Nishihara, Okinawa 903-0213, Japan

Hidenori Tani
Research Institute for Environmental Management Technology, National Institute of Advanced Industrial Science and Technology (AIST), 16-1 Onogawa, Tsukuba, Ibaraki 305-8569, Japan

Atsuya Yamashita and Kohji Moriishi
Department of Microbiology, Division of Medicine, Graduate School of Medicine and Engineering, University of Yamanashi, 1110 Shimokato, Chuo-shi, Yamanashi 409-3898, Japan

Masamichi Nakakoshi and Masayoshi Tsubuki
Institute of Medical Chemistry, Hoshi University, 2-4-41 Ebara, Shinagawa-ku, Tokyo 142-8501, Japan

Poh Wee Peng, Youichi Suzuki and Naoki Yamamoto
Department of Microbiology, Yong Loo Lin School of Medicine, National University of Singapore, Center for Translational Medicine, 14 Medical Drive, #15-02, Level 15, Singapore 117599, Singapore

Fatih Karadeniz
Marine Biotechnology Center for Pharmaceuticals and Foods, Silla University, Busan 617-736, Korea

Myeong Sook Kwon and Chang-Suk Kong
Department of Food and Nutrition, College of Medical and Life Science, Silla University, Busan 617-736, Korea

Jung-Ae Kim
Marine Biotechnology Center for Pharmaceuticals and Foods, Silla University, Busan 617-736, Korea
Department of Food and Nutrition, College of Medical and Life Science, Silla University, Busan 617-736, Korea

Byul-Nim Ahn
Department of Organic Material Science and Engineering, Pusan National University, Busan 609-735, Korea

Isabel Rodríguez Amado, José Antonio Vázquez and Pilar González
Group of Recycling and Valorisation of Waste Materials (REVAL), Marine Research Institute (IIM-CSIC), r/ Eduardo Cabello, 6. Vigo, Galicia 36208, Spain

Diego Esteban-Fernández
Department of Chemistry, Humboldt-Universitaet zu Berlin, Brook-Taylor Strasse 2, Berlin 12489, Germany

Mónica Carrera
Institute of Molecular Systems Biology (IMSB), ETH Zürich, Zürich 8093, Switzerland

Carmen Piñeiro
Scientific Instrumentation Service (SICIM), Marine Research Institute (IIM-CSIC), r/ Eduardo Cabello, 6. Vigo, Galicia 36208, Spain

Index

www.ingramcontent.com/pod-product-compliance
Lightning Source LLC
LaVergne TN
LVHW081013210926
848485LV00004B/852